Springer Series in Statistics

Advisors:
P. Bickel, P. Diggle, S. Fienberg, K. Krickeberg,
I. Olkin, N. Wermuth, S. Zeger

Springer
New York
Berlin
Heidelberg
Hong Kong
London
Milan
Paris
Tokyo

Springer Series in Statistics

Andersen/Borgan/Gill/Keiding: Statistical Models Based on Counting Processes.
Atkinson/Riani: Robust Diagnotstic Regression Analysis.
Berger: Statistical Decision Theory and Bayesian Analysis, 2nd edition.
Borg/Groenen: Modern Multidimensional Scaling: Theory and Applications
Brockwell/Davis: Time Series: Theory and Methods, 2nd edition.
Chan/Tong: Chaos: A Statistical Perspective.
Chen/Shao/Ibrahim: Monte Carlo Methods in Bayesian Computation.
David/Edwards: Annotated Readings in the History of Statistics.
Devroye/Lugosi: Combinatorial Methods in Density Estimation.
Efromovich: Nonparametric Curve Estimation: Methods, Theory, and Applications.
Eggermont/LaRiccia: Maximum Penalized Likelihood Estimation, Volume I: Density Estimation.
Fahrmeir/Tutz: Multivariate Statistical Modelling Based on Generalized Linear Models, 2nd edition.
Farebrother: Fitting Linear Relationships: A History of the Calculus of Observations 1750-1900.
Federer: Statistical Design and Analysis for Intercropping Experiments, Volume I: Two Crops.
Federer: Statistical Design and Analysis for Intercropping Experiments, Volume II: Three or More Crops.
Glaz/Naus/Wallenstein: Scan Statistics.
Good: Permutation Tests: A Practical Guide to Resampling Methods for Testing Hypotheses, 2nd edition.
Gouriéroux: ARCH Models and Financial Applications.
Gu: Smoothing Spline ANOVA Models.
Györfi/Kohler/Krzyżak/ Walk: A Distribution-Free Theory of Nonparametric Regression.
Haberman: Advanced Statistics, Volume I: Description of Populations.
Hall: The Bootstrap and Edgeworth Expansion.
Härdle: Smoothing Techniques: With Implementation in S.
Harrell: Regression Modeling Strategies: With Applications to Linear Models, Logistic Regression, and Survival Analysis
Hart: Nonparametric Smoothing and Lack-of-Fit Tests.
Hastie/Tibshirani/Friedman: The Elements of Statistical Learning: Data Mining, Inference, and Prediction
Hedayat/Sloane/Stufken: Orthogonal Arrays: Theory and Applications.
Heyde: Quasi-Likelihood and its Application: A General Approach to Optimal Parameter Estimation.
Huet/Bouvier/Gruet/Jolivet: Statistical Tools for Nonlinear Regression: A Practical Guide with S-PLUS Examples.
Ibrahim/Chen/Sinha: Bayesian Survival Analysis.
Jolliffe: Principal Component Analysis.
Kolen/Brennan: Test Equating: Methods and Practices.
Knottnerus: Sample Survey Theory.

(continued after index)

Jun S. Liu

Monte Carlo Strategies in Scientific Computing

With 56 Figures

 Springer

Jun S. Liu
Department of Statistics
Harvard University
Cambridge, MA 02138
jliu@hustat.harvard.edu
USA

Library of Congress Cataloging-in-Publication Data
Liu, Jun S.
  Monte Carlo strategies in scientific computing / Jun S. Liu
    p. cm. — (Springer series in statistics)
  Includes bibliographical references and index.
  ISBN 0-387-95230-6 (hc. : alk. paper)
    1. Science—Statistical methods.   2. Monte Carlo method.   I. Title.   II. Series
  Q180.55.S7 L58 2001
  501′519282—dc21   00-069243

ISBN 0-387-95230-6   Printed on acid-free paper.

© 2001 Springer-Verlag New York, Inc.
All rights reserved. This work may not be translated or copied in whole or in part without the written permission of the publisher (Springer-Verlag New York, Inc., 175 Fifth Avenue, New York, NY 10010, USA), except for brief excerpts in connection with reviews or scholarly analysis. Use in connection with any form of information storage and retrieval, electronic adaptation, computer software, or by similar or dissimilar methodology now known or hereafter developed is forbidden.
The use in this publication of trade names, trademarks, service marks, and similar terms, even if they are not identified as such, is not to be taken as an expression of opinion as to whether or not they are subject to proprietary rights.

Printed in the United States of America.

9 8 7 6 5 4 3 2 (Corrected printing, 2003)   SPIN 10886636

Typesetting: Pages created by the author using a Springer L$_A$T$_E$X macro package.

www.springer-ny.com

Springer-Verlag   New York Berlin Heidelberg
A member of BertelsmannSpringer Science+Business Media GmbH

To my wife Wei

Preface

An early experiment that conceives the basic idea of Monte Carlo computation is known as "Buffon's needle" (Dörrie 1965), first stated by Georges Louis Leclerc Comte de Buffon in 1777. In this well-known experiment, one throws a needle of length l onto a flat surface with a grid of parallel lines with spacing D ($D > l$). It is easy to compute that, under ideal conditions, the chance that the needle will intersect one of the lines is $2l/\pi D$. Thus, if we let p_N be the proportion of "intersects" in N throws, we can have an estimate of π as

$$\hat{\pi} = \lim_{N \to \infty} \frac{2l}{p_N D},$$

which will "converge" to π as N increases to infinity. Numerous investigators actually used this setting to estimate π. The idea of simulating random processes so as to help evaluate certain quantities of interest is now an essential part of scientific computing.

A systematic use of the Monte Carlo method for real scientific problems appeared in the early days of electronic computing (1945-55) and accompanied the development of the world's first programmable "super" computer, MANIAC (Mathematical Analyzer, Numerical Integrator and Computer), at Los Alamos during World War II. In order to make a good use of these fast computing machines, scientists (Stanislaw Ulam, John von Neumann, Nicholas Metropolis, Enrico Fermi, etc.) invented a statistical sampling-based method for solving numerical problems concerning random neutron diffusion in fissile material in atomic bomb designs and for estimating eigenvalues of the Schrödinger equation. The basic idea underlying the method was first brought up by Ulam and deliberated between him and von Neumann in a car when they drove together from Los Alamos to Lamy. Allegedly, Nick Metropolis coined the name "Monte Carlo," which played an essential role in popularizing the method.

In the early 1950s, statistical physicists (N. Metropolis, A. Rosenbluth, M. Rosenbluth, A. Teller, and E. Teller) introduced a Markov-chain-based dynamic Monte Carlo method for the simulation of simple fluids. This method was later extended to cover more and more complex physical systems, including spin glass models, harmonic crystal, polymer models, etc. In the 1980s, statisticians and computer scientists developed Monte-Carlo-based methods for a wide variety of tasks such as combinatorial optimizations, nonparametric statistical inference (e.g., jackknife and bootstrap), likelihood computation with missing observations, statistical genetics analysis, Bayesian modeling and computations, and others. In the 1990s, the method began to play an important role in computational biology and was used to solve problems in sequence motif identification and the analysis of complex pedigree. Now, the list of application areas of Monte Carlo methods includes biology (Leach 1996, Karplus and Petsko 1990, Lawrence, Altschul, Boguski, Liu, Neuwald and Wootton 1993), chemistry (Alder and Wainwright 1959), computer science (Kirkpatrick, Gelatt and Vecchi 1983), economics and finance (Gouriérourx and Monfort 1997); engineering (Geman and Geman 1984), material science (Frenkel and Smit 1996), physics (Metropolis, Rosenbluth, Rosenbluth, Teller and Teller 1953, Goodman and Sokal 1989, Marinari and Parisi 1992), statistics (Efron 1979, Gelfand and Smith 1990, Rubin 1987, Tanner and Wong 1987), and many others. Among all Monte Carlo methods, *Markov chain Monte Carlo* (MCMC) provides an enormous scope for dealing with very complicated stochastic systems and has been the central pillar in the study of macromolecules and other physical systems. Recently, the MCMC methodology has drawn much attention from statisticians because the method enables them to entertain more sophisticated and realistic statistical models.

Being attracted by the extreme flexibility and power of the Monte Carlo method, many researchers in different scientific areas have contributed to its development. However, because a substantial amount of domain-specific knowledge is required in order to understand problems in any of these fields, communications among researchers in these fields are very limited. Many efforts have been devoted to the reinvention of techniques that have been developed in other fields. It is therefore desirable to develop a relatively general framework in which scientists in every field — e.g., theoretical chemists, statistical physicists, structural biologists, statisticians, econometricians, and computer scientists — can compare their Monte Carlo techniques and learn from each other. For a large number of scientists and engineers who employ Monte Carlo simulation and related global optimization techniques (such as simulated annealing) as an essential tool in their work, there is also a need to keep up to date with recent advances in Monte Carlo methodologies and to understand the nature and connection of various proposed methods. The aim of this book is to provide a self-contained, unified, and up-to-date treatment of the Monte Carlo method.

This book is intended to serve three audiences: researchers specializing in the study of Monte Carlo algorithms; scientists who are interested in using advanced Monte Carlo techniques; and graduate students in statistics, computational biology, and computer sciences who want to learn about Monte Carlo computations. The prerequisites for understanding most of the methods described in this book are rather minimal: a one-semester course on probability theory (Pitman 1993) and a one-semester course on theoretical statistics (Rice 1994), both at the undergraduate level. However, it would be more desirable if the reader has some background in a specific scientific field such as artificial intelligence, computational biology, computer vision, engineering, or Bayesian statistics in which heavy computations are involved. This book is most suitable for a second-year graduate-level course on Monte Carlo methods, with an emphasis on their relevance to scientific and statistical research.

The author is most grateful to his mentor and friend Wing Hung Wong for his many important suggestions, his overwhelming passion for Monte Carlo and scientific problems, and his continuous encouragement. The author is also grateful to Persi Diaconis for teaching him many things including Markov chain theory, group theory, and nonparametric Bayes methods, to both Susan Holmes and Persi for numerous enlightening conversations on Markov chain Monte Carlo and other related problems, to Donald B. Rubin for insights on the missing data formulation and the Bayesian thinking, to Jonathan Goodman for helpful comments on multigrid Monte Carlo, to Yingnian Wu and Songchun Zhu for their materials on pattern simulations and thoughts on conditional sampling, to Faming Liang for his supply of many examples and figures, and to Minghui Chen and David van Dyk for helpful comments. Several former graduate students in the statistics departments of Stanford and Harvard universities — Yuguo Chen, Lingyu Chen, Chiara Sabatti, Tanya Logvinenko, Zhaohui Qin and Juni Zhang — have contributed in many ways to the development of this book. Ms. Helen Tombropoulos has provided editorial assistance to the author both for this book and for many articles published earlier. Finally, the author is greatly indebted to his wife Wei for her love and her continuous support of his research activities these years. Part of the book was written when the author was on the faculty of the Statistics Department of Stanford University. This work was also partially supported by the National Science Foundation Grants DMS-9803649 and DMS-0094613.

Cambridge, Massachusetts *Jun Liu*
March 2001

Contents

Preface   vii

1 Introduction and Examples   1
    1.1 The Need of Monte Carlo Techniques 1
    1.2 Scope and Outline of the Book 3
    1.3 Computations in Statistical Physics 7
    1.4 Molecular Structure Simulation 9
    1.5 Bioinformatics: Finding Weak Repetitive Patterns 10
    1.6 Nonlinear Dynamic System: Target Tracking 14
    1.7 Hypothesis Testing for Astronomical Observations 16
    1.8 Bayesian Inference of Multilevel Models 18
    1.9 Monte Carlo and Missing Data Problems 19

2 Basic Principles: Rejection, Weighting, and Others   23
    2.1 Generating Simple Random Variables 23
    2.2 The Rejection Method 24
    2.3 Variance Reduction Methods 26
    2.4 Exact Methods for Chain-Structured Models 28
       2.4.1 Dynamic programming 29
       2.4.2 Exact simulation 30
    2.5 Importance Sampling and Weighted Sample 31
       2.5.1 An example 31
       2.5.2 The basic idea 33
       2.5.3 The "rule of thumb" for importance sampling 34

xii   Contents

		2.5.4	Concept of the weighted sample	36
		2.5.5	Marginalization in importance sampling	37
		2.5.6	Example: Solving a linear system	38
		2.5.7	Example: A Bayesian missing data problem	40
	2.6	Advanced Importance Sampling Techniques		42
		2.6.1	Adaptive importance sampling	42
		2.6.2	Rejection and weighting	43
		2.6.3	Sequential importance sampling	46
		2.6.4	Rejection control in sequential importance sampling	48
	2.7	Application of SIS in Population Genetics		49
	2.8	Problems		51

3 Theory of Sequential Monte Carlo   53

	3.1	Early Developments: Growing a Polymer		55
		3.1.1	A simple model of polymer: Self-avoid walk	55
		3.1.2	Growing a polymer on the square lattice	56
		3.1.3	Limitations of the growth method	59
	3.2	Sequential Imputation for Statistical Missing Data Problems		60
		3.2.1	Likelihood computation	60
		3.2.2	Bayesian computation	62
	3.3	Nonlinear Filtering		64
	3.4	A General Framework		67
		3.4.1	The choice of the sampling distribution	69
		3.4.2	Normalizing constant	69
		3.4.3	Pruning, enrichment, and resampling	71
		3.4.4	More about resampling	72
		3.4.5	Partial rejection control	75
		3.4.6	Marginalization, look-ahead, and delayed estimate	76
	3.5	Problems		77

4 Sequential Monte Carlo in Action   79

	4.1	Some Biological Problems		79
		4.1.1	Molecular Simulation	79
		4.1.2	Inference in population genetics	81
		4.1.3	Finding motif patterns in DNA sequences	84
	4.2	Approximating Permanents		90
	4.3	Counting 0-1 Tables with Fixed Margins		92
	4.4	Bayesian Missing Data Problems		94
		4.4.1	Murray's data	94
		4.4.2	Nonparametric Bayes analysis of binomial data	95
	4.5	Problems in Signal Processing		98
		4.5.1	Target tracking in clutter and mixture Kalman filter	98
		4.5.2	Digital signal extraction in fading channels	102
	4.6	Problems		103

5 Metropolis Algorithm and Beyond — 105
- 5.1 The Metropolis Algorithm … 106
- 5.2 Mathematical Formulation and Hastings's Generalization … 111
- 5.3 Why Does the Metropolis Algorithm Work? … 112
- 5.4 Some Special Algorithms … 114
  - 5.4.1 Random-walk Metropolis … 114
  - 5.4.2 Metropolized independence sampler … 115
  - 5.4.3 Configurational bias Monte Carlo … 116
- 5.5 Multipoint Metropolis Methods … 117
  - 5.5.1 Multiple independent proposals … 118
  - 5.5.2 Correlated multipoint proposals … 120
- 5.6 Reversible Jumping Rule … 122
- 5.7 Dynamic Weighting … 124
- 5.8 Output Analysis and Algorithm Efficiency … 125
- 5.9 Problems … 127

6 The Gibbs Sampler — 129
- 6.1 Gibbs Sampling Algorithms … 129
- 6.2 Illustrative Examples … 131
- 6.3 Some Special Samplers … 133
  - 6.3.1 Slice sampler … 133
  - 6.3.2 Metropolized Gibbs sampler … 133
  - 6.3.3 Hit-and-run algorithm … 134
- 6.4 Data Augmentation Algorithm … 135
  - 6.4.1 Bayesian missing data problem … 135
  - 6.4.2 The original DA algorithm … 136
  - 6.4.3 Connection with the Gibbs sampler … 137
  - 6.4.4 An example: Hierarchical Bayes model … 138
- 6.5 Finding Repetitive Motifs in Biological Sequences … 139
  - 6.5.1 A Gibbs sampler for detecting subtle motifs … 140
  - 6.5.2 Alignment and classification … 141
- 6.6 Covariance Structures of the Gibbs Sampler … 143
  - 6.6.1 Data Augmentation … 143
  - 6.6.2 Autocovariances for the random-scan Gibbs sampler … 144
  - 6.6.3 More efficient use of Monte Carlo samples … 146
- 6.7 Collapsing and Grouping in a Gibbs Sampler … 146
- 6.8 Problems … 151

7 Cluster Algorithms for the Ising Model — 153
- 7.1 Ising and Potts Model Revisit … 153
- 7.2 The Swendsen-Wang Algorithm as Data Augmentation … 154
- 7.3 Convergence Analysis and Generalization … 155
- 7.4 The Modification by Wolff … 157
- 7.5 Further Generalization … 157
- 7.6 Discussion … 158

8 General Conditional Sampling — 161
- 8.1 Partial Resampling — 161
- 8.2 Case Studies for Partial Resampling — 163
  - 8.2.1 Gaussian random field model — 163
  - 8.2.2 Texture synthesis — 165
  - 8.2.3 Inference with multivariate t-distribution — 169
- 8.3 Transformation Group and Generalized Gibbs — 171
- 8.4 Application: Parameter Expansion for Data Augmentation — 174
- 8.5 Some Examples in Bayesian Inference — 176
  - 8.5.1 Probit regression — 176
  - 8.5.2 Monte Carlo bridging for stochastic differential equation — 178
- 8.6 Problems — 181

9 Molecular Dynamics and Hybrid Monte Carlo — 183
- 9.1 Basics of Newtonian Mechanics — 184
- 9.2 Molecular Dynamics Simulation — 185
- 9.3 Hybrid Monte Carlo — 189
- 9.4 Algorithms Related to HMC — 192
  - 9.4.1 Langevin-Euler moves — 192
  - 9.4.2 Generalized hybrid Monte Carlo — 193
  - 9.4.3 Surrogate transition method — 194
- 9.5 Multipoint Strategies for Hybrid Monte Carlo — 195
  - 9.5.1 Neal's window method — 195
  - 9.5.2 Multipoint method — 197
- 9.6 Application of HMC in Statistics — 198
  - 9.6.1 Indirect observation model — 199
  - 9.6.2 Estimation in the stochastic volatility model — 201

10 Multilevel Sampling and Optimization Methods — 205
- 10.1 Umbrella Sampling — 206
- 10.2 Simulated Annealing — 209
- 10.3 Simulated Tempering — 210
- 10.4 Parallel Tempering — 212
- 10.5 Generalized Ensemble Simulation — 215
  - 10.5.1 Multicanonical sampling — 216
  - 10.5.2 The $1/k$-ensemble method — 217
  - 10.5.3 Comparison of algorithms — 218
- 10.6 Tempering with Dynamic Weighting — 219
  - 10.6.1 Ising model simulation at sub-critical temperature — 221
  - 10.6.2 Neural network training — 222

(7.7 Problems — 159)

11 Population-Based Monte Carlo Methods — 225

- 11.1 Adaptive Direction Sampling: Snooker Algorithm — 226
- 11.2 Conjugate Gradient Monte Carlo — 227
- 11.3 Evolutionary Monte Carlo — 228
  - 11.3.1 Evolutionary movements in binary-coded space — 230
  - 11.3.2 Evolutionary movements in continuous space — 231
- 11.4 Some Further Thoughts — 233
- 11.5 Numerical Examples — 235
  - 11.5.1 Simulating from a bimodal distribution — 235
  - 11.5.2 Comparing algorithms for a multimodal example — 236
  - 11.5.3 Variable selection with binary-coded EMC — 237
  - 11.5.4 Bayesian neural network training — 239
- 11.6 Problems — 242

12 Markov Chains and Their Convergence — 245

- 12.1 Basic Properties of a Markov Chain — 245
  - 12.1.1 Chapman-Kolmogorov equation — 247
  - 12.1.2 Convergence to stationarity — 248
- 12.2 Coupling Method for Card Shuffling — 250
  - 12.2.1 Random-to-top shuffling — 250
  - 12.2.2 Riffle shuffling — 251
- 12.3 Convergence Theorem for Finite-State Markov Chains — 252
- 12.4 Coupling Method for General Markov Chain — 254
- 12.5 Geometric Inequalities — 256
  - 12.5.1 Basic setup — 257
  - 12.5.2 Poincaré inequality — 257
  - 12.5.3 Example: Simple random walk on a graph — 259
  - 12.5.4 Cheeger's inequality — 261
- 12.6 Functional Analysis for Markov Chains — 263
  - 12.6.1 Forward and backward operators — 264
  - 12.6.2 Convergence rate of Markov chains — 266
  - 12.6.3 Maximal correlation — 267
- 12.7 Behavior of the Averages — 269

13 Selected Theoretical Topics — 271

- 13.1 MCMC Convergence and Convergence Diagnostics — 271
- 13.2 Iterative Conditional Sampling — 273
  - 13.2.1 Data augmentation — 273
  - 13.2.2 Random-scan Gibbs sampler — 275
- 13.3 Comparison of Metropolis-Type Algorithms — 277
  - 13.3.1 Peskun's ordering — 277
  - 13.3.2 Comparing schemes using Peskun's ordering — 279
- 13.4 Eigenvalue Analysis for the Independence Sampler — 281
- 13.5 Perfect Simulation — 284
- 13.6 A Theory for Dynamic Weighting — 287
  - 13.6.1 Definitions — 287

		13.6.2	Weight behavior under different scenarios	288

           13.6.2 Weight behavior under different scenarios 288
           13.6.3 Estimation with weighted samples 291
           13.6.4 A simulation study 292

A Basics in Probability and Statistics   295

  A.1 Basic Probability Theory 295
        A.1.1 Experiments, events, and probability 295
        A.1.2 Univariate random variables and their properties . . 296
        A.1.3 Multivariate random variable 298
        A.1.4 Convergence of random variables 300
  A.2 Statistical Modeling and Inference 301
        A.2.1 Parametric statistical modeling 301
        A.2.2 Frequentist approach to statistical inference 302
        A.2.3 Bayesian methodology 304
  A.3 Bayes Procedure and Missing Data Formalism 306
        A.3.1 The joint and posterior distributions 306
        A.3.2 The missing data problem 308
  A.4 The Expectation-Maximization Algorithm 310

References   313

Author Index   333

Subject Index   338

1
Introduction and Examples

1.1 The Need of Monte Carlo Techniques

An essential part of many scientific problems is the computation of integral

$$I = \int_D g(\mathbf{x})d\mathbf{x},$$

where D is often a region in a high-dimensional space and $g(\mathbf{x})$ is the target function of interest. If we can draw independent and identically distributed (i.i.d.) random samples $\mathbf{x}^{(1)}, \ldots, \mathbf{x}^{(m)}$ uniformly from D (by a computer), an approximation to I can be obtained as

$$\hat{I}_m = \frac{1}{m}\{g(\mathbf{x}^{(1)}) + \cdots + g(\mathbf{x}^{(m)})\}.$$

The *law of large numbers* states that the average of many independent random variables with common mean and finite variances tends to stabilize at their common mean (see the Appendix); that is,

$$\lim_{m \to \infty} \hat{I}_m = I, \text{ with probability 1.}$$

Its convergence rate can be assessed by the *central limit theorem* (CLT):

$$\sqrt{m}(\hat{I}_m - I) \to N(0, \sigma^2), \text{ in distribution,}$$

where $\sigma^2 = \text{var}\{g(\mathbf{x})\}$. Hence, the "error term" of this Monte Carlo approximation is $O(m^{-1/2})$, regardless of the dimensionality of $\mathbf{x}$. This basic setting underlies the potential role of the Monte Carlo methodology in science and statistics.

In the simplest case when $D = [0,1]$ and $I = \int_0^1 g(x)dx$, one can approximate I by

$$\tilde{I}_m = \frac{1}{m}\{g(b_1) + \cdots + g(b_m)\},$$

where $b_j = j/m$. This method can be called the *Riemann approximation*. When g is reasonably smooth, the Riemann approximation gives us an error rate of $O(m^{-1})$, better than that of the Monte Carlo method. More sophisticated methods such as Simpson's rule and the Newton-Cotes rules give better numerical approximations (Thisted 1988). However, a fatal defect of these deterministic methods is that they do not scale well as the dimensionality of D increases. For example, in a 10-dimensional space with $D = [0,1]^{10}$, we will have to evaluate $O(m^{10})$ grid points in order to achieve an accuracy of $O(m^{-1})$ in the Riemann approximation of I. In contrast, the naive Monte Carlo approach, which draws $x^{(1)}, \ldots, x^{(m)}$ uniformly from D, has an error rate $O(m^{-1/2})$ regardless of the dimensionality of D, at least theoretically.

Although the "error rate" of a Monte Carlo integration scheme remains the same in high-dimensional problems, two intrinsic difficulties arise: (a) when the region D is large in high-dimensional space, the variance σ^2, which measures how "uniform" the function g is in region D, can be formidably large; (b) one may not be able to produce uniform random samples in an arbitrary region D. To overcome these difficulties, researchers often employ the idea of *importance sampling* in which one generates random samples $\mathbf{x}^{(1)}, \ldots, \mathbf{x}^{(m)}$ from a nonuniform distribution $\pi(\mathbf{x})$ that puts more probability mass on "important" parts of the state space D. One can estimate integral I as

$$\hat{I} = \frac{1}{m}\sum_{j=1}^m \frac{g(\mathbf{x}^{(j)})}{\pi(\mathbf{x}^{(j)})},$$

which has a variance $\sigma_\pi^2 = \text{var}_\pi\{g(\mathbf{x})/\pi(\mathbf{x})\}$. In the most fortunate case, we may choose $\pi(\mathbf{x}) \propto g(\mathbf{x})$ when g is non-negative and I is finite, which results in an exact estimate of I. But in no known application of the Monte Carlo method has this "luckiest situation" ever occurred. More realistically, we may hope to find a good "candidate" π which will explore more in regions where the value of g is high. In such a situation, generating random draws from π can be a challenging problem.

Demands for sampling from a nonuniform distribution π are also seen from another set of problems in bioinformatics, computational chemistry, physics, structural biology, statistics, etc. In these problems, the desired probability distribution $\pi(\mathbf{x})$ of a complex system, where $\mathbf{x}$ is often called a *configuration* of the system, arises from basic laws in physics and statistical inference. For example, in the study of a macromolecule, **x** may represent the *structure* of a molecule in the form of three-dimensional coordinates of all the atoms in the molecule. The target probability distribution is defined by the Boltzmann distribution $\pi(\mathbf{x}) = Z(T)e^{-h(\mathbf{x})/kT}$, where k is the Boltzmann constant, T is the system's temperature, $h(\mathbf{x})$ is the energy function, and $Z(T)$ is the *partition function* which is difficult to compute. Scientists are often interested in certain "average characteristics" of the system, many of which can be expressed mathematically as $E_\pi[g(\mathbf{x})]$ for a suitable function g. In Bayesian statistical inference, **x** often represents the joint configuration of missing data and parameter values and $\pi(\mathbf{x})$ is usually the posterior distribution of these variables. One has to integrate out nuisance parameters and the missing data so as to make a proper inference on the parameter of interest and to make valid predictions for future observations. These tasks can, once again, be expressed as computing the expectation of a function of the configuration space.

Sometimes, an optimization problem can also be formulated as a Monte Carlo sampling problem. Suppose we are interested in finding the minimum of a target function, $h(\mathbf{x})$, defined on a possibly complex configuration space. The problem is equivalent to finding the maximum of another function, $q_T(\mathbf{x}) = e^{-h(\mathbf{x})/T}$ (as long as $T > 0$). In the case when $q_T(\mathbf{x})$ is integrable for all $T > 0$, which is most common in practice, we can make up a family of probability distributions:

$$\pi_T(\mathbf{x}) \propto e^{-h(\mathbf{x})/T}, \quad T > 0.$$

If we can sample from $\pi_T(\mathbf{x})$ when T is sufficiently small, resulting random draws will most likely be located in the vicinity of the global minimum of $h(\mathbf{x})$. This consideration is the basis of the well-known simulated annealing algorithm (Kirkpatrick et al. 1983) and is also key to the *tempering* techniques for designing more efficient Monte Carlo algorithms (Chapter 10).

1.2 Scope and Outline of the Book

A fundamental step in all Monte Carlo methods is to generate (pseudo-) random samples from a probability distribution function $\pi(\mathbf{x})$, often known only up to a normalizing constant. The variable of interest **x** usually takes value in $\mathbb{R}^k$, but occasionally can take value in other spaces such as a permutation or transformation group (Diaconis 1988, Liu and Wu 1999). In most applications, directly generating independent samples from the target distribution π is not feasible. It is often the case that either the generated samples have to be dependent or the distribution used to generate the samples is different from π, or both. The rejection method (von Neumann 1951), importance sampling (Marshall 1956), and sampling-importance-resampling (SIR) (Rubin 1987) are schemes that make use of samples generated from a *trial distribution* $p(\mathbf{x})$, which differs from, but should be similar to, the target distribution π. The Metropolis algorithm (Metropolis et al. 1953) which, together with Hastings's (1970) generalizations, serves as the basic building block of Markov chain Monte Carlo (MCMC), is the one that generates dependent samples from a Markov chain with π as its equilibrium distribution. In other words, MCMC is essentially a Monte Carlo integration procedure in which the random samples are produced by evolving a Markov chain.

Because of the great potential of Monte Carlo methodology, various techniques have been developed by researchers in their respective fields. Recent advances in Monte Carlo techniques include the cluster method, data augmentation, parameter expansion, multicanonical sampling, multigrid Monte Carlo (MGMC), umbrella sampling, density-scaling Monte Carlo, simulated tempering, parallel tempering, hybrid Mont Carlo (HMC), multiple try Metropolis (MTM), sequential Monte Carlo, particle filtering, etc. There is also a trend in moving toward a population-based approach. These advances in one way or another were all motivated by the need to sample from very complex probability distributions for which the standard Metropolis method tends to be trapped in a local "energy" well. Many of these methods are related, and some are even identical. For example, the configurational bias Monte Carlo (Siepmann and Frenkel 1992) is equivalent to a sequential importance sampling combined with a Metropolized independence sampler (Chapters 2 & 3); the exchange Monte Carlo (Hukushima and Nemoto 1996) is reminiscent of *parallel tempering* (Geyer 1991); the multiple-try Metropolis (Liu, Liang and Wong 2000) generalizes a method described by Frenkel and Smit (1996); the parameter expansion (Liu and Wu 1999) recently developed is a special case of the *partial resampling* technique (Goodman and Sokal 1989); and the bootstrap filter and sequential imputation (Gordon, Salmond and Smith 1993, Kong, Liu and Wong 1994) can be traced back to the "growth method" (Hammersley and Morton 1954, Rosenbluth and Rosenbluth 1955). By providing a systematic account of these methods, this book focuses on the following aspects: understanding the properties and characteristics of these methods, revealing their connections and differences, comparing their performances and proposing generalizations, and demonstrating their use in scientific and statistical problems.

The remaining part of this chapter presents motivating examples in statistical physics, molecular simulation, bioinformatics, dynamic system analysis, statistical hypothesis testing, Bayesian inference for hierarchical models, and other statistical missing data problems.

Chapter 2 covers basic Monte Carlo techniques including the inversion method, rejection sampling, antithetic sampling, control variate method, stratified sampling, importance sampling, and the exact sampling method for chain-structured models. The last method is usually not covered by the standard Monte Carlo books but is becoming increasingly important in modern statistical analysis, artificial intelligence, and computational biology. Special attention is given to methods related to importance sampling (e.g., that for solving linear equations, for phylogenetic analysis, and for Bayesian inference with missing data).

Chapter 3 explains in detail the origin and the theoretical framework of sequential Monte Carlo. Started with the Monte Carlo treatment of a self-avoid random walk by Morton, Hammersley, Rosenbluth, and Rosenbluth, in the 1950s, this chapter shows the reader an important common structure in seemingly unrelated problems, such as polymer simulation and Bayesian missing data problems. A general methodology built upon sequential importance sampling, resampling (or pruning and enrichment), and rejection sampling is described to generalize the methods used in the polymer simulation and Bayesian missing data problems. Chapter 4 illustrates how sequential Monte Carlo methods can be used in different problems such as molecular simulation, population genetics, computational biology, nonparametric Bayes analysis, approximating permanents, target tracking, and digital communications.

The later chapters focus primarily on Markov chain based dynamic Monte Carlo strategies. Chapter 5 introduces the basic building block of almost all Markov chain Monte Carlo strategies — the Metropolis-Hastings transition rule. A few recent generalizations of the rule, such as the multipoint rule, the reversible jumping rule, and the dynamic weighting rule are described so that the reader can be equipped with a full array of basic tools in designing a MCMC sampler. The basic method for analyzing efficiency of a MCMC algorithm is described at the end of this chapter.

Chapters 6-8 analyze and generalize another main class of Monte Carlo Markov chains — those built upon iterative sampling from conditional distributions. A prominent special case is the Gibbs sampler, which is now a standard tool for statistical computing. Data augmentation, which was originally designed for solving statistical missing data problems, is recasted as a strategy for improving the ease of computation and the convergence speed of a MCMC sampler. The cluster algorithm for the Ising model (Swendsen and Wang 1987) is treated under this unified view. Another important generalization is the view of partial resampling (Goodman and Sokal 1989), which can be used to conduct more global moves based on a transformation group formulation (Liu and Sabatti 2000, Liu and Wu 1999). The analytical form of the required conditional distributions in this general setting, as well as its applications in Gaussian random field, texture modeling, probit regression, and stochastic differential equations, are given in Chapter 8.

Starting with the basic Newtonian mechanics, Chapter 9 introduces the method of hybrid Monte Carlo (HMC), a means to construct Markov chain moves by evolving Hamiltonian equations. This method reveals the close connection between Monte Carlo and molecular dynamics algorithms, the latter having been one of the most widely used tools in structural biology and theoretical chemistry. A few strategies for improving the efficiency of an

HMC or MC algorithm are discussed; these include the surrogate transition method, the window method, and the multipoint method. We also want to draw the reader's attention to unconventional uses of HMC, especially its application in statistical problems.

Chapters 10 and 11 discuss a few recent developments for efficient Monte Carlo sampling. Incidentally, many of these new methods rely on the idea of running multiple Monte Carlo Markov chains in parallel. Mechanisms that enable communications among the multiple chains are incorporated into the samplers so as to speed up their exploration of the configuration space. These techniques can be grouped into three main classes: temperature-based methods (simulated tempering, parallel tempering, and simulated annealing), reweighting methods (umbrella sampling, multicanonical sampling, and $1/k$-ensemble method), and evolution-based methods (adaptive direction sampling, conjugate gradient Monte Carlo, and evolutionary Monte Carlo). Some of these approaches can be combined so as to produce a more efficient sampler. For many important scientific problems, these new techniques are indispensable.

Chapter 12 provides a basic theory for the general Markov chain and a few analytical techniques for studying its convergence rate. The basic theory includes the Chapman-Kolmogorov equation and the geometric convergence theorem for finite-state Markov chains. For advanced techniques, we show how to use the coupling method, the Poincaré inequalities, and Cheeger's inequality to bound the second largest eigenvalue of the Markov chain transition matrix and how to use the basic functional analysis tools to get a qualitative understanding of the Markov chain's convergence.

Chapter 13 selects a few theoretical topics in the analysis of Monte Carlo Markov chain. The chapter starts with short discussion of the general convergence issue in MCMC sampling and proceeds to an analysis of the covariance structures of data augmentation and the random-scan Gibbs sampler, showing that these structures can be used to gain insight into the design of an efficient Gibbs sampler. Peskun's theorem is described and used to compare different Metropolis samplers. A complete eigenstructure analysis for the Metropolized independence sampler is provided. One of the most exciting recent advances in Markov chain theory, the so-called *perfect simulation*, is briefly described and the related literature mentioned. Finally, a theoretical analysis of the dynamic weighting method is given.

In the Appendix, we outline the basics of the probability theory and statistical inference procedures. The interested reader can also find there a rather short description of the popular expectation-maximization (EM) algorithm (Dempster, Laird and Rubin 1977) and a brief discussion of its property.

1.3 Computations in Statistical Physics

Scientists are often interested in simulating from a *Boltzmann distribution* (or Gibbs distribution) which is of the form

$$\pi(\mathbf{x}) = \frac{1}{Z} e^{-U(\mathbf{x})/kT}, \qquad (1.1)$$

where $\mathbf{x}$ is a particular configuration of a physical system, $U(\mathbf{x})$ is its potential energy, T is the temperature, and k is the Boltzmann constant. The function $Z = Z(T)$ is called the *partition function* (also called the normalizing constant in non-physics literature).

The Ising model serves to model the behavior of a magnet and is perhaps the best known and the most thoroughly researched model in statistical physics. The intuition behind the model is that the magnetism of a piece of material is the collective contribution of dipole moments of many atomic spins within the material. A simple 2-D Ising model places these atomic spins on a $N \times N$ lattice space, $\mathcal{L} = \{(i,j), i = 1, \ldots, N; j = 1, \ldots, N\}$, as shown in Figure 1.1.

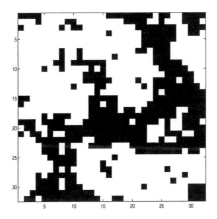

FIGURE 1.1. A configuration of the Ising model on a 32 × 32 grid space, denoted as $\mathcal{L}$, with a temperature slightly higher than the critical temperature.

In the model, each site $\sigma \in \mathcal{L}$ hosts a particle that has either a positive or a negative spin. Abstractly, the state of each particle can be represented by a random variable x_σ which is either +1 or −1. A configuration of the whole system is then $\mathbf{x} = \{x_\sigma, \sigma \in \mathcal{L}\}$, whose potential energy is defined as

$$U(\mathbf{x}) = -J \sum_{\sigma \sim \sigma'} x_\sigma x_{\sigma'} + \sum_\sigma h_\sigma x_\sigma, \qquad (1.2)$$

where the symbol $\sigma \sim \sigma'$ means that they are a neighboring pair, J is called the interaction strength, and h_σ the external magnetic field.

Several important quantities regarding a physical system are often of interest. First, the *internal energy* is defined as

$$\langle U \rangle = E_\pi \{U(\mathbf{x})\}.$$

The notation on the left-hand side is frequently used by physicists to refer to the "state average" of the potential energy, whereas the right-hand side notation is employed mostly by mathematicians and statisticians and is read as "the mathematical expectation of $U(\mathbf{x})$ with respect to π." They are, of course, the same thing and are equal to the integral

$$I = \int_D U(\mathbf{x})\pi(\mathbf{x})d\mathbf{x},$$

where D is the set of all possible configurations of $\mathbf{x}$. Clearly, estimating $\langle U \rangle$ based on random samples drawn uniformly from D is disastrous. A much better estimate would have been resulted if we could simulate random draws from the Boltzmann distribution $\pi(\mathbf{x})$.

If we let $\beta = 1/kT$, an interesting relationship between the internal energy and the *partition function* can be derived:

$$\frac{\partial \log Z}{\partial \beta} = -\langle U \rangle.$$

This implies that we can use Monte Carlo methods to estimate the temperature derivative of the partition function, which can then be used to estimate the partition function itself (up to a multiplicative constant). The *free energy* of the system, defined as $F = -kT \log Z$, can also be estimated, up to an additive constant. To date, problems related to the estimation of partition functions of various probabilistic systems still present a significant challenge to researchers in different fields (Meng and Wong 1996, Chen and Shao 1997).

The *specific heat* of the system is defined as

$$C = \frac{\partial \langle U \rangle}{\partial T} = \frac{1}{kT^2} \text{var}_\pi U(\mathbf{x}),$$

and the system's *entropy* is

$$S = (\langle U \rangle - F)/T.$$

For the Ising model, one is also interested in the *mean magnetization per spin*, defined as

$$\langle m \rangle = E_\pi \left\{ \frac{1}{N^2} \left| \sum_{\sigma \in S} x_\sigma \right| \right\},$$

which can, again, be estimated by Monte Carlo sample averages. Generally, many physical quantities of interest correspond to taking expectations with respect to the Boltzmann distribution and can be estimated by Monte Carlo simulations.

1.4 Molecular Structure Simulation

Simple Liquids Model. In this model, the configuration space is a compact subset in $\mathbb{R}^{3k}$: $\mathbf{x} = \{x_i;\ i = 1, \ldots, k\}$, where $x_i = (x_{i1}, x_{i2}, x_{i3})^T$ represents the position vector of the ith particle. A simple energy function is of the form

$$U(\mathbf{x}) = \sum_{\text{all } i,j} \Phi(|x_i - x_j|) = \sum_{\text{all } i,j} \Phi(r_{ij}),$$

where

$$\Phi(r) = 4\epsilon \left[\left(\frac{\sigma}{r}\right)^{12} - \left(\frac{\sigma}{r}\right)^{6} \right]$$

is called the *Lennard-Jones pair potential*. Its shape is depicted in Figure 1.2.

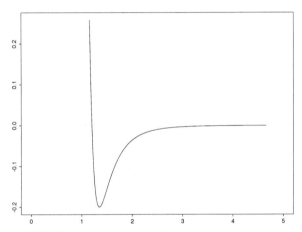

FIGURE 1.2. A plot of the Lennard-Jones function.

A more complicated model for macromolecules, which is widely used in protein structural simulations (Creighton 1993), has a potential energy function of the form

$$U(\mathbf{x}) = \sum_{\text{bonds}} (\text{bond terms}) + \sum_{i,j} \left\{ \Phi(r_{ij}) + \frac{q_i q_j}{4\pi\epsilon_0 r_{ij}} \right\},$$

where the last term on the right-hand side represents electrostatic interaction between two atoms. In macromolecule simulations, one often uses three bond terms that have the form

$$\text{bond terms} = \sum_{\text{bonds}} \frac{k_i}{2}(l_i - l_{i,0})^2 + \sum_{\text{angles}} \frac{k_i}{2}(\theta_i - \theta_{i,0})^2 + \sum_{\text{torsions}} v(\omega_i),$$

where l_i is the bond length, θ_i is the *bond angle*[1], and ω_i is the *torsion angle*[2]. The torsional term $v(\omega)$ has the form

$$v(\omega) = \frac{V_n}{2}(1 + \cos(n\omega - \gamma)). \qquad (1.3)$$

Also see Leach (1996) for more details. A water molecule is shown in Figure 1.3.

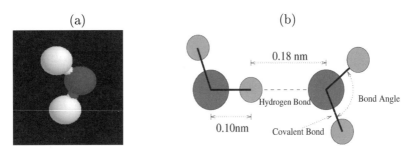

FIGURE 1.3. (a) A water molecule generated by Rasmol, a molecular visualization software (http://www.umass.edu/microbio/rasmol/). There are two covalent bonds in H_2O, one between each hydrogen atom and the oxygen atom. (b) A schematic plot of the interactions between two water molecules.

Figure 1.4 shows how a protein molecule interacts with the double-helix structure of a DNA molecule at the atomic level. This interaction is important for gene regulation. In this class of problems, one is often interested in seeing the likely structures of a stable macromolecule and estimating several basic physical quantities such as free energy and specific heat. In protein folding problems, one is sometimes more interested in finding the "minimal-energy" configuration of the system [i.e., finding the configuration $\mathbf{x}_0$ that minimized $U(\mathbf{x})$].

1.5 Bioinformatics: Finding Weak Repetitive Patterns

The linear biopolymers, DNA, RNA, and proteins, are the three central molecular building blocks of life. DNA is an information storage molecule.

---

[1] The *bond angle* is defined as the angle between the lines connecting the nucleus of one atom to the nuclei of two other atoms that are bonded to it.

[2] The *torsion angle* is the angle of rotation about the bond B-C in a series of bonded atoms A-B-C-D needed to make all the four atoms be on the same plane (remember that they are in a three-dimensional space). The positive sense is clockwise. If the torsion angle is 180°, the four atoms lie in a planar zigzag (Z-shaped).

1.5 Bioinformatics: Finding Weak Repetitive Patterns

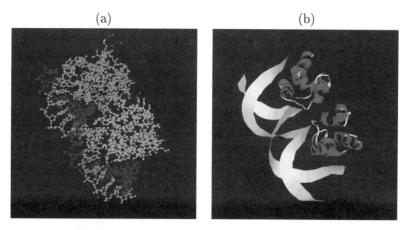

FIGURE 1.4. (a) A ball-and-stick plot of the interaction between 3CRO, a regulatory protein in bacteriophage 434, and the DNA segment to which it binds. (b) The same structure as in (a), but expressed by a ribbon representation widely used in the protein structure modeling community.

All of the hereditary information of an individual organism is contained in its genome, which consists of sequences of the four DNA bases (nucleotides), A, T, C, and G. RNA has a wide variety of roles, including a small but important set of functions. Proteins, which are chains of 20 different amino acid residues, are the action molecules of life, being responsible for nearly all of the functions of all living beings and forming many of life's structures. All protein sequences are coded by segments of the genome called genes. The universal genetic code is used to translate triplets of DNA bases, called *codons*, to the 20-letter alphabet of proteins (Campbell 1999). For example, codons "CCA" and "CCG" are both translated into the amino acid "Proline" (abbreviated as Pro or P). How genetic information flows from DNA to RNA and then to protein is regarded as the central paradigm of molecular biology.

The human genome and many other genome sequencing projects have resulted in rapidly growing and publicly available databases of DNA and protein sequences (e.g., http://www.ncbi.nlm.nih.gov). The data in these databases are sequences of letters using d-letter ($d = 4$ for DNA or $d = 20$ for proteins) alphabets without punctuation or space characters. One of the most interesting questions scientists are concerned with is how to get any useful biological information from "looking" at these databases. This task is often termed "data mining" for other types of data. The recent announcement of the near-completion of the human genome makes this interesting question more an urgent task for all interested scientists. However, "mining" a biopolymer database is noticeably different from mining other types of databases because (i) many sophisticated structures have

12   1. Introduction and Examples

been built in well-organized biopolymer databases [one can take a look at NCBI's GeneBank (whose web address is *http://www.ncbi.nlm.nih.gov*) to have a rough idea], (ii) there is an enormous amount of biological knowledge, and (iii) fundamental laws in physics and chemistry can be applied. Consequently, more sophisticated mathematical/statistical models are often critical in developing a "mining" strategy.

One important problem in analyzing biological sequence data is to find patterns shared by multiple protein or DNA sequences. It is closely related to the task of *local sequence alignment*. The fact that a common pattern appears in several otherwise dissimilar protein sequences often indicates that these proteins may be functionally or structurally related. For example, in Figure 1.4(b), one can see that a helical part of protein 3CRO is inside the major groove of a DNA double-helix structure. This helical part of the protein, often referred to as the *helix-turn-helix* motif, plays an important role in the binding of 3CRO to a DNA segment and turns out to be a rather conserved part in a large family of proteins responsible for gene regulation. In another example, as shown in Figure 1.5, the helix in the light color is the pattern shared by a number of dinucleotide binding proteins; this helix segment is important in interacting with the ligand, dinucleotide. This pattern was discovered by a stochastic search algorithm, the *Gibbs motif sampler* (Liu, Neuwald and Lawrence 1995) to be described later.

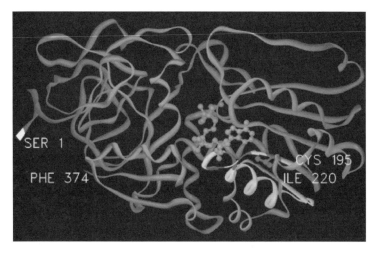

FIGURE 1.5. A ribbon model of the trace of the backbone chain of a dinucleotide binding protein, ADHE. The helical segment in light color (residues 195-220) corresponds to the pattern common to a number of such proteins. The cofactor (NAD) is represented by a stick-and-ball model.

Common patterns in DNA sequences also have important biological implications. For example, the existence of a common short sequence motif

1.5 Bioinformatics: Finding Weak Repetitive Patterns 13

in the upstream untranslated DNA regions (5' UTR) of a set of candidate genes may suggest that these genes are regulated in a similar fashion. This motif pattern, whose occurrences are often very close to the start of a gene (less than a few hundreds base pairs), may correspond to the sites bound by a common regulatory protein, called the transcription factor,[3] which regulates the expression level of these genes (Stormo and Hartzell 1989, Lawrence and Reilly 1990, Lawrence et al. 1993, Liu 1994a). These short patterns are often called *binding motifs.* In Figure 1.4(b), the segment of DNA (double-helix structure) that are interacting with protein 3CRO is a *binding site* whose pattern is conserved among the 5' UTRs of a number of related genes.

The above pattern discovery problem can be abstracted as follows: Given a set of K sequences $R_1, \ldots, R_K$, we seek within each sequence mutually similar segments of a specified width w. An analogy is that each R_i is a sentence and our task is to find some "word" (or words) that is most "common" to all the sentences in consideration. If this word occurs in each sequence exactly without any misspellings, one can find it without too much difficulties. But in the biology world, there is seldom anything as good as "exact" matches, implying that we have to describe both the common pattern and the remaining parts of the sequences probabilistically. The simplest statistical model is the *block-motif* model as depicted in Figure 1.6:

FIGURE 1.6. A schematic plot of the block-motif model used for our pattern finding.

In this model, the letter at the ith position within the "motif" (pattern) is assumed to be drawn independently from a multinomial distribution with parameter $\boldsymbol{\theta}_i$, $i = 1, \ldots, w$; letters elsewhere follow a common background multinomial model with parameter $\boldsymbol{\theta}_0$, where $\boldsymbol{\theta}_i$ is a probability vector of length d, the size of the letter alphabet ($d = 4$ for DNA and $d = 20$ for protein). In other words, the residues (or base pairs) we see outside the motif pattern are treated as i.i.d. observations from the multinomial distribution with parameter $\boldsymbol{\theta}_0$; a residue observed at position i within the motif pattern is generated from probability vector $\boldsymbol{\theta}_i$. [It is possible to use a more sophisticated background model to improve the algorithm's efficiency, see Liu, Brutlag and Liu (1995) and McCue et al. (2001).] What makes this problem difficult is that we do not know the location of the "word"

---

[3] A transcription factor is a protein which binds to certain site of a DNA molecule to either enhance or repress the gene expression.

(pattern) in each sequence. Based on this simple statistical model and the Gibbs sampling principle, Liu (1994a) and Lawrence et al. (1993) derived the following simple, yet effective, Monte Carlo algorithm (Chapter 6).

In what they called the *site sampler*, the motif locations (sites) are initialized at random; that is, position $a_k^{(0)}$ (for $k = 1, \ldots, K$) is a randomly chosen position of the kth sequence. For $t = 1, \ldots, m$, the algorithm iterates the following steps:

- Select a sequence, say the kth sequence, either deterministically or at random.

- Draw a new motif location a_k according to the predictive distribution

$$P(a_k \mid a_1^{(t)}, \ldots, a_{k-1}^{(t)}, a_{k+1}^{(t)}, \ldots, a_K^{(t)}) \qquad (1.4)$$

and update the current motif location $a_k^{(t)}$ to $a_k^{(t+1)} = a_k$.

- Let $a_j^{(t+1)} = a_j^{(t)}$ for $j \neq k$.

Although there are many choices of the predictive distribution function used for updating the alignment [e.g., one can let $P(a_k \mid \ldots)$ be proportional to certain fitness measure of the segment indexed by a_k to the current multinomial profile resulting from the $a_j^{(t)}$, $j \neq k$], those that make the foregoing iteration consistent have to be the ones derived from a complete Bayesian statistical model. More detailed derivations will be given in Chapter 6.

1.6 Nonlinear Dynamic System: Target Tracking

Dynamic modeling is widely used in areas such as computer vision, economical data analysis, feedback control systems, mobile communication, radar or sonar surveillance systems, etc. An important problem in such a dynamical system is the on-line (instantaneously in real time) processing of information (such as estimation and prediction) regarding the system characteristics. These tasks are generally termed "filtering" in engineering and statistical literature. A main challenge to researchers in these fields is to find efficient filtering algorithms.

Target tracking in a clutter environment as shown in Figure 1.7 is a typical example of dynamic modeling. To facilitate tracking, one often uses a linear Gaussian *state-space model* [also called *dynamic linear model* by West and Harrison (1989)] to describe the movement of the target object. A typical 2-D tracking model is as follows.

1.6 Nonlinear Dynamic System: Target Tracking

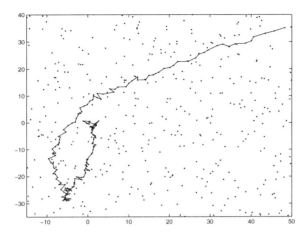

FIGURE 1.7. A simulation of the target tracking problem in a 2-D space. The dots connected by the line represent the signals (y_t) generated by the true positions of the target; the dots elsewhere represent the nearby confusing objects (simulated from a Poisson point process with rate 8%).

State Equation:

$$\begin{pmatrix} v_{t,1} \\ v_{t,2} \end{pmatrix} = \begin{pmatrix} v_{t-1,1} \\ v_{t-1,2} \end{pmatrix} + \begin{pmatrix} \epsilon_{t,1} \\ \epsilon_{t,2} \end{pmatrix},$$

$$\begin{pmatrix} s_{t,1} \\ s_{t,2} \end{pmatrix} = \begin{pmatrix} s_{t-1,1} \\ s_{t-1,2} \end{pmatrix} + \begin{pmatrix} v_{t-1,1} \\ v_{t-1,2} \end{pmatrix} + \frac{1}{2}\begin{pmatrix} \epsilon_{t,1} \\ \epsilon_{t,2} \end{pmatrix},$$

where $s_t = (s_{t,1}, s_{t,2})^T$ is the object's position vector at time t and $v_t = (v_{t,1}, v_{t,2})^T$ is its current velocity vector. The state equation innovation $\epsilon_t = (\epsilon_{t,1}, \epsilon_{t,2})^T$ is distributed as $\mathcal{N}(\mathbf{0}, \sigma_a^2 \mathbf{I})$. This model says that the speed (vector) of the object evolves like a Gaussian random walk and the position of the object follows the change of its speed. We assume, however, that a noisy version of the object's true position $(y_{t,1}, y_{t,2})^T$ is observable.

Observation Equation:

$$\begin{pmatrix} y_{t,1} \\ y_{t,2} \end{pmatrix} = \begin{pmatrix} s_{t,1} \\ s_{t,2} \end{pmatrix} + \begin{pmatrix} e_{t,1} \\ e_{t,2} \end{pmatrix},$$

where the observation noise $e_t = (e_{t,1}, e_{t,2})^T$ follows $\mathcal{N}(\mathbf{0}, \sigma_b^2 \mathbf{I})$.

By writing $x_t = (s_{t,1}, s_{t,2}, v_{t,1}, v_{t,2})^T$ and $y_t = (y_{t,1}, y_{t,2})^T$, we can rewrite the foregoing system more briefly as

$$\begin{aligned} x_t &= G x_{t-1} + \epsilon_t, \quad \epsilon_t \sim \mathcal{N}(\mathbf{0}, \sigma_a^2 A); \\ y_t &= H x_t + e_t, \quad e_t \sim \mathcal{N}(\mathbf{0}, \sigma_b^2 \mathbf{I}), \end{aligned}$$

where

$$G = \begin{pmatrix} 1 & 0 & 1 & 0 \\ 0 & 1 & 0 & 1 \\ 0 & 0 & 1 & 0 \\ 0 & 0 & 0 & 1 \end{pmatrix}, \; A = \begin{pmatrix} \frac{1}{4} & 0 & \frac{1}{2} & 0 \\ 0 & \frac{1}{4} & 0 & \frac{1}{2} \\ \frac{1}{2} & 0 & 1 & 0 \\ 0 & \frac{1}{2} & 0 & 1 \end{pmatrix}, \; \text{and} \; H = \begin{pmatrix} 1 & 0 \\ 0 & 1 \\ 0 & 0 \\ 0 & 0 \end{pmatrix}^T.$$

Mathematically, the tracking task is accomplished by *on-line* estimation of the object's position, $(s_{t,1}, s_{t,2})$, based on all information available until time t. If the y_t are always observable at time t (i.e., there is no clutter), this estimation task can be achieved rather efficiently via the *Kalman filter* (Kalman 1960) because of the linear Gaussian structures being employed.

If we consider a clutter environment, however, then at time t we observe instead a clutter of points, $z_t = \{z_{t,1}, \ldots, z_{t,k_t}\}$, in a 2-D detection region with area Δ in which the number of false signals follow a spatial Poisson process with rate λ. The set z_t includes the true measurement $y_t = (y_{t,1}, y_{t,2})^T$ with probability p_d. Other z's are treated as uniform within the detection range. This model for tracking in clutter is no longer a linear Gaussian system. To date, there has not been a universally effective algorithm for dealing with nonlinear and non-Gaussian systems. Depending on the features of individual problems, some generalizations of the Kalman filter can be effective. A few well-known generalizations are the extended Kalman filters (Gelb 1974), the Gaussian sum filters (Anderson and Moore 1979), and the iterated extended Kalman filters (Jazwinski 1970). Most of these methods are based on local linear approximations of the nonlinear system. More recently, researchers are attracted to a new class of filtering methods based on the sequential Monte Carlo approach. In Section 4.5.1, we will show that the method based on sequential Monte Carlo performs very well for the target tracking problem.

1.7 Hypothesis Testing for Astronomical Observations

Efron and Petrosian (1999) described an interesting problem in astronomy. Figure 1.8 shows us a set of doubly truncated astronomical data in which log-luminosities y_i (or the boundary point of its truncation interval) is plotted against redshifts z_i for $n = 210$ quasars. In other words, due to experimental constraint, we are able to observe y_i when it is within a known interval R_i depending on z_i, otherwise we only know that y_i is outside of the interval R_i. Two questions are of interest: (a) Are the y_i independent of the z_i? (b) Assuming independence, can we estimate the marginal distribution of the y_i?

The real data actually consist of independently collected quadruplets

$$(z_i, m_i, a_i, b_i), \quad i = 1, \ldots, n,$$

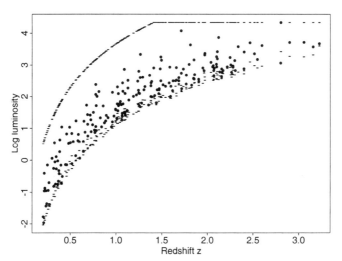

FIGURE 1.8. Doubly truncated data from an astronomical study of Efron and Petrosian (1999). Points represent (redshift, log-luminosity) for 210 quasars. Luminosity subject to upper and lower truncation is indicated by "−" in the figure.

where a_i and b_i are lower and upper truncation limits on m_i, respectively, and m_i is the apparent magnitude of the ith quasar. Quasars with apparent magnitude above b_i were too dim to produce usable redshift observations. The lower limit a_i was used to avoid confusion with nonquasar stellar objects. In the dataset, $a_i = 16.08$ for all i, and b_i is between 18.494 and 18.934. Since further quasars appear dimmer (and have larger values of m_i), one needs to use Hubble's law, which states that distance is proportional to redshift, to transform m_i into a luminosity measurement that should be independent of distance. The plotted y_i is thus derived as

$$y_i = 19.894 - 2.303\frac{m_i}{2.5} + 2\log(Z_i - \sqrt{Z_i}) - \frac{1}{2}\log(Z_i),$$

where $Z_i = 1 + z_i$ [more details are given in Efron and Petrosian (1999)].

To answer question (a), a simple permutation test was considered by Efron and Petrosian (1999). In other words, if we could observe the y_i without truncation, then under the null hypothesis that the y_i and z_i are independent, the permuted y_i should "look" the same as the old y_i. Thus, if we permute the y_i many times and compute the test statistic, say $T = t(\mathbf{y}^*)$, repeatedly, where $\mathbf{y}^* = (y_1^*, \ldots, y_n^*)$ is a permutation of the original y's, we can obtain the *"null distribution"* of T: that is, if the independence hypothesis were true, we would expect T to behave like what we see in the repeated permutations. Because of truncations, however, the $\mathbf{y}^*$ is observable only if all the $y_i^* \in R_i$ (i.e., satisfying the double truncation requirement).

Let $\mathcal{Y}$ be the set of all permutations that satisfy the truncation requirement. Then to compute the p-values for the statistical test, we need to be able to generate permutations of $(y_1,\ldots,y_n)$ uniformly in $\mathcal{Y}$. In order to achieve this sampling with a reasonable amount of computing time, one needs to employ the *Metropolis algorithm* (Chapter 5). Briefly, to obtain an approximately uniform sample from the set of all allowable permutations $\mathcal{Y}$, we can start with the identity permutation $\sigma(i) = i$, $i = 1,\ldots,n$. At each iteration step, we do the following:

- Randomly pick a pair of elements, i and j, say, and propose a new permutation σ' which differs from σ only by transposing $\sigma(i)$ and $\sigma(j)$.

- If the new permutation resulting from the transposition is still in $\mathcal{Y}$, then we accept the new configuration; otherwise, we stay put.

The validity of this algorithm can be understood in part by the Markov invariance property explained in Chapter 5. Diaconis, Graham and Holmes (2001) showed that the space $\mathcal{Y}$ is "connected" by the foregoing Markov moves. Thus, the equilibrium state of the algorithm follows the target distribution (i.e., uniform in $\mathcal{Y}$). In other words, if we carry out a large number (m=1,000,000, say) of iterations of the above steps, the latter m_0 (m_0=900,000, say) correlated permutations may be regarded as being drawn uniformly from $\mathcal{Y}$. These samples can be used to estimate the null distribution of T. An importance sampling approach can also be applied to achieve the same goal.

1.8 Bayesian Inference of Multilevel Models

A basic regression model used for the prediction of a student's first-year average (FYA) from the Law School Aptitude Test (LSAT) score and undergraduate grade point average (UGPA) is

$$\widehat{\text{FYA}} \propto \text{LSAT} + (?) \times \text{UGPA}.$$

However, the important question for each law school is how to choose its own multiplier of UGPA in this equation (Rubin 1980). In the past, a number of law schools simply chose 200 as the multiplier. But due to recent "grade inflation," some recommended smaller multipliers such as 130. Many law schools favored estimating the multiplier empirically, using recent data from attending students. Because different law schools may have a different education effect on its students, we do not expect that the multipliers used by different schools to be the same. On the other hand, one may also feel that all law schools have certain common characteristics and the multiplier used in one school should provide information for other schools to determine their own.

The dataset in Rubin (1980) consisted of records from 82 law schools. A standard linear regression was carried out for each school to obtain an initial estimate of the multiplier, $\hat{L}_i$. Then Rubin (1980) proceeded by assuming that $Y_i = \arctan(\hat{L}_i/200)$ follows a Gaussian distribution with mean θ_i and variance s_i^2, where s_i^2 is estimated from the ith law school's past record. A hierarchical structure is further imposed on the θ_i:

$$\theta_i \stackrel{\text{i.i.d.}}{\sim} N(\mu, \sigma^2).$$

The $\hat{\theta}_i$ estimated from this model can be used to calculate the multiplier [i.e., $\hat{L}_i = 200\tan(\hat{\theta}_i)$] for the ith school in its FYA prediction.

A more general form of the model Rubin used can be stated as follows:

$$Y_i \mid \theta_i \stackrel{\text{i.i.d.}}{\sim} f_i(y_i \mid \theta_i), \qquad (1.5)$$

$$\theta_i \stackrel{\text{i.i.d.}}{\sim} G(\theta \mid \lambda), \qquad (1.6)$$

for $i = 1, \ldots, k$. This model is often referred to as a "hierarchical model." Of interest in this model are the estimation of all the unknown parameters, θ_i and λ, and the quantification of uncertainties in these estimates. A *hierarchical Bayes model* is completed by giving a prior distribution $f_0(\mu, \sigma^2)$ to the hyper-parameter λ.

With notation $\boldsymbol{\theta} = (\theta_1, \ldots, \theta_{82})$, we can write down the joint distribution of the data and the parameters in Rubin's study:

$$p(Y, \boldsymbol{\theta}, \mu, \sigma^2) = f_0(\mu, \sigma) \prod_{i=1}^{82} \phi(Y_i; \theta_i, s_i) \phi(\theta_i; \mu, \sigma),$$

where $\phi(x; \mu, \sigma)$ is the density function for $N(\mu, \sigma^2)$. By the Bayes theorem, we derive the *posterior distribution* of all the unknown variables:

$$\pi(\boldsymbol{\theta}, \mu, \sigma^2) = \frac{p(Y, \boldsymbol{\theta}, \mu, \sigma^2)}{p(Y)} \propto p(Y, \boldsymbol{\theta}, \mu, \sigma^2).$$

The *Bayes estimator* of θ_i is its posterior mean:

$$\hat{\theta}_i = E(\theta_i \mid Y) = \int \pi(\boldsymbol{\theta}, \mu, \sigma^2) d\boldsymbol{\theta}_{[-i]} d\mu d\sigma^2,$$

where $\boldsymbol{\theta}_{[-i]}$ is all but the ith component of $\boldsymbol{\theta}$. This quantity is not analytically available. Its numerical approximation needs high-dimensional integration and can be most effectively solved by Monte Carlo techniques (Gelfand and Smith 1990).

1.9 Monte Carlo and Missing Data Problems

A central theme of statistics is to infer unobserved parameters from observed data. Let $\boldsymbol{\theta}$ be the parameter vector of interest. In a *parametric* inference problem, the observed vector **y** is regarded as the realized value of a random vector whose distribution is $f(\mathbf{y} \mid \boldsymbol{\theta})$, known up to a finite-dimensional parameter $\boldsymbol{\theta}$. Two of the most popular approaches for inferring $\boldsymbol{\theta}$ are the *maximum likelihood estimation* (MLE) method and the *Bayes* method. In the MLE, the unknown parameter $\boldsymbol{\theta}$ is estimated by the $\hat{\boldsymbol{\theta}}$ that maximizes $f(\mathbf{y} \mid \boldsymbol{\theta})$, which is also called the *likelihood*. Computationally, this becomes an optimization problem. In the Bayesian methods, one relies on the posterior distribution of $\boldsymbol{\theta}$, $p(\boldsymbol{\theta} \mid \mathbf{y})$, to make inferential statement, which can be formulated as a high-dimensional integration problem. (See Section A.2 of the Appendix for more details.)

Many statistical problems do not naturally suggest "nice" models enabling simple analytical solutions to the posterior calculation. However, a tractable structure can often be obtained if some auxiliary parts are augmented to the system. In the field of statistics, these auxiliary components can often be viewed as "missing data" (Tanner and Wong 1987). For example, the hierarchical model discussed in the previous section can be treated as a "missing data problem" in which the individual effects θ_i are viewed as missing data. In a state-space model (Sections 1.6, 3.3, and 4.5), the unobserved state variable x_t can be viewed as missing data. In the biological motif discovery problem (Section 1.5), the motif pattern location a_k in each sequence can be treated as missing data.

A Bayesian missing data problem can be formulated as follows: Suppose the "complete–data" model $f(\mathbf{y} \mid \boldsymbol{\theta})$ has a nice clean form from which we can obtain the analytical form of the posterior distribution. However, only part of **y**, denoted as $\mathbf{y}_{\text{obs}}$, is observed, and the remaining part, $\mathbf{y}_{\text{mis}}$, is missing. Let $\mathbf{y} = (\mathbf{y}_{\text{obs}}, \mathbf{y}_{\text{mis}})$. The *joint* posterior distribution of $\mathbf{y}_{\text{mis}}$ and $\boldsymbol{\theta}$ is

$$\pi(\boldsymbol{\theta}, \mathbf{y}_{\text{mis}}) \propto f(\mathbf{y}_{\text{mis}}, \mathbf{y}_{\text{obs}} \mid \boldsymbol{\theta}) f_0(\boldsymbol{\theta}), \tag{1.7}$$

and, marginally, the *observed-data posterior* is

$$p(\boldsymbol{\theta} \mid \mathbf{y}_{\text{obs}}) = \pi(\boldsymbol{\theta}) = \int \pi(\boldsymbol{\theta}, \mathbf{y}_{\text{mis}}) d\mathbf{y}_{\text{mis}}.$$

If we can draw Monte Carlo samples $(\boldsymbol{\theta}^{(1)}, \mathbf{y}_{\text{mis}}^{(1)}), \ldots, (\boldsymbol{\theta}^{(m)}, \mathbf{y}_{\text{mis}}^{(m)})$ from the joint posterior distribution π, then the histogram based on $\boldsymbol{\theta}^{(1)}, \ldots, \boldsymbol{\theta}^{(m)}$ can serve as an approximation to the observed-data posterior distribution $p(\boldsymbol{\theta} \mid \mathbf{y}_{\text{obs}})$. Furthermore, if we are interested in estimation some posterior expectation of $\boldsymbol{\theta}$, say $E\{h(\boldsymbol{\theta}) \mid \mathbf{y}_{\text{obs}}\}$, we can estimate it by

$$\hat{h} = \frac{1}{m} \left[h\left(\boldsymbol{\theta}^{(1)}\right) + \cdots + h\left(\boldsymbol{\theta}^{(m)}\right) \right].$$

Note that the form of π in (1.7) is no different, at least in principle, from that in (1.1). Hence, statisticians and physicists are indeed facing a similar computational problem.

1.9 Monte Carlo and Missing Data Problems

To a broader scientific audience, the concept of "missing data" is perhaps a little odd, for many scientists may not believe that they have any missing data. In the most general and abstract form, the "missing data" can refer to any augmented *component* of the probabilistic system under consideration and the inclusion of this component often results in a simpler structure and easier computation (there are more examples to demonstrate this need in the later chapters). However, this component needs to be marginalized (integrated) out in the final inference. Indeed, it is the scientist's desire to marginalize part or whole of a probabilistic system under investigation that has been the main impetus to the development of Monte Carlo techniques (Gilks, Richardson and Spiegelhalter 1998).

2
Basic Principles: Rejection, Weighting, and Others

2.1 Generating Simple Random Variables

To generate random variables that follow a general probability distribution function π, we need first to generate random variables *uniformly* distributed in [0,1]. These random variables are often called *random numbers* for simplicity. However, this "simple-sounding" task is not easily achievable on a computer. But even if it were possible, it might not be desirable to use authentic random numbers because of the need to debug computer programs. In debugging a program, we often have to repeat the same computation many times; this require us to reproduce the same sequence of random numbers repeatedly. What becomes an accepted alternative in the community of scientific computing is to generate *pseudo-random* numbers. More formally, we can define a *uniform pseudo-random number generator* as an algorithm which, starting from an initial value u_0 (i.e., the *seed*), produces a sequence $(u_i) = (D^i(u_0))$ of values in [0,1]. For all n, the values $(u_1, \ldots, u_n)$ should reproduce the behavior of an i.i.d. sample $(V_1, \ldots, V_n)$ of uniform random variables. A few very good pseudo-random number generators are available; we refer the reader to Marsaglia and Zaman (1993) and Knuth (1997) for further reference. Consequently, we *assume* from now on that uniform random variables can be satisfactorily produced on the computer. The following simple lemma enables us to produce nonuniform random variables. Its proof is left as an exercise for the reader.

Lemma 2.1.1 *Suppose $U \sim Uniform[0,1]$ and F is a one-dimensional cumulative distribution function (cdf). Then, $X = F^{-1}(U)$ has the distribution F. Here we define $F^{-1}(u) = \inf\{x;\ F(x) \geq u\}$.*

This lemma suggests to us an explicit way (i.e., the inversion method) of generating a one-dimensional random variable when its cdf is available. However, because many distributions (e.g., Gaussian) do not have a closed-form cdf, it is often difficult to directly apply the above inversion procedure. For distributions with nice mathematical properties, special methods are often available for drawing random samples from them. For example, a fast way of generating standard Gaussian random variables is to use the property that a standard *bivariate* Gaussian random vector (X,Y) (with zero mean and identity covariance matrix) can be generated by first uniformly choosing an angle in $\mathbb{R}^2$ (two-dimensional Euclidean space) and then generating the square distance from an Exponential distribution (Devroye 1986). A Beta random variable can be constructed as the ratio $X_1/(X_1 + X_2)$, where X_1 and X_2 are independent Gamma random variables. For mathematically less convenient distributions, von Neumann (1951) proposed a very general algorithm, the *rejection method*, which can — at least in principle — be applied to draw from any probability distribution with a density function given up to a normalizing constant, regardless of dimensions.

2.2 The Rejection Method

Suppose $l(\mathbf{x}) = c\pi(\mathbf{x})$ is computable, where π is a probability distribution function or density function and c is unknown. If we can find a sampling distribution $g(\mathbf{x})$ and "covering constant" M so that the envelope property [i.e., $Mg(\mathbf{x}) \geq l(\mathbf{x})$] is satisfied for all $\mathbf{x}$, then we can apply the following procedure.

Rejection sampling [von Neumann (1951)]:

(a) Draw a sample $\mathbf{x}$ from $g(\ )$ and compute the ratio

$$r = \frac{l(\mathbf{x})}{Mg(\mathbf{x})} \quad (\leq 1).$$

(b) Flip a coin with success probability r;
- if the head turns up, we accept and return the $\mathbf{x}$;
- otherwise, we reject the $\mathbf{x}$ and go back to (a).

The accepted sample follows the target distribution π.

To show that the foregoing procedure is correct, we let I be the indicator function so that $I = 1$ if the sample X drawn from $g(\ )$ is accepted, and

$I = 0$, otherwise. Then, we observe that

$$P(I=1) = \int P(I=1 \mid \mathbf{X} = \mathbf{x}) g(\mathbf{x}) d\mathbf{x} = \int \frac{c\pi(\mathbf{x})}{Mg(\mathbf{x})} g(\mathbf{x}) d\mathbf{x} = \frac{c}{M}.$$

Hence,

$$p(\mathbf{x} \mid I = 1) = \frac{c\pi(\mathbf{x})}{Mg(\mathbf{x})} g(\mathbf{x}) / P(I=1) = \pi(\mathbf{x}).$$

Because the expected number of "operations" for obtaining one accepted sample is M, The key to a successful application of the algorithm is to find a good trial distribution $g(\mathbf{x})$ which gives rise to a small M. It is usually very difficult to apply the simple rejection method for a high-dimensional Monte Carlo simulation problem.

Example: *Truncated Gaussian distribution.* Suppose we want to draw random samples from $\pi(x) \propto \phi(x) I_{\{x>c\}}$, where $\phi(x)$ is the standard normal density and I is the indicator function. A simple strategy can be applied when $c < 0$: We continue to generate random samples from a standard Gaussian distribution until a sample satisfying $X > c$ is obtained. In the worst case, the efficiency of this method is 50%.

For $c > 0$, especially when c is large, the above strategy is very inefficient. we can use the rejection method with an exponential distribution as the envelope function. Suppose the density of this exponential distribution has the form $\lambda_0 e^{-\lambda_0 x}$. We want to find the smallest constant b such that

$$\frac{\phi(x+c)}{1 - \Phi(c)} \leq b\lambda_0 e^{-\lambda_0 x}, \quad \forall \, x \geq 0.$$

The optimal choice of b is

$$b = \frac{\exp\{(\lambda_0^2 - 2\lambda_0 c)/2\}}{\sqrt{2\pi} \lambda_0 (1 - \Phi(c))}.$$

The acceptance rate for using the posited exponential distribution as the envelope function is then $1/b$. To achieve the minimum rejection rate, we further find that the best choice for λ_0 is

$$\lambda_0 = (c + \sqrt{c^2 + 4})/2.$$

With this choice of λ_0 and b, we can implement the rejection method. The rejection rate for this scheme decreases as c increases, and this rate becomes very small for moderate to large c. For example, for $c = 0, 1,$ and 2, the rejection rates are 0.24, 0.12, and 0.07, respectively.

2.3 Variance Reduction Methods

Here we briefly describe a few techniques commonly used for variance reduction in Monte Carlo computations. More detailed descriptions can be found in standard Monte Carlo books (Hammersley and Handscomb 1964, Rubinstein 1981).

Stratified Sampling. It is a powerful and commonly used technique in population survey and is also very useful in Monte Carlo computations. Mathematically, this method can be viewed as a special importance sampling method with its trial density $g(x)$ constructed as a piecewise constant function. Suppose we are interested in estimating $\int_{\mathcal{X}} f(x)dx$. If possible, we want to break the region $\mathcal{X}$ into the union of k disjoint subregions, $D_1, \ldots, D_k$, so that within each subregion, the function $f(x)$ is relatively "homogeneous" (e.g., close to being a constant). Then, we can spend m_i random samples, $X^{(i,1)}, \ldots, X^{(i,m_i)}$, in the subregion D_i, and approximate the subregional integral $\int_{D_i} f(x)dx$ by

$$\hat{\mu}_i = \frac{1}{m_i}[f(X^{(i,1)}) + \cdots + f(X^{(i,m_i)})].$$

The overall integral μ can be approximated by

$$\hat{\mu} = \hat{\mu}_1 + \cdots + \hat{\mu}_k,$$

whose variance is easily calculated as

$$\text{var}(\hat{\mu}) = \frac{\sigma_1^2}{m_1} + \cdots + \frac{\sigma_m^2}{m_k},$$

where σ_i^2 is the variation of $f(x)$ in region D_i. In contrast, if we use all of the $m = m_1 + \cdots + m_k$ samples to do a plain uniform sampling in the region $\mathcal{X}$, the variance of the estimate would be σ^2/n, with σ^2 being the overall variation of $f(x)$ in $\mathcal{X}$.

Clearly, if we fail to have relatively homogeneous $f(x)$ in each region D_i (in other words, if σ_i^2 is not much different from σ^2), stratified sampling actually makes the computation less accurate than a plain Monte Carlo. The moral is this: There is no free lunch, and one needs to think carefully before adopting any advanced techniques.

Control Variates Method. In this method, one uses a control variate C, which is correlated with the sample X, to produce a better estimate. Suppose the estimation of $\mu = E(X)$ is of interest and $\mu_C = E(C)$ is known. Then, we can construct Monte Carlo samples of the form

$$X(b) = X + b(C - \mu_C),$$

which have the same mean as X, but a new variance

$$\text{var}\{X(b)\} = \text{var}(X) - 2b\text{cov}(X,C) + b^2\text{var}(C).$$

If the computation of $\text{cov}(X,C)$ and $\text{var}(C)$ is easy, then we can let $b = \text{cov}(X,C)/\text{var}(C)$, in which case

$$\text{var}\{X(b)\} = (1 - \rho_{XC}^2)\text{var}(X) < \text{var}(X).$$

Another situation is when we know only that $E(C)$ is equal to μ. Then, we can form $X(b) = bX + (1-b)C$. It is easy to show that if C is correlated with X, we can always choose a proper b so that $X(b)$ has a smaller variance than X. Extensions to more than one control variate are also useful in Monte Carlo computations, but are omitted in this book.

Antithetic Variates Method. This method is due to Hammersley and Morton (1956), where they describe a way of producing negatively correlated samples. Suppose U is the random number used in the production of a sample X that follows a distribution with cdf F [i.e., $X = F^{-1}(U)$ according to Lemma 2.1.1]. Then, $X' = F^{-1}(1-U)$ also follows distribution F. More generally, if g is a monotonic function, then

$$\{g(u_1) - g(u_2)\}\{g(1-u_1) - g(1-u_2)\} \leq 0$$

for any $u_1, u_2 \in [0,1]$. For two independent uniform random variables U_1 and U_2 (in fact, it is only required that the two are i.i.d. a with symmetric density in $[0,1]$), we have

$$E[\{g(U_1) - g(U_2)\}\{g(1-U_1) - g(1-U_2)\}] = \operatorname{cov}(X, X') \leq 0,$$

where $X = g(U)$ and $X' = g(1-U)$. Thus, $\operatorname{var}[(X+X')/2] \leq \operatorname{var}(X)/2$, implying that using the pair X and X' is better than using two independent Monte Carlo draws for estimating $E(X)$.

Rao-Blackwellization. This method reflects a basic principle (or rule of thumb) in Monte Carlo computation: One should carry out analytical computation as much as possible. The problem can be formulated as follows: Suppose we have drawn independent samples $\mathbf{x}^{(1)}, \ldots, \mathbf{x}^{(m)}$ from the target distribution $\pi(\mathbf{x})$ and are interested in evaluating $I = E_\pi h(\mathbf{x})$. A straightforward estimator is

$$\hat{I} = \frac{1}{m}\left\{h(\mathbf{x}^{(1)}) + \cdots + h(\mathbf{x}^{(m)})\right\}.$$

Suppose, in addition, that $\mathbf{x}$ can be decomposed into two parts (x_1, x_2) and that the conditional expectation $E[h(\mathbf{x}) \mid x_2]$ can be carried out analytically. An alternative estimator of I is

$$\tilde{I} = \frac{1}{m}\left\{E[h(\mathbf{x}) \mid x_2^{(1)}] + \cdots + E[h(\mathbf{x}) \mid x_2^{(m)}]\right\}.$$

Clearly, both $\hat{I}$ and $\tilde{I}$ are unbiased[1] because of the simple fact that

$$E_\pi h(\mathbf{x}) = E_\pi[E\{h(\mathbf{x}) \mid x_2\}].$$

---

[1] An estimator $\hat{\mu}$ is called an unbiased estimator of μ if $E_\mu \hat{\mu} = \mu$. In words, this means that the average behavior of $\hat{\mu}$ is "on target."

If the computational effort for obtaining the two estimates are the same, then $\tilde{I}$ should be preferred because

$$\mathrm{var}\{h(\mathbf{x})\} = \mathrm{var}\{E[h(\mathbf{x}) \mid x_2]\} + E\{\mathrm{var}[h(\mathbf{x}) \mid x_2]\},$$

which implies that

$$\mathrm{var}(\hat{I}) = \frac{\mathrm{var}\{h(\mathbf{x})\}}{m} \geq \frac{\mathrm{var}\{E[h(\mathbf{x}) \mid x_2]\}}{m} = \mathrm{var}(\tilde{I}).$$

In statistics, $\hat{I}$ is often called the "histogram estimator" or the empirical estimator and $\tilde{I}$ the "mixture estimator." Statisticians find that by conditioning an inferior estimator on the value of sufficient statistics, one can obtain the optimal estimator. This procedure is often referred to as *Rao-Blackwellization* (Bickel and Doksum 2000). Some other uses of Rao-Blackwellization in Monte Carlo estimations can be found in Casella and Robert (1996) More discussions on this issue can be found in Section 2.5.5 and Chapter 6.

2.4 Exact Methods for Chain-Structured Models

An important probability distribution used in many applications has the following form:

$$\pi(\mathbf{x}) \propto \exp\left\{-\sum_{i=1}^{d} h_i(x_{i-1}, x_i)\right\}, \tag{2.1}$$

where $\mathbf{x} = (x_0, x_1, \ldots, x_d)$. This type of model can be seen as having a "Markovian structure" because the conditional distribution $\pi(x_i \mid \mathbf{x}_{[-i]})$, where $\mathbf{x}_{[-i]} = (x_1, \ldots, x_{i-1}, x_{i+1}, \ldots, x_d)$, depends only on the two neighboring variables x_{i-1} and x_{i+1}; that is,

$$\pi(x_i \mid \mathbf{x}_{[-i]}) \propto \exp\left\{-h(x_{i-1}, x_i) - h(x_i, x_{i+1})\right\}.$$

The unobserved state variables in a state-space model (Section 1.6) can clearly be represented in this form, which can also be depicted by the following graph:

When $x_0, \ldots, x_d$ are discrete random variables taking values in a finite set $S = \{s_1, \ldots, s_k\}$, this structure is often referred to as the *hidden Markov model* (HMM) and we can do many things with it. First, the "dynamic programming" (DP) method (Bellman 1957) can be used to find the global maximum of $\pi(\mathbf{x})$ and its maximizer $\hat{\mathbf{x}}$ with $O(dk^2)$ operations. Second,

2.4 Exact Methods for Chain-Structured Models

an algorithm of the same order as the DP exists for finding the marginal distribution of each x_i and drawing "exact" random samples from $\pi(\mathbf{x})$. Clearly, these exact algorithms are only practical when k, the number of distinctive values that x_i can take, is not too large.

2.4.1 Dynamic programming

Suppose each x_i in $\mathbf{x}$ only takes values in set $\mathcal{S} = \{s_1, \ldots, s_k\}$. Maximizing the distribution $\pi(\mathbf{x})$ in (2.1) is equivalent to minimizing its exponent

$$H(\mathbf{x}) = h_1(x_0, x_1) + \cdots + h_d(x_{d-1}, x_d).$$

A recursive procedure can be carried out:

- Define
$$m_1(x) = \min_{s_i \in \mathcal{S}} h_1(s_i, x), \text{ for } x = s_1, \ldots, s_k.$$

- Recursively compute the function
$$m_t(x) = \min_{s_i \in \mathcal{S}} \{m_{t-1}(s_i) + h_t(s_i, x)\}, \text{ for } x = s_1, \ldots, s_k.$$

- The optimal value $H(\mathbf{x})$ is attained by $\min_{s \in \mathcal{S}} m_d(s)$.

It is not difficult to see that the minimum of $m_1(x)$ is equal to the minimum of $h_1(x_0, x_1)$. By induction, one can easily argue that

$$\min_{x \in \mathcal{S}} m_t(x) = \min_{x_0, \ldots, x_t \in \mathcal{S}} [h_1(x_0, x_1) + \cdots + h_t(x_{t-1}, x_t)].$$

Thus, the above procedure indeed minimizes the target function $H(\mathbf{x})$.

To find out which $\mathbf{x}$ gives rise to the global minimum of $H(\mathbf{x})$, we can trace backward as follows:

- Let $\hat{x}_d$ be the minimizer of $m_d(x)$; that is,
$$\hat{x}_d = \arg \min_{s_i \in \mathcal{S}} m_d(s_i).$$

Break ties arbitrarily.

- For $t = d-1, d-2, \ldots, 1$, we let
$$\hat{x}_t = \arg \min_{s_i \in \mathcal{S}} \{m_t(s_i) + h_{t+1}(s_i, \hat{x}_{t+1})\}.$$

Break ties arbitrarily.

Configuration $\hat{\mathbf{x}} = (\hat{x}_1, \ldots, \hat{x}_d)$ obtained by this method is the minimizer of $H(\mathbf{x})$.

2.4.2 Exact simulation

The first step for simulating from $\pi(\mathbf{x})$ in (2.1) is to draw x_d from its marginal distribution. This requires us to marginalize $x_1, \ldots, x_{d-1}$ in the joint distribution $\pi(\mathbf{x})$. After we have drawn x_d, we can work our way backward recursively; that is, sampling x_{d-1} from $\pi(x_{d-1}|x_d)$; x_{d-2} from $\pi(x_{d-2}|x_{d-1})$; and so on. The principle behind the marginalization step is based on the observation that the overall summation can be decomposed into recursive steps; that is,

$$Z \equiv \sum_{\mathbf{x}} \exp\{-H(\mathbf{x})\} = \sum_{x_d}\left[\cdots\left[\sum_{x_1}\left\{\sum_{x_0} e^{-h_1(x_0,x_1)}\right\} e^{-h_2(x_1,x_2)}\right]\cdots\right].$$

More precisely, the following recursions similar to those in DP can be carried out by a computer:

- Define $V_1(x) = \sum_{x_0 \in \mathcal{S}} e^{-h_1(x_0,x)}$.
- Compute recursively for $t = 2, \ldots, d$:

$$V_t(x) = \sum_{y \in \mathcal{S}} V_{t-1}(y) e^{-h_t(y,x)}. \tag{2.2}$$

Then, the partition function is $Z = \sum_{x \in \mathcal{S}} V_d(x)$ and the marginal distribution of x_d is $\pi(x_d) = V_d(x_d)/Z$. To simulate $\mathbf{x}$ from π, we can do the following:

- Draw x_d from $\mathcal{S}$ with probability $V_d(x_d)/Z$;
- For $t = d-1, d-2, \ldots, 1$, we draw x_t from distribution

$$p_t(x) = \frac{V_t(x) e^{-h_{t+1}(x,x_{t+1})}}{\sum_{y \in \mathcal{S}} V_t(y) e^{-h_{t+1}(y,x_{t+1})}}.$$

The random sample $\mathbf{x} = (x_1, \ldots, x_d)$ obtained in this way follows the distribution $\pi(\mathbf{x})$.

As an example, we can use the forward-recursion formula (2.2) to compute the partition function for a one-dimensional Ising model

$$\pi(\mathbf{x}) = Z^{-1} \exp\{\beta(x_0 x_1 + \cdots + x_{d-1} x_d)\},$$

where x_i takes value in $\mathcal{S} = \{-1, 1\}$. First, we have

$$V_1(x) = e^{\beta x} + e^{-\beta x} \equiv e^{\beta} + e^{-\beta},$$

which is a constant. Applying the recursion, we easily obtain that

$$V_t(x) = (e^{-\beta} + e^{\beta})^t$$

and $Z = 2(e^{-\beta}+e^{\beta})^d$. Details for an exact simulation from this distribution are left to the reader.

An important feature of the target distribution $\pi(\mathbf{x})$ treated in this section is that it can be written as

$$\pi(\mathbf{x}) \propto \exp\left\{-\sum_{C \in \mathcal{C}} h(\mathbf{x}_C)\right\},$$

where $\mathcal{C}$ is the set of some subsets of $\{1,\ldots,d\}$ and $\mathbf{x}_C = (x_i,\ i \in C)$. In Lauritzen and Spiegelhalter (1988), each subset C in $\mathcal{C}$ is called a "clique." Any two subsets C_1 and C_2 are said to "connected" if they share at least one common component. Any probability distribution that possesses this dependency structure is termed a *graphical model*. Our model in this section can be seen as a special graphical model in which the set of cliques is $\mathcal{C} = \{C_1,\ldots,C_d\}$, where $C_i = \{i-1,i\}$. When $\mathcal{C}$ forms a "tree" and each clique does not have too many vertices (components), one can derive efficient algorithms for optimization and exact simulation similar to the algorithms described in this section (Lauritzen and Spiegelhalter 1988). There does not seem to be a common name for this sampling method. Some people call it the *peeling algorithm* because this method was first developed for a genetic linkage problem (Cannings, Thompson and Skolnick 1978). Some others refer it as the *forward-summation-backward-sampling* method, a rather awkward name. We prefer to use the name *propagation method* for the reason that both the forward and the backward steps can be thought of as propagating information along the chain graph.

2.5 Importance Sampling and Weighted Sample

2.5.1 An example

Suppose we wish to evaluate the quantity

$$\theta = \int_{\mathcal{X}} h(x)\pi(x)dx = E_\pi[h(X)],$$

where the support of the random variable X is denoted as $\mathcal{X}$ and $h(x) \geq 0$. In standard numerical methods, we discretize the domain $\mathcal{X}$ by regular grids, evaluate $h(x)\pi(x)$ on each of the grid points, and then use the Riemann sum as an approximation. Consider the target function given by Figure 2.1(a):

$$f(x,y) = 0.5e^{-90(x-0.5)^2-45(y+0.1)^4} + e^{-45(x+0.4)^2-60(y-0.5)^2},$$

where $(x,y) \in [-1,1]\times[-1,1]$. More than two-thirds of computing time are wasted on evaluating those grid points on which the function is virtually

32   2. Basic Principles: Rejection, Weighting, and Others

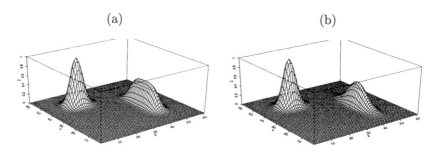

FIGURE 2.1. (a) The target function whose integral is of interest. (b) A possible trial distribution $g(x,y)$ for applying importance sampling.

zero. It is easy to imagine that the situation deteriorates very rapidly as the dimension of space $\mathcal{X}$ increases.

By taking m random samples, $(x^{(1)}, y^{(1)}), \ldots, (x^{(m)}, y^{(m)})$, uniformly in $[-1,1] \times [-1,1]$, we implemented a vanilla Monte Carlo algorithm to estimate the integral $\mu = \int \int f(x,y) dx dy$. Because the density for the sampling distribution is a constant, $1/4$, in the region, the estimate of the integral was produced as

$$\hat{\mu} = \frac{4}{m} \left\{ f^{(1)} + \cdots + f^{(m)} \right\},$$

where $f^{(i)} = f(x^{(i)}, y^{(i)})$. With $m=2{,}500$, we obtained $\hat{\mu}= 0.1307$, with a standard deviation 0.009, which was estimated by

$$\text{std}(\hat{\mu}) = \sqrt{\frac{1}{n(n-1)} \sum_{i=1}^{m} (f_i - \hat{\mu})^2}.$$

Clearly, this vanilla Monte Carlo algorithm suffers a similar problem as its deterministic counterpart: It wastes a lot of effort in evaluating random samples located in regions where the function value is almost zero. Although the theoretical convergence rate is $m^{-1/2}$ for essentially *all* Monte Carlo methods, it is the constant in front of this rate that makes a huge difference in a real problem.

2.5.2   The basic idea

The *importance sampling* idea (Marshall 1956) suggests that one should focus on the region(s) of "importance" so as to save computational resources. Although it may not seem so important in the toy example shown earlier, the idea of biasing toward "importance" regions of the sample space becomes essential for Monte Carlo computation with high-dimensional models such as those in statistical physics, molecular simulation, and Bayesian

2.5 Importance Sampling and Weighted Sample

statistics. In high-dimensional problems, the region in which the target function is meaningfully nonzero compared with the whole space $\mathcal{X}$ is just like a needle compared with a haystack. Vanilla Monte Carlo schemes (e.g., sampling uniformly from a regular region) are bound to fail in these problems.

Suppose one is interested in evaluating

$$\mu = E_\pi\{h(\mathbf{x})\} = \int h(\mathbf{x})\pi(\mathbf{x})d\mathbf{x}.$$

The following procedure is a simple form of the *importance sampling algorithm*:

(a) Draw $\mathbf{x}^{(1)}, \ldots, \mathbf{x}^{(m)}$ from a *trial distribution* $g(\cdot)$.

(b) Calculate the *importance weight*

$$w^{(j)} = \pi(\mathbf{x}^{(j)})/g(\mathbf{x}^{(j)}), \quad \text{for } j = 1, \ldots, m.$$

(c) Approximate μ by

$$\hat{\mu} = \frac{w^{(1)}h(\mathbf{x}^{(1)}) + \cdots + w^{(m)}h(\mathbf{x}^{(m)})}{w_1 + \cdots + w_m}. \tag{2.3}$$

Thus, in order to make the estimation error small, one wants to choose $g(\mathbf{x})$ as "close" in shape to $\pi(\mathbf{x})h(\mathbf{x})$ as possible. A major advantage of using (2.3) instead of the unbiased estimate,

$$\tilde{\mu} = \frac{1}{m}\left\{w^{(1)}h(\mathbf{x}^{(1)}) + \cdots + w^{(m)}h(\mathbf{x}^{(m)})\right\}, \tag{2.4}$$

is that in using the former, we need *only* to know the ratio $\pi(\mathbf{x})/g(\mathbf{x})$ up to a multiplicative constant; whereas in the latter, the ratio needs to be known exactly. Additionally, although inducing a small bias, (2.3) often has a smaller mean squared error than the unbiased one $\tilde{\mu}$.

Another scenario for resorting to importance sampling is when we want to generate i.i.d. random samples from π but doing so directly is infeasible. In this case, we may generate random samples from a different, but similar, trivial distribution $g(\ )$, and then correct the bias by using the importance sampling procedure. Similar to the rejection method, a successful application of importance sampling in this case requires that the sampling distribution g is reasonably close to π; in particular, that g has a longer tail than π. Note that finding a good trial distribution g can be a major — and sometimes impossible — undertaking in high-dimensional problems.

Alternatively, we can opt for *correlated* samples produced by running a Markov chain whose stationary distribution is π. This methodology is referred to as *Markov chain Monte Carlo* (MCMC) throughout the book. A

34   2. Basic Principles: Rejection, Weighting, and Others

very general recipe for designing a proper Markov chain was first proposed by Metropolis et al. (1953) and has been subject to active research in the past few decades. We will discuss this class of methods in the latter part of this book (Chapters 5–11).

Let us illustrate the *importance sampling* method with the toy example shown in Figure 2.1. After a visual examination of function $f(x,y)$, we decided to use a distribution $g(x,y)$, which is of the form

$$g(x,y) \propto 0.5 e^{-90(x-0.5)^2 - 10(y+0.1)^2} + e^{-45(x+0.4)^2 - 60(y-0.5)^2},$$

with $(x,y) \in [-1,1] \times [-1,1]$. This is a truncated mixture of Gaussian distributions:

$$0.46 \mathcal{N}\left[\begin{pmatrix} 0.5 \\ -0.1 \end{pmatrix}, \begin{pmatrix} \frac{1}{180} & 0 \\ 0 & \frac{1}{20} \end{pmatrix}\right] + 0.54 \mathcal{N}\left[\begin{pmatrix} -0.4 \\ 0.5 \end{pmatrix}, \begin{pmatrix} \frac{1}{90} & 0 \\ 0 & \frac{1}{120} \end{pmatrix}\right]$$

We can sample from this mixture distribution by a two-step procedure: (a) a biased coin (with probability 0.464 of showing heads) is first tossed; (b) if the head turns up, we draw a random vector from the first Gaussian distribution, otherwise, we draw from the second Gaussian distribution. The integral can then be estimated by averaging the ratios $w(x,y) = f(x,y)/g(x,y)$, with $w = 0$ when (x,y) falls out of the region $\mathcal{X}$. With $m=2500$, our estimate of μ is 0.1259, with a standard error 0.0005.

2.5.3   The "rule of thumb" for importance sampling

Importance sampling suggests estimating $\mu = E_\pi\{h(\mathbf{x})\}$ by first generating independent samples $\mathbf{x}^{(1)}, \ldots, \mathbf{x}^{(m)}$ from an easy-to-sample trial distribution, $g(\ )$, and then correcting the bias by incorporating the importance weight $w^{(j)} \propto \pi(\mathbf{x}^{(j)})/g(\mathbf{x}^{(j)})$ in estimation using either (2.3) or 2.4. By properly choosing $g(\cdot)$, one can reduce the variance of the estimate substantially. A good candidate for $g(\cdot)$ is one that is close to the shape of $h(\mathbf{x})\pi(\mathbf{x})$. Therefore, the importance sampling method can be super-efficient; that is, the resulting variance of $\hat{\mu}$ can be smaller than that obtained using independent samples from π. The method is generalized to the case of, say, evaluating $E_\pi\{h(\mathbf{x})\}$ when sampling from $\pi(\cdot)$ directly is difficult but generating from $g(\cdot)$ and computing the importance ratio $w(\mathbf{x}) = \pi(\mathbf{x})/g(\mathbf{x})$ (up to a multiplicative constant) are easy. The efficiency of such a method is then difficult to measure. A useful "rule of thumb" is to use the *effective sample size* (ESS) to measure how different the trial distribution is from the target distribution. Suppose m independent samples are generated from $g(\mathbf{x})$; then, the ESS of this method is defined as

$$\text{ESS}(m) = \frac{m}{1 + \text{var}_g[w(\mathbf{x})]}.$$

Since the target distribution π is known only up to a normalizing constant in many problems, the variance of the *normalized* weight needs to be estimated by the *coefficient of variation* of the unnormalized weight:

$$\text{cv}^2(w) = \frac{\sum_{j=1}^m (w^{(j)} - \bar{w})^2}{(m-1)\bar{w}^2}, \qquad (2.5)$$

where $\bar{w}$ is the sample average of the $w^{(j)}$. The ESS measure of efficiency can be partially justified by the delta method as follows (Kong et al. 1994, Liu 1996a).

Note that $E_p\{w(\mathbf{x})\} = 1$; hence, both (2.3) and (2.4) are proper estimates of μ. In particular, the two estimates are related to each other in the following form:

$$\hat{\mu} = \frac{\frac{1}{m}\sum_{j=1}^m h(\mathbf{x}^{(j)}) w(\mathbf{x}^{(j)})}{\frac{1}{m}\sum_{j=1}^m w(\mathbf{x}^{(j)})} \equiv \frac{\bar{Z}}{\bar{W}}. \qquad (2.6)$$

Let $Z = h(\mathbf{x})w(\mathbf{x})$, $W = w(\mathbf{x})$, and let $\bar{Z}$ and $\bar{W}$ be the corresponding sample averages. As we have mentioned in Section 2.5.2, there are two advantages for choosing $\hat{\mu}$ over $\bar{Z}$ for estimation: The importance sampling ratios need only to be evaluated up to an unknown constant; and $\hat{\mu}$ may have smaller mean squared error than $\tilde{\mu} \equiv \bar{Z}$. By the delta method, we see that

$$E_g(\hat{\mu}) \approx E_g\{\bar{Z}[1 - (\bar{W}-1) + (\bar{W}-1)^2 + \cdots]\}$$
$$\approx \mu - \frac{\text{cov}_g(W,Z)}{m} + \frac{\mu \text{var}_g W}{m}$$

The variance of $\hat{\mu}$ can be explored by using the standard delta method for ratio statistics:

$$\text{var}_g(\hat{\mu}) \approx \frac{1}{m}[\mu^2 \text{var}_g(W) + \text{var}_g(Z) - 2\mu \text{cov}_g(W,Z)]. \qquad (2.7)$$

In contrast, $E_g(\tilde{\mu}) = \mu$ and $\text{var}_g(\tilde{\mu}) = \text{var}_g(Z)/m$. Hence, the mean squared error (MSE) of $\tilde{\mu}$ is

$$\text{MSE}(\tilde{\mu}) = E_g(\tilde{\mu}-\mu)^2 = \text{var}_g(Z)/m,$$

and that for $\hat{\mu}$ is

$$\begin{aligned}\text{MSE}(\hat{\mu}) &= [E_g(\hat{\mu}) - \mu]^2 + \text{var}_g(\hat{\mu}) \\ &= \frac{1}{m}\text{MSE}(\tilde{\mu}) + \frac{1}{m}[\mu^2 \text{var}_g(W) - 2\mu\text{cov}_g(W,Z)] + O(m^{-2})\end{aligned}$$

Without loss of generality, we let $\mu > 0$. Then, MSE$(\hat{\mu})$ is smaller in comparison with MSE$(\tilde{\mu})$ when $\mu^2 - 2\mu\text{cov}_g(W,Z) < 0$ (i.e., when W and Z are strongly correlated).

Denoting $H = h(\mathbf{x})$, we observe that $Z = WH$, $\mu = E_g(WH)$, and
$$\text{cov}_g(W, Z) = E_\pi(HW) - \mu = \text{cov}_\pi(W, H) + \mu E_\pi(W) - \mu.$$
Similarly
$$\begin{aligned}\text{var}_g(Z) &= E_\pi(WH^2) - \mu^2 \\ &\approx E_\pi(W)E_\pi^2(H) + \text{var}_\pi(H)E_\pi(W) + 2\mu\text{cov}_\pi(W, H) - \mu^2,\end{aligned}$$
where the approximation is made based on the delta method involving the first two moments of W and H. It is easy to show that the remainder term in the above approximation is
$$E_\pi[\{W - E_\pi(W)\}(H - \mu)^2], \tag{2.8}$$
which is not necessarily small. By reformulating (2.7), we find that
$$\text{var}_g(\tilde{\mu}) \approx \text{var}_\pi(H)\{1 + \text{var}_p(W)\}/m.$$
Roughly speaking, if μ were estimated by $\hat{\mu} = \sum_{j=1}^m h(\mathbf{y}^{(j)})/m$ where the $\mathbf{y}^{(j)}$ are i.i.d. draws from π, then the efficiency of $\hat{\mu}$ relative to $\tilde{\mu}$ is
$$\frac{\text{var}_\pi\{h(\mathbf{y})\}}{\text{var}_g\{h(\mathbf{x})w(\mathbf{x})\}} \approx \frac{1}{1 + \text{var}_g\{w(\mathbf{x})\}}.$$
This can be interpreted as that the m weighted samples is worth of $m/\{1 + \text{var}_g[w(\mathbf{x})]\}$ i.i.d. samples drawn from the target distribution. Obviously, the "rule of thumb" approximation can be substantially off if the remainder term (2.8) is large. The advantage of the "rule" is that it does not involve $h(\mathbf{x})$, which makes it particularly useful as a measure of the relative efficiency of the method when many different h's are of potential interest.

2.5.4 Concept of the weighted sample

The concept of a *properly weighted sample* is a useful generalization to the foregoing importance sampling procedure. Suppose we are interested in Monte Carlo estimation of $\mu = E_\pi h(\mathbf{x})$ for some arbitrary function h. The importance sampling principle suggests that the random sample $\mathbf{x}^{(1)}, \ldots, \mathbf{x}^{(m)}$ used to estimate μ need not be drawn from π — they can be drawn from almost any distribution provided that a proper set of weights are associated with the sample and the weights are not too skewed.

Definition 2.5.1 *A set of weighted random samples $\{(\mathbf{x}^{(j)}, w^{(j)})\}_{j=1}^m$ is called proper with respect to π if for any square integrable function $h(\cdot)$,*
$$E[h(\mathbf{x}^{(j)})w^{(j)}] = cE_\pi h(\mathbf{x}), \quad \text{for } j = 1, \ldots, m,$$
where c is a normalizing constant common to all the m samples.

With this set of weighted samples, we can estimate μ as

$$\hat{\mu} = \frac{1}{W}\sum_{j=1}^{m} w^{(j)}h(\mathbf{x}^{(j)}), \qquad (2.9)$$

where $W = \sum_{j=1}^{m} w^{(j)}$. Mathematically, this says that the joint distribution $g(w,\mathbf{x})$ for both the weight and the sample satisfies the relationship: For any square integrable $h(\cdot)$, $E_g\{h(\mathbf{x})w\}/E_g(w) = E_\pi\{h(\mathbf{x})\}$. This equality also implies that

$$\frac{E_g(w\mid \mathbf{x})}{E_g(w)}g(\mathbf{x}) = \pi(\mathbf{x}), \qquad (2.10)$$

where $g(\mathbf{x})$ is the marginal distribution of $\mathbf{x}$ under $g(\mathbf{x},w)$. Thus, a *necessary and sufficient* condition for $\mathbf{x}$ to be properly weighted by w with respect to π is (2.10).

The whole point of this generalization is to emphasize that there are many possible choices of the weighting function w for any given $\mathbf{x}$. In the context of importance sampling, the importance weight w is a deterministic function of the corresponding sample $\mathbf{x}$ [i.e., $w = \pi(\mathbf{x})/g(\mathbf{x})$]. Thus, in this case the joint distribution of $(w, \mathbf{x})$ is a degenerate one. It is also possible that conditional on $\mathbf{x}$, the weight variable w has a proper distribution and this flexibility can be useful for combining importance sampling with MCMC algorithms (more details in later chapters).

2.5.5 Marginalization in importance sampling

As we explained in Section 2.3, the method of Rao-Blackwellization is useful for improving estimation in a vanilla Monte Carlo scheme. Here, we show that the same technique takes the form of *marginalization* in importance sampling and is useful for reducing the variance of the importance weight.

Theorem 2.5.1 *Let $f(\mathbf{z}_1,\mathbf{z}_2)$ and $g(\mathbf{z}_1,\mathbf{z}_2)$ be two probability densities, where the support of f is a subset of the support of g. Then,*

$$\mathrm{var}_g\{\frac{f(\mathbf{Z}_1,\mathbf{Z}_2)}{g(\mathbf{Z}_1,\mathbf{Z}_2)}\} \geq \mathrm{var}_g\{\frac{f_1(\mathbf{Z}_1)}{g_1(\mathbf{Z}_1)}\},$$

where $f_1(\mathbf{z}_1) = \int f(\mathbf{z}_1,\mathbf{z}_2)d\mathbf{z}_2$ and $g_1(\mathbf{z}_1) = \int g(\mathbf{z}_1,\mathbf{z}_2)d\mathbf{z}_2$ are marginal densities. The variances are taken with respect to g.

Proof: It is easy to see that

$$\begin{aligned}\frac{f_1(\mathbf{z}_1)}{g_1(\mathbf{z}_1)} &= \int \frac{f(\mathbf{z}_1,\mathbf{z}_2)}{g_1(\mathbf{z}_1)g_{2\mid 1}(\mathbf{z}_2\mid \mathbf{z}_1)}g_{2\mid 1}(\mathbf{z}_2\mid \mathbf{z}_1)d\mathbf{z}_2 \\ &= E_g\left\{\frac{f(\mathbf{Z}_1,\mathbf{Z}_2)}{g(\mathbf{Z}_1,\mathbf{Z}_2)}\;\middle|\; \mathbf{Z}_1 = \mathbf{z}_1\right\}.\end{aligned}$$

Hence,

$$\text{var}_g\left\{\frac{f(Z_1,Z_2)}{g(Z_1,Z_2)}\right\} \geq \text{var}_g\left\{E_g\left[\frac{f(Z_1,Z_2)}{g(Z_1,Z_2)}\,\bigg|\,Z_1\right]\right\} = \text{var}_g\left\{\frac{f_1(Z_1)}{g_1(Z_1)}\right\}.$$

We can also obtain an explicit expression of the variance reduction:

$$\text{var}_g\left\{\frac{f(Z_1,Z_2)}{g(Z_1,Z_2)}\right\} - \text{var}_g\left\{\frac{f_1(Z_1)}{g_1(Z_1)}\right\} = E_g\left\{\text{var}\left[\frac{f(Z_1,Z_2)}{g(Z_1,Z_2)}\,\bigg|\,Z_1\right]\right\},$$

which, in the Analysis of Variance (ANOVA) terminology, is the average "within-group" variation with the group indexed by Z_1. ◇

The moral of the theorem is, again, that in Monte Carlo computations, one is encouraged to do as much analytical work as possible. Bringing down dimensionality is almost surely a good practice, although some examples exist in which one actually wants to *increase* the dimension of the space to improve the efficiency of the Monte Carlo algorithms (Section 7). This theorem was used in MacEachern, Clyde and Liu (1999) to justify a new importance sampling algorithm for a nonparametric Bayesian inference problem.

Another place to use Rao-Blackwellization is in estimation. For example, if the sample $\mathbf{x}^{(j)}$ can be decomposed as $(x_1^{(j)}, x_2^{(j)})$ and if $E_\pi[h(\mathbf{x}) \mid x_2]$ is available in closed-form, then an often more efficient estimator of $\mu = E_\pi h(\mathbf{x})$ is

$$\breve{\mu} = \frac{1}{W}\sum_{j=1}^{m} w^{(j)} E_\pi[h(\mathbf{x}) \mid x_2^{(j)}], \quad W = \sum_{j=1}^{m} w^{(j)},$$

whose asymptotic unbiasedness is easily shown. However, it is no longer as trivial as in Section 2.3 to prove its optimality — it in fact can not be proved that the new estimator $\breve{\mu}$ is always better than the plain estimator in (2.3).

2.5.6 Example: Solving a linear system

It has been noted that many deterministic systems can be solved by Monte Carlo methods (Hammersley and Handscomb 1964, Ripley 1987). Such systems include the boundary problems of partial differential equations, general high-dimensional linear equations, and other fixed-point problems. We here follow the general formulation in Section 4 of Griffiths and Tavare (1994). Suppose a system can be written in a recursive form as

$$q(\mathbf{x}) = \sum_{\mathbf{y}\in\mathcal{A}} r(\mathbf{x},\mathbf{y})q(\mathbf{y}) + \sum_{\mathbf{z}\in\mathcal{B}} r(\mathbf{x},\mathbf{z})q(\mathbf{z}), \quad \text{for } \mathbf{x}\in\mathcal{B}, \tag{2.11}$$

where $q(\mathbf{x})$ is known for $\mathbf{x} \in \mathcal{A}$ and is unknown for $\mathbf{x} \in \mathcal{B}$ (this happens in solving a difference equation with a given boundary condition). By repeatedly substituting the unknown $q(\mathbf{z})$ in the right-hand side of (2.11) by relationship (2.11), we have

$$\begin{aligned} q(\mathbf{x}) &= \sum_{\mathbf{y} \in \mathcal{A}} r(\mathbf{x}, \mathbf{y}) q(\mathbf{y}) + \sum_{\mathbf{y}_1 \in \mathcal{B}} \sum_{\mathbf{y} \in \mathcal{A}} r(\mathbf{x}, \mathbf{y}_1) r(\mathbf{y}_1, \mathbf{y}) q(\mathbf{y}) \\ &\quad + \sum_{\mathbf{y}_1 \in \mathcal{B}} \sum_{\mathbf{y}_2 \in \mathcal{B}} \sum_{\mathbf{y} \in \mathcal{A}} r(\mathbf{x}, \mathbf{y}_1) r(\mathbf{y}_1, \mathbf{y}_2) r(\mathbf{y}_2, \mathbf{y}) q(\mathbf{y}) + \cdots \\ &= \sum_{k=0}^{\infty} \left\{ \sum_{\mathbf{y}_1 \in \mathcal{B}} \cdots \sum_{\mathbf{y}_k \in \mathcal{B}} \sum_{\mathbf{y} \in \mathcal{A}} r(\mathbf{y}_1, \mathbf{y}_2) \cdots r(\mathbf{y}_k, \mathbf{y}) q(\mathbf{y}) \right\}. \quad (2.12) \end{aligned}$$

Suppose we can construct a Markov transition function $A(\mathbf{x}, \mathbf{y})$ that satisfies the following conditions: (a) $A(\mathbf{x}, \mathbf{y}) > 0$ whenever $r(\mathbf{x}, \mathbf{y}) > 0$ and (b) the chain visits $\mathcal{A}$ with probability 1 starting from any $\mathbf{x} \in \mathcal{B}$. Then, for any given $\mathbf{x}_0 \in \mathcal{B}$, we can simulate this Markov chain (some basics of Markov chain theory is given in Chapter 12) with $\mathbf{x}_0$ as its initial state (i.e. $\mathbf{X}_0 = \mathbf{x}_0$) and run the chain until it hits $\mathcal{A}$ for the first time. With this construction, expression (2.12) can be rewritten probabilistically as

$$q(\mathbf{x}_0) = E_{\mathbf{x}_0} \left\{ q(\mathbf{X}_\tau) \prod_{k=1}^{\tau} \frac{r(\mathbf{X}_{k-1}, \mathbf{X}_k)}{A(\mathbf{X}_{k-1}, \mathbf{X}_k)} \right\},$$

where τ is the first time the chain visits $\mathcal{A}$ (the hitting time). Thus, $q(\mathbf{x})$ can be estimated as follows. We run m independent Markov chains with the transition function $A(\cdot, \cdot)$ and the starting value $\mathbf{x}_0$ until hitting $\mathcal{A}$. Let these chains be $\mathbf{X}_1^{(j)}, \ldots, \mathbf{X}_{\tau_j}^{(j)}$, where τ_j is the hitting time of the jth chain; then,

$$\hat{q} = \frac{1}{m} \left\{ \sum_{j=1}^{m} q(\mathbf{X}_{\tau_j}^{(j)}) \prod_{k=1}^{\tau_j} \frac{r(\mathbf{X}_{k-1}^{(j)}, \mathbf{X}_k^{(j)})}{A(\mathbf{X}_{k-1}^{(j)}, \mathbf{X}_k^{(j)})} \right\}.$$

However, this method is usually inferior to its deterministic counterpart except for a few special cases (Ripley 1987). For example, this Monte Carlo approach might be attractive when one is only interested in estimating a few values of $q(\mathbf{x})$. We will discuss in Section 4.1.2 and Chapter 3 several techniques (e.g., resampling, rejection control, etc.) for improving efficiencies of the importance sampling method [see also Liu and Chen (1998)]. A proper implementation of these new techniques might lead to a Monte Carlo method that is more appealing than the corresponding deterministic approach (Chen and Liu 2000b).

2.5.7 Example: A Bayesian missing data problem

In statistics, it is often the case that part of the data is missing which renders the standard likelihood computation difficult (see the Appendix). The current example is constructed by Murray (1977) in the discussion of Dempster et al. (1977). Table 1 contains 12 observations assumed to be drawn from a bivariate normal distribution with known mean vector (0,0) and unknown covariance matrix.

1	1	−1	−1	2	2	−2	−2	*	*	*	*
1	−1	1	−1	*	*	*	*	2	2	−2	−2

TABLE 2.1. An artificial dataset of bivariate Gaussian observations with missing parts (Murray, 1977). The symbol * indicates that the value is missing.

Let ρ denote the correlation coefficient and let σ_1 and σ_2 denote the marginal variances. The complete data are $y_1, \ldots, y_{12}$, where $y_t = (y_{t,1}, y_{t,2})$ for $t = 1, \ldots, 12$. So the $y_{t,2}$ are missing for $t = 5, \ldots, 8$, whereas $y_{t,1}$ are missing for the $t = 9, \ldots, 12$. We are interested in the posterior distribution of ρ given the incomplete data. Note that the information on σ_1 and σ_2 provided by the eight incomplete observations cannot be ignored in drawing likelihood-based inference about ρ.

For simplicity, the covariance matrix of the bivariate normal is assigned the Jeffreys' noninformative prior distribution (Box and Tiao 1973)

$$\pi(\Sigma) \propto |\Sigma|^{-\frac{d+1}{2}}, \tag{2.13}$$

where d is the dimensionality and is equal to 2 in this example. The posterior distribution of Σ given t i.i.d. observed complete data $y_1, \ldots, y_t$ is

$$p(\Sigma \mid y_1, \ldots, y_t) \propto |\Sigma|^{-\frac{t+d+1}{2}} \exp\left\{-\frac{1}{2} \operatorname{tr}[\Sigma \cdot S]\right\},$$

where $S = (s_{ij})_{2 \times 2}$ is the uncorrected sum of squares matrix (i.e., $s_{ij} = \sum_{s=1}^{t} y_{s,i} y_{s,j}$). This distribution is called the *inverse Wishart* distribution. Box and Tiao (1973) and Gelman, Carlin, Stern and Rubin (1995) give more details on the standard Bayes inference with multivariate Gaussian observations. An introduction on the general Bayesian inference can be found in the Appendix.

By letting $\mathcal{Z} = \Sigma^{-1}$, we see that $\mathcal{Z}$ follows a Wishart(t, S) distribution:

$$p(\mathcal{Z} \mid \text{complete data}) \propto \mathcal{Z}^{\frac{t-d-1}{2}} \exp\left\{-\frac{1}{2} \operatorname{tr}[\mathcal{Z} \cdot S]\right\}. \tag{2.14}$$

See Johnson and Kotz (1972) for a detailed derivation. In this distribution, n is often referred as its *degree of freedom* and S its scale matrix (required to be positive definite). Sampling from this Wishart distribution when $t \geq d+1$ can be accomplished as follows: Simulate t independent samples $\epsilon_1, \ldots, \epsilon_t$ from a d-dimensional multivariate Gaussian distribution, $N(0, S)$, and then let $Z = \sum_{i=1}^{t} \epsilon_i \epsilon_i^T$. Suppose S is decomposed as $S = CC^T$, a more efficient algorithm proposed by Odell and Feiveson (1966) is as follows:

(a) Simulate independent random samples $V_j \sim \chi^2(t-j)$, $j = 1, \ldots, d$.

(b) Simulate independent random variables $N_{ij} \sim N(0, 1)$, for $i < j \leq d$.

(c) Construct a symmetric matrix $B = (b_{ij})_{d \times d}$ as follows:

$$b_{11} = V_1, \quad b_{1j} = N_{1j}\sqrt{V_1};$$

$$b_{jj} = V_j + \sum_{i=1}^{j-1} N_{ij}^2, \quad j = 2, \ldots, d;$$

$$b_{ij} = N_{ij}\sqrt{V_i} + \sum_{k=1}^{i-1} N_{ki}N_{kj}, \quad i < j \leq d.$$

(d) Then, $Z = CBC^T$ follows the Wishart(t, S) distribution in (2.14).

Now let us go back to the original problem of making inference on ρ with incomplete data. In this case, we can write down the joint posterior distribution of Σ and missing data $\mathbf{y}_{\text{mis}} = (y_{5,2}, \ldots, y_{8,2}, y_{9,1}, \ldots, y_{12,1})$ as

$$p(\Sigma, \mathbf{y}_{\text{mis}} \mid \mathbf{y}_{\text{obs}}) \propto p(\mathbf{y}_{\text{mis}}, \mathbf{y}_{\text{obs}} \mid \Sigma) p(\Sigma)$$

$$\propto |\Sigma|^{-\frac{12+3}{2}} \exp\left\{-\frac{1}{2}\text{tr}[\Sigma^{-1} \cdot S(\mathbf{y}_{\text{mis}})]\right\},$$

where $S(\mathbf{y}_{\text{mis}})$ emphasizes that its value depends on the value of $\mathbf{y}_{\text{mis}}$. Thus, the posterior distribution of Σ can be derived from the above joint distribution with $\mathbf{y}_{\text{mis}}$ integrated out. An importance sampling algorithm for achieving this goal can be implemented as follows:

- Sample Σ from some trial distribution $g_0(\Sigma)$,
- Draw $\mathbf{y}_{\text{mis}}$ from $p(\mathbf{y}_{\text{mis}} \mid \mathbf{y}_{\text{obs}}, \Sigma)$.

It should be noted that given Σ, the predictive distribution of $\mathbf{y}_{\text{mis}}$ is simply the Gaussian distribution. For example,

$$[y_{5,2} \mid \Sigma, \mathbf{y}_{\text{obs}}] = [y_{5,2} \mid \Sigma, y_{5,1}] \sim N(\mu_*, \sigma_*^2)$$

where $\mu_* = \rho y_{5,1}\sqrt{\sigma_{11}/\sigma_{22}}$ and $\sigma_*^2 = (1-\rho^2)\sigma_{11}$. Thus, the key question is how to draw Σ. A simple idea is to draw Σ from its posterior distribution conditional only on the first four complete observations:

$$g_0(\Sigma) \propto |\Sigma|^{-7/2} \exp\{-\text{tr}[\Sigma^{-1} S_4]/2\},$$

where S_4 refers to the sum of the square matrix computed from the first four observations. With g_0 so chosen, the estimated coefficient of variation of the importance weights is 2.25 with 5000 Monte Carlo samples. Of course, other choices of Σ are also possible.

For a comparison, we obtained the analytical form of the observed-data posterior distribution of ρ up to a normalizing constant:

$$p(\rho \mid \text{data in Table 2.1}) \propto \frac{[(1-\rho^2)^{4.5}]}{[(1.25-\rho^2)^8]}.$$

Figure 2.2 displays the Monte Carlo estimates of the density versus the true posterior density of ρ.

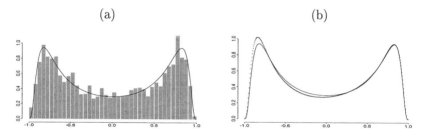

FIGURE 2.2. Importance sampling estimate of the posterior density of the correlation coefficient ρ (with 5000 iterations) for a bivariate Gaussian model with Murray's (1977) data. (a) the usual estimate with $m=5000$ overlaid by the "true" density; (b) the estimate resulting from Rao-Blackwellization (Section 2.5.5) with $m=1000$ (by courtesy of Mr. Yuguo Chen).

2.6 Advanced Importance Sampling Techniques

2.6.1 Adaptive importance sampling

It is often a good idea to "learn" about the target distribution of interest along with Monte Carlo sampling. A simple way of achieving this is to start with a trial density, say $g_0(\mathbf{x}) = t_\alpha(\mathbf{x}; \mu_0, \Sigma_0)$, where t_α represents a t-distribution with α degrees of freedom. With weighted Monte Carlo samples, one can estimate the mean and covariance matrix, denoted as μ_1 and Σ_1, respectively, of the target distribution. Then, a new trial density can be constructed as $g_1(\mathbf{x}) = t_\alpha(\mathbf{x}, \mu_1, \Sigma_1)$ (Oh and Berger 1992). This procedure can be iterated until a certain measure of discrepancy between the trial distribution and the target distribution, such as the coefficient of variation of the importance weights, does not improve any more.

Another way of doing an adaptation (Oh and Berger 1993) is to assume a parametric form for the new trial density $g_1(\mathbf{x})$ [e.g., suppose it is of the form $g(\mathbf{x}; \lambda)$]. Then, we try to find the optimal choice of λ, defined as the one that minimizes, say, the estimated coefficient of variation of the importance weights, based on the current sample. Let $w(\mathbf{x}; \lambda) = \pi(\mathbf{x})/g_1(\mathbf{x})$. We note that

$$\mathrm{var}_{g_1}(w) = \int \frac{\pi^2(\mathbf{x})}{g_1(\mathbf{x})g_0(\mathbf{x})} g_0(\mathbf{x}) d\mathbf{x} - 1.$$

Suppose we have drawn $\mathbf{x}^{(1)}, \ldots, \mathbf{x}^{(m)}$ from the trial distribution $g_0(\mathbf{x})$. The coefficient of variation of $\pi(\mathbf{x})/g_1(\mathbf{x})$ can be approximated as

$$\widehat{cv}^2(\lambda) = \bar{H}(\lambda) - 1,$$

where $\bar{H}(\lambda) = \sum_{j=1}^m H^{(j)}(\lambda)/m$ and

$$H^{(j)}(\lambda) = \frac{\pi^2(\mathbf{x}^{(j)})}{g_1(\mathbf{x}^{(j)})g_0(\mathbf{x}^{(j)})} = \frac{\{\pi(\mathbf{x}^{(j)})/g_0(\mathbf{x}^{(j)})\}^2}{g(\mathbf{x}^{(j)}; \lambda)/g_0(\mathbf{x}^{(j)})}.$$

When $\pi(\mathbf{x})$ can only be evaluated up to a normalizing constant, we need to use the estimate

$$\widehat{cv}^2(\lambda) = \frac{\bar{H}(\lambda)}{\bar{W}_0^2} - 1,$$

where $\bar{W}_0$ is the sample average of the un-normalized importance ratio $\pi(\mathbf{x}^{(j)})/g_0(\mathbf{x}^{(j)})$. Then, we can implement a Newton-Raphson method to find the optimal λ. Of particular interest is that for $\lambda = (\epsilon, \mu, \Sigma)$,

$$g(\mathbf{x}; \lambda) = \epsilon g_0(\mathbf{x}) + (1 - \epsilon) t_\nu(\mathbf{x}; \mu, \Sigma);$$

that is, the "improved version" is the mixture of the previous trial distribution g_0 with a new parametric component.

The reader should be cautious in using these adaptive methods since they are typically unstable. Perhaps a less greedy but more robust approach is to minimize a more robust distance measure between the trial and the target densities (e.g., the Hellinger or the Kullback-Leibler distance).

2.6.2 Rejection and weighting

When implementing the rejection method, one needs to find a trial density $g(\ )$ and an envelope constant M such that $\pi(\mathbf{x}) \leq Mg(\mathbf{x})$ for all $\mathbf{x}$. Its efficiency is determined as $1/M$; that is, on the average, M random samples have to be generated in order to produce an accepted one. Thus, finding a good M is crucial, but is nontrivial. Suppose one uses a reasonable M but is unsure whether the envelope inequality holds in the entire support of $\pi(\cdot)$; one can, in fact, accept those $\mathbf{x}$'s that lie in the region $\{\mathbf{x} : \pi(\mathbf{x}) > Mg(\mathbf{x})\}$, and adjust the bias by giving these samples appropriate weights. In this way, we may achieve faster computation and better efficiency.

When applying importance sampling, one often produces random samples with very small importance weights because of a less than ideal trial density. Suppose we are interested in estimating $E_\pi[h(\mathbf{x})]$, but the evaluation of $h(\mathbf{x})$ is expensive. In this case, we would like to evaluate as few samples as possible but without losing much information or creating a bias. The following simple technique for combining rejection and importance weighting can be used.

Suppose we have drawn samples $\mathbf{x}^{(1)},\ldots,\mathbf{x}^{(m)}$ from $g(\mathbf{x})$. Let $w^{(j)} = \pi(\mathbf{x}^{(j)})/g(\mathbf{x}^{(j)})$. We can conduct the the following operation for any given threshold value $c > 0$:

Rejection Control (RC)

- For $j = 1,\ldots,m$, accept $\mathbf{x}^{(j)}$ with probability

$$r^{(j)} = \min\left\{1, \frac{w^{(j)}}{c}\right\}.$$

- If the jth sample $\mathbf{x}^{(j)}$ is accepted, its weight is updated to $w^{(*j)} = q_c w^{(j)}/r^{(j)}$, where

$$q_c = \int \min\left\{1, \frac{w(\mathbf{x})}{c}\right\} g(\mathbf{x}) d\mathbf{x}.$$

Constant q_c is maintained only for conceptual clarity instead of computational need in estimating a expectation with respect to π. This is because q_c, the same for all the accepted samples, is not needed for the evaluation of the ratio estimate in (2.3). But in cases where one is interested in estimating the normalizing constant of the target distribution (also called the *partition function*), one may need a good estimate of q_c.

The above RC scheme can be viewed as a technique for adjusting the trial density g in light of current importance weights. The new trial density $g^*(\mathbf{x})$ resulting from this adjustment is expected to be closer to the target function $\pi(\mathbf{x})$. In fact, it can be seen that

$$g^*(\mathbf{x}) = q_c^{-1} \min\{g(\mathbf{x}), \pi(\mathbf{x})/c\}. \tag{2.15}$$

Because of the relationship

$$q_c = \int \min\{g(\mathbf{x}), \pi(\mathbf{x})/c\} d\mathbf{x} = E_g \min\left\{1, \frac{w(\mathbf{x})}{c}\right\} d\mathbf{x},$$

the normalizing constant q_c can be unbiasedly estimated from the sample as

$$\hat{p}_c = \frac{1}{m} \sum_{j=1}^{m} \min\left\{1, \frac{w^{(j)}}{c}\right\}.$$

2.6 Advanced Importance Sampling Techniques 45

After applying rejection control, we will typically have fewer than N samples. More samples can be drawn from either $g(x)$ or $g^*(x)$ (via rejection control) to make up for the rejected samples. The usefulness of the method in sequential importance sampling as shown by Liu, Chen and Wong (1998) will be discussed in the next chapter. Theoretically, one can show the following:

Theorem 2.6.1 *The rejection control method indeed reduces the χ^2 distance between the target distribution and the modified trial distribution; that is,*

$$\mathrm{var}_{g*}[\pi(\mathbf{x})/g^*(\mathbf{x})] \leq \mathrm{var}_g[\pi(\mathbf{x})/g(\mathbf{x})]. \qquad (2.16)$$

Proof: With $w(\mathbf{x}) = \pi(\mathbf{x})/g(\mathbf{x})$, the rejection probability q_c in (2.15) can be expressed as

$$q_c = \int \min\left\{g(\mathbf{x}), \frac{\pi(\mathbf{x})}{c}\right\} d\mathbf{x} = \frac{1}{c} E_g[\min\{w(\mathbf{x}), c\}]. \qquad (2.17)$$

On the other hand, we have

$$1 + \mathrm{var}_{g*}\left\{\frac{\pi(\mathbf{x})}{g^*(\mathbf{x})}\right\} = \int \frac{\pi^2(\mathbf{x})}{g^*(\mathbf{x})} d\mathbf{x} = q_c \int \frac{\pi(\mathbf{x})}{\min\{g(\mathbf{x}), \pi(\mathbf{x})/c\}} \pi(\mathbf{x}) d\mathbf{x}$$

$$= \int q_c \max\{w(\mathbf{x}), c\} \pi(\mathbf{x}) d\mathbf{x} \qquad (2.18)$$

$$= q_c E_g[\max\{w(\mathbf{x}), c\} w(\mathbf{x})]. \qquad (2.19)$$

Now we show that for any $w_1 > 0, w_2 > 0$,

$$h(w_1, w_2) = [\min\{w_1, c\} - \min\{w_2, c\}][w_1 \max\{w_1, c\} - w_2 \max\{w_2, c\}] \geq 0.$$

There are three scenarios for value c: (i) $\min(w_1, w_2) > c$, then $h(w_1, w_2) = 0$; (ii) $c > \max(w_1, w_2)$, then $h(w_1, w_2) = c(w_1 - w_2)^2 \geq 0$; and (iii) c is between w_1 and w_2, in which case we assume without loss of generality that $w_1 \leq c \leq w_2$, then

$$h(w_1, w_2) = (c - w_1)(w_2^2 - cw_1) \geq 0.$$

Hence, the two random variables $\min\{w(\mathbf{x}), c\}$ and $w(\mathbf{x}) \max\{w(\mathbf{x}), c\}$ are positively correlated. Together with the fact that

$$\min\{w(\mathbf{x}), c\} \max\{w(x), c\} = cw(x),$$

and formulas (2.17) and (2.19), we have

$$c\left[1 + \mathrm{var}_{g*}\left\{\frac{\pi(\mathbf{x})}{g^*(\mathbf{x})}\right\}\right] = E_g[\min\{w(\mathbf{x}), c\}] E_g[\max\{w(\mathbf{x}), c\} w(\mathbf{x})]$$

$$\leq E_g[\min\{w(\mathbf{x}), c\} \max\{w(\mathbf{x}), c\} w(\mathbf{x})]$$

$$= cE_g[w^2(\mathbf{x})] = c\left[1 + \mathrm{var}_g\left\{\frac{\pi(\mathbf{x})}{g(\mathbf{x})}\right\}\right].$$

Hence, we have proved the result (2.16). ◇

2.6.3 Sequential importance sampling

It is nontrivial to design a good trial distribution for doing importance sampling in high-dimensional problems. One of the most useful strategies in these problems is to build up the trial density sequentially. Suppose we can decompose $\mathbf{x}$ as $\mathbf{x} = (x_1, \ldots, x_d)$ where each of the x_j may be multidimensional. Then, our trial density can be constructed as

$$g(\mathbf{x}) = g_1(x_1) g_2(x_2 \mid x_1) \cdots g_d(x_d \mid x_1, \ldots, x_{d-1}), \quad (2.20)$$

by which we hope to obtain some guidance from the target density while building up the trial density. Corresponding to the decomposition of $\mathbf{x}$, we can rewrite the target density as

$$\pi(\mathbf{x}) = \pi(x_1) \pi(x_2 \mid x_1) \cdots \pi(x_d \mid x_1, \ldots, x_{d-1}) \quad (2.21)$$

and the importance weight as

$$w(\mathbf{x}) = \frac{\pi(x_1) \pi(x_2 \mid x_1) \cdots \pi(x_d \mid x_1, \ldots, x_{d-1})}{g_1(x_1) g_2(x_2 \mid x_1) \cdots g_d(x_d \mid x_1, \ldots, x_{d-1})}. \quad (2.22)$$

Equation (2.22) suggests a recursive way of computing and monitoring the importance weight; that is, by denoting $\mathbf{x}_t = (x_1, \ldots, x_t)$ (thus, $\mathbf{x}_d \equiv \mathbf{x}$), we have

$$w_t(\mathbf{x}_t) = w_{t-1}(\mathbf{x}_{t-1}) \frac{\pi(x_t \mid \mathbf{x}_{t-1})}{g_t(x_t \mid \mathbf{x}_{t-1})}.$$

At the end, w_d is equal to $w(\mathbf{x})$ in (2.22). Potential advantages of this recursion and (2.21) are the following: (a) We can stop generating further components of $\mathbf{x}$ if the *partial weight* derived from the sequentially generated *partial sample* is too small and (b) we can take advantage of $\pi(x_t \mid \mathbf{x}_{t-1})$ in designing $g_t(x_t \mid \mathbf{x}_{t-1})$. In other words, the marginal distribution $\pi(\mathbf{x}_t)$ can be used to guide the generation of $\mathbf{x}$.

Although the above "idea" sounds interesting, the trouble is that the decomposition of π as in (2.21) and that of w as in (2.22) are impractical at all! The reason is that in order to get (2.21), one needs to have the marginal distribution

$$\pi(\mathbf{x}_t) = \int \pi(x_1, \ldots, x_d) dx_{t+1} \cdots dx_d,$$

whose computation involves integrating out components $x_{t+1}, \ldots, x_d$ in $\pi(\mathbf{x})$ and is as difficult as — or even more difficult than — the original problem.

In order to carry out the sequential sampling idea, we need to introduce another layer of complexity. Suppose we can find a sequence of "auxiliary distributions," $\pi_1(x_1), \pi_2(\mathbf{x}_2), \ldots, \pi_d(\mathbf{x})$, so that $\pi_t(\mathbf{x}_t)$ is a reasonable approximation to the marginal distribution $\pi(\mathbf{x}_t)$, for $t = 1, \ldots, d-1$ and

$\pi_d = \pi$. We want to emphasize that the π_t are only required to be known up to a normalizing constant and they *only* serve as "guides" to our construction of the whole sample $\mathbf{x} = (x_1, \ldots, x_d)$. The *sequential importance sampling* (SIS) method can then be defined as the following recursive procedure (for $t = 2, \ldots, d$).

SIS Step:

(A) Draw $X_t = x_t$ from $g_t(x_t|\mathbf{x}_{t-1})$, and let $\mathbf{x}_t = (\mathbf{x}_{t-1}, x_t)$.

(B) Compute
$$u_t = \frac{\pi_t(\mathbf{x}_t)}{\pi_{t-1}(\mathbf{x}_{t-1}) g_t(x_t \mid \mathbf{x}_{t-1})}, \qquad (2.23)$$
and let $w_t = w_{t-1} u_t$.

In the SIS step, we call u_t an "incremental weight." It is easy to show that $\mathbf{x}_t$ is properly weighted by w_t with respect to π_t provided that $\mathbf{x}_{t-1}$ is properly weighted by w_{t-1} with respect to π_{t-1}. Thus, the whole sample $\mathbf{x}$ obtained in this sequential fashion is properly weighted by the final importance weight, w_d, with respect to the target density $\pi(\mathbf{x})$. One reason for the sequential buildup of the trial density is that it breaks a difficult task into manageable pieces. The SIS framework is particularly attractive, as it can use the sequence of "auxiliary distributions" $\pi_1, \pi_2, \ldots, \pi_d$ to help construct more efficient trial distribution:

- We can build g_t in light of π_t. For example, one can choose (if possible)
$$g_t(x_t \mid \mathbf{x}_{t-1}) = \pi_t(x_t \mid \mathbf{x}_{t-1}).$$
Then, the incremental weight becomes
$$u_t = \pi_t(\mathbf{x}_t)/\pi_{t-1}(\mathbf{x}_{t-1}).$$

- When we observe that w_t is getting too small, we can choose to reject the sample halfway and restart again. In this way, we avoid wasting time on generating samples that are doomed to have little effect in the final estimation. However, as an outright rejection incurs bias, the rejection control technique described in Section 2.6.2 can be used to correct such bias (Section 2.6.4)

In configurational bias Monte Carlo (Siepmann and Frenkel 1992), the SIS is used as a proposal (independent) transition in a Metropolis-Hastings algorithm (see Section 5.4.3).

The most important unanswered question in the SIS framework is how to find a reasonable set of "auxiliary distributions." This issue will be illustrated through several practical examples in Chapters 3 and 4. For example, in a nonlinear filtering problem, the "auxiliary distributions" often correspond to the "current" posterior distributions of the true signals.

48   2. Basic Principles: Rejection, Weighting, and Others

2.6.4 Rejection control in sequential importance sampling

Although the rejection control method (Section 2.6.2) was described in a "static" form, it can be applied dynamically to improve an SIS scheme. Suppose a sequence of "check points," $0 < t_1 < t_2 < \cdots < t_k \leq d$, and a sequence of threshold values $c_1, \ldots, c_k$, are given in advance. The following procedure can be implemented:

1. At each check point t_j, start RC(t_k) as described in Section 2.6.2 with the threshold value $c = c_j$. If the partial sample $(x_1, \ldots, x_{t_j})$ has a weight w_{t_j}, then we accept this partial sample with probability $\min\{1, w_{t_j}/c_j\}$ and, if accepted, replace its weight by $w^*_{t_j} = \max\{w_{t_j}, c_j\}$.

2. For each rejected partial sample, restart from the beginning again and let it pass through all the check points at $t_1, \ldots, t_j$, with threshold values $c_1, \ldots, c_j$, respectively. If rejected in any middle check point, start again.

Note that after the first rejection control at stage t_1, the sampling distribution $g^*_t(\mathbf{x}_t)$ for $\mathbf{X}_t$ is no longer the same as the one described in (2.20). It is shown by Liu, Chen and Wong (1998) that for any time t, partial sample $\mathbf{x}_t$ resulting from the above procedure is properly weighted with respect to π_t by their modified weights w^*_t. To retain a proper estimate of the normalizing constant for π, one has to estimate p_c, the probability of acceptance, and adjust the weight to $p_c w^*_t$. Since this method requires that each rejected sample be restarted from stage 0, it tends to be impractical when the number of components d is large. An interesting way to combine the RC operation with resampling is described in Section 3.4.5.

2.7 Application of SIS in Population Genetics

Evolutionary theory holds that stochastic mutational events may alter the genome of an individual and that these changes may be passed to its progeny. Thus, comparing homologous DNA regions (segments) of a random sample of individuals taken from a certain population can shed light on the evolutionary process of this population. This comparison can also yield important information for locating genes that are responsible for genetic diseases. Recent advances in the biotechnology revolution have provided a wealth of DNA sequence data for which meaningful studies on the evolution process can be made and biologically verified.

Following Griffiths and Tavare (1994) and Stephens and Donnelly (2000), we consider the simplest demographic model focusing on populations of constant size N which evolve in non-overlapping generations. Each individual in a population is a sufficiently small chromosomal region in which no recombination is allowed (in reality, recombination can happen with a very small probability). Thus, each chromosomal segment seen in the dataset can be treated as a descendent of a single parental segment in the previous generation — and it is sufficient to consider the haploid model (i.e., each "individual" only has one parent). Each segment has a genetic type and the set of all possible types is denoted as E. If a parental segment is of type $\alpha \in E$, then its progeny is of type $\alpha \in E$ with probability $1 - \mu$ and of type $\beta \in E$ with probability $\mu P_{\alpha\beta}$. Thus, μ can be seen as the mutation rate per chromosome per generation. The mutation transition matrix $\boldsymbol{P} = (P_{\alpha\beta})$ is assumed to have a unique stationary distribution.

Suppose we observe a random sample from the current population assumed to be at stationarity. The ancestral relationships among the individuals in the sample — when being traced back to their most recent common ancestor (MRCA) — can be described by a tree. Figure 2.3 shows a genealogical tree for an example when the segment has only two genetic types, C and T (i.e., $E = \{C, T\}$), and the sample consists of five observations.

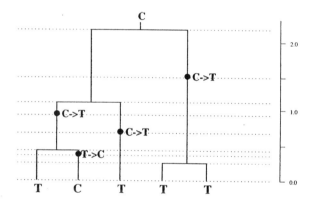

FIGURE 2.3. Illustration of a genealogical tree. The set of five observed individuals at the current time is $\{T, C, T, T, T\}$ (the bottom of the tree). The plotted tree illustrates a possible route for these five individuals to descend from a common ancestor of type C. Ancestral lineages are jointed by horizontal lines (and are said to coalesce) when they share a common ancestor. The dots represent mutations and the horizontal dotted lines indicate the times at which events (coalescence and mutations) occur. The history $\mathcal{H} = (H_{-k}, H_{-(k-1)}, \ldots, H_{-1}, H_0)$ in this case is $(\{C\}, \{C, C\}, \{C, T\}, \{C, C, T\}, \{C, T, T\}, \{T, T, T\}, \{T, T, T, T\}, \{C, T, T, T\}, \{C, T, T, T, T\})$. Reproduced from Stephens and Donnelly (2000).

Stephens and Donnelly (2000) used $\mathcal{H} = (H_{-m}, \ldots, H_{-1}, H_0)$ to denote the whole ancestral history (unobserved) of the observed individuals at the present time, where k is the first time when all the individuals in the sample coalesce (i.e., the first time they have a common ancestor). Each H_{-i} is an unordered list of genetic types of the ancestors i generations ago. Thus, the history $\mathcal{H}$ has a one-to-one correspondence with the tree topology (evolution time is not reflected in $\mathcal{H}$). Note that only H_0 is observable. For any given $\mathcal{H}$ that is compatible with H_0, however, we can compute the likelihood function $p_\theta(\mathcal{H})$ as

$$p_\theta(\mathcal{H}) \propto p_\theta(H_{-k}) p_\theta(H_{-k+1} \mid H_{-k}) \cdots p_\theta(H_0 \mid H_{-1}) p_\theta(\text{stop} \mid H_0).$$

Here, $p_\theta(H_{-k}) = \pi_0(H_{-k})$, with π_0 being the stationary distribution of $\boldsymbol{P}$. The coalescence theory (Griffiths and Tavare 1994, Stephens and Donnelly 2000) tells us that

$$p_\theta(H_i \mid H_{i-1}) = \begin{cases} \dfrac{n_\alpha}{n} \dfrac{\theta}{n-1+\theta} P_{\alpha\beta} & \text{if } H_i = H_{i-1} - \alpha + \beta \\ \dfrac{n_\alpha}{n} \dfrac{n-1}{n-1+\theta} & \text{if } H_i = H_{i-1} + \alpha \\ 0 & \text{otherwise,} \end{cases}$$

for $i = -(k-1), \ldots, 0$ and the process is stopped just before a new genetic type is produced:

$$p_\theta(\text{stop}|H_0) = \sum_\alpha \frac{n_\alpha}{n} \frac{n-1}{n-1+\theta}. \qquad (2.24)$$

Here, n is the sample size at generation H_{i-1} and n_α is the number of chromosomes of type α in the sample. The notation $H_i = H_{i-1} + \alpha$ indicates that the new generation H_i is obtained from H_{i-1} by a split of a line of type α, and the notation $H_i = H_{i-1} - \alpha + \beta$ means that H_i is obtained from H_{i-1} by a mutation from a type α to a type β. The parameter $\theta = 2N\mu/\nu$, with N being the population size (assumed to be constant throughout the history) and ν^2 being the variance of the number of progeny of a random chromosome.

To infer the value of θ, one can use the MLE method (see the Appendix for more descriptions), which requires us to compute for each given θ the likelihood value

$$p_\theta(H_0) = \sum_{\mathcal{H}:\text{compatible with } H_0} p_\theta(\mathcal{H}).$$

This computation cannot be solved analytically and we have to resort to some approximation methods — Monte Carlo appears to be a natural choice. In a naive Monte Carlo, one may randomly choose the generation number k and then simulate forward from H_{-k}, which only has a single individual, to H_0. However, except for trivial dataset, such simulated history $\mathcal{H}$ has little chance to be compatible with the observed H_0. An alternative strategy is to simulate $\mathcal{H}$ backward starting from H_0 and then use weight to correct bias. In a sequential importance sampling strategy (equivalent to the method of Griffiths and Tavare (1994)), we can simulate $H_{-1}, H_{-2}, \ldots$

2.7 Application of SIS in Population Genetics

from a trial distribution built up sequentially by reversing the forward sampling probability at a fixed θ_0; that is, for $t = 1, \ldots, k$, we have

$$g_t(H_{-t}|H_{-t+1}) = \frac{p_{\theta_0}(H_{-t+1}|H_{-t})}{\sum_{\text{all } H'_{-t}} p_{\theta_0}(H_{-t+1}|H'_{-t})},$$

and the final trial distribution

$$g(\mathcal{H}) = g_1(H_{-1} \mid H_0) \cdots g_k(H_{-k} \mid H_{-k+1}).$$

In other words, each g_t is the "local" posterior distribution of H_{-t}, under a uniform prior, conditional on H_{-t+1}. By simulating from $g(\ )$ multiple copies of the history, $\mathcal{H}^{(j)}, j = 1, \ldots, m$, we can approximate the likelihood function as

$$\hat{p}_\theta(H_0) = \frac{1}{m} \sum_{j=1}^m \frac{p_\theta(\mathcal{H}^{(j)})}{g(\mathcal{H}^{(j)})}.$$

In this approach, the choice of θ_0 can influence the final result. We tested this importance sampling method on a small test dataset in Stephens and Donnelly (2000), $\{8, 11, 11, 11, 11, 12, 12, 12, 12, 13\}$, with $E = \{0, 1, \ldots, 19\}$ and a simple random walk mutation transition on E. Figure 2.4 shows the likelihood curve of θ estimated from m=1,000,000 Monte Carlo samples and $\theta_0 = 10$. Stephens and Donnelly (2000) recently proposed a new SIS construction of the trial distribution and is significantly better than the simple construction described in this section. We will discuss a resampling method in Section 4.1.2 that can improve both algorithms.

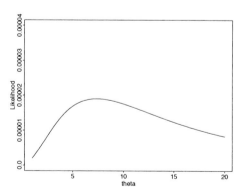

FIGURE 2.4. The estimation of the likelihood function for a dataset in Stephens and Donnelly (2000), with $\theta_0 = 10$ and m=1,000,000 iterations.

2.8 Problems

1. Evaluate integral $\int_0^1 \sin^2(1/x)dx$ by both a deterministic method and a Monte Carlo method. Comment on relative advantages and disadvantages.

2. Prove that the rejection method does produce random variables that follow the target distribution π. Show that the expected acceptance rate is $1/c$, where c is the "envelope constant."

3. Implement an adaptive importance sampling algorithm to evaluate mean and variance of a density

$$\pi(\mathbf{x}) \propto N(\mathbf{x}; \mathbf{0}, 2I_4) + 2N(\mathbf{x}; 3e, I_4) + 1.5N(\mathbf{x}; -3e, D_4),$$

where $e = (1, 1, 1, 1)$, $I_4 = \text{diag}(1, 1, 1, 1)$, and $D_4 = \text{diag}(2, 1, 1, .5)$. A possible procedure is as follows:

   - Start with a trial density $g_0 = t_\nu(0, \Sigma)$;
   - Recursively, we build

$$g_k(\mathbf{x}) = (1 - \epsilon)g_{k-1}(\mathbf{x}) + \epsilon t_\nu(\mu, \Sigma),$$

   in which one chooses (ϵ, μ, Σ) to minimize the variation of coefficient of the importance weights.

4. Describe the process of simulating from a Wishart distribution and prove that the proposed method is correct.

5. Formulate the method described in Section 2.5.6 for the continuous state space and use it to solve a differential equation.

3
Theory of Sequential Monte Carlo

In the previous chapter, we introduced the basic framework of sequential importance sampling (SIS), in which one builds up the trial sampling distribution sequentially and computes the importance weights recursively. More precisely, we first decompose the random variable $\mathbf{x}$ into a number of components, $\mathbf{x} = (x_1, \ldots, x_d)$, and then build up the trial distribution as

$$g(\mathbf{x}) = g_1(x_1)g_2(x_2 \mid x_1) \cdots g_d(x_d \mid x_1, \ldots, x_{d-1}), \qquad (3.1)$$

where the g_t are chosen by the investigator so as to make the joint sampling distribution of $\mathbf{x}$ as close to the target distribution, $\pi(\mathbf{x})$, as possible. A technical implication of decomposition (3.1) is that the simulation is carried out *sequentially* on a computer: The first component x_1 is drawn first, and then x_2 is drawn conditional on x_1, and so on. When this recursion is repeated independently for m times, we obtain m i.i.d. random samples $\mathbf{x}^{(1)}, \ldots, \mathbf{x}^{(m)}$ from the trial distribution $g(\mathbf{x})$, which can then be used to estimate integrals related to the target distribution π. A mathematically, but not computationally, equivalent view of this process is to start m independent sequential sampling processes *in parallel*. More precisely, at stage 1, we generate m i.i.d. samples $\{x_1^{(1)}, \ldots, x_1^{(m)}\}$ from g_1; at stage 2, we generate $x_2^{(j)}$ from $g_2(\cdot \mid x_1^{(j)})$ for $j = 1, \ldots, m$, and so on. The ideas described in this chapter will be most easily understood under this view of *parallel* operation.

The importance weight of the sample simulated from $g(\mathbf{x})$ can be evaluated either at the end when all the parts of $\mathbf{x}$ are in place, or recursively as

$$w_t = w_{t-1}\frac{\pi_t(x_1,\ldots,x_t)}{g_t(x_t \mid x_1,\ldots,x_{t-1})\pi_{t-1}(x_1,\ldots,x_{t-1})}, \quad t=1,\ldots,d,$$

in which the sequence of distributions $\pi_1,\ldots,\pi_d$ are referred to as *a sequence of auxiliary distributions*. A necessary condition for these distributions is that $\pi_d(x_1,\ldots,x_d) = \pi(x_1,\ldots,x_d)$. The π_t can be thought of as an approximation to the marginal distribution

$$\pi(x_1,\ldots,x_t) = \int \pi(\mathbf{x})dx_{t+1}\cdots dx_d$$

of the partial sample $\mathbf{x}_t = (x_1,\ldots,x_t)$.

If we have a good sequence of auxiliary distributions (i.e., those π_t's that track the marginal distributions of π), we would be able to judge the "quality" of a partially generated sample. Based on such a judgment, we can consider stopping before finishing the generation of the whole sample $\mathbf{x}$. For example, after seeing $w_2 = 0$ for the first two generated components, we would certainly want to stop simulating further along the line. Similarly, a very small weight of w_k at stage k would suggest to us a possible early stop. To avoid introducing bias, the rejection control procedure described in Section 2.6.2 should be used (Liu, Chen and Wong 1998). Since early stopping will decrease the total number of samples being produced at the end, one would like to make up these lost samples by repeating the same sampling procedure from the beginning. We show in the later part of this chapter that a surprisingly powerful strategy for making up those lost samples due to early stopping is by *resampling* from the currently available "good" partial samples.

A good sequence of auxiliary distributions $\{\pi_t\}$ can also help us choose good trial densities. As will be shown in Section 3.4, a convenient choice of g_t is

$$g_t(x_t \mid x_1,\ldots,x_{t-1}) = \pi_t(x_t \mid x_1,\ldots,x_{t-1}).$$

We will show in this and the next chapters that in many simulation and optimization problems, especially for those models of dynamic nature, the auxiliary distribution sequence $\pi_1,\pi_2,\ldots$ and the sequential sampling distributions $g_1,g_2,\ldots$ often emerge naturally.

The SIS-based Monte Carlo methods have been invented independently in at least three main research areas. The first invention dated back to the 1950s and was motivated by the problems of simulating macromolecules (Hammersley and Morton 1954, Rosenbluth and Rosenbluth 1955). The other two incidences are more recent: one was motivated by statistical missing data problems (Kong et al. 1994, Liu and Chen 1995), and the other motivated by nonlinear filtering problems in radar tracking (Gordon et al. 1993). In the remaining part of this chapter, we will trace these independent developments, summarize and abstract special techniques in each of the areas, and describe a general *sequential Monte Carlo* framework which can be used as a basis to contemplate newer techniques.

3.1 Early Developments: Growing a Polymer

The sequential Monte Carlo methodology can be traced back to Hammersley and Morton (1954) and Rosenbluth and Rosenbluth (1955) who invented a method to estimate the average squared extension of a self-avoiding random walk of length N on a lattice space. Since its inception, the Rosenbluth's method has received attentions from structural simulation and minimization community and various modifications have been proposed (Grassberger 1997, Siepmann and Frenkel 1992).

3.1.1 A simple model of polymer: Self-avoid walk

Simulation of chain polymers is perhaps one of the first serious scientific endeavors with the help of electronic computers. Although the history goes back to the 1950s, the problem of simulating a long chain of biopolymer still presents a major challenge to the scientific community because of both its difficult nature and its extreme importance in biology and chemistry (Kremer and Binder 1988, Grassberger 1997).

The self-avoiding random walk (SAW) in a two- or three-dimensional lattice space is often used as a simple model for chain polymers such as polyester and polyethylene. For the sake of a concise description of the basic method, we content ourselves with the two-dimensional lattice model. In such a model, the realization of a chain polymer of length N is fully characterized by the positions of all of its molecules, $\mathbf{x} = (x_1, \ldots, x_N)$, where x_i is a point in the 2-D lattice space [i.e., $x_i = (a,b)$, where a and b are integers]. The distance between x_i and x_{i+1} has to be exactly 1, and $x_{i+1} \neq x_k$ for all $k < i$. The line connecting x_i and x_{i+1} is called a (covalent) *bond*. Figure 3.1 shows two realizations of a simple random walk on a 2-D lattice space, of which the first one is a self-avoiding walk (i.e., the chain does not cross itself), whereas the second one is not. It is not difficult to understand why it is appealing to use the trace of an SAW to imitate a chain polymer — they do look similar.

The force interaction between a pair of nonadjacent monomers (with no covalent bonds) in a macromolecule is usually very weak when they are certain distance apart and their mutual repulsion becomes huge once they are closer than a distance limit. A good approximation to this type of polymers is the hard-shell model. An even simpler one is the lattice model. A possible Boltzmann distribution for the chain polymer modeled by an SAW is the *uniform* distribution. More formally, the target distribution of

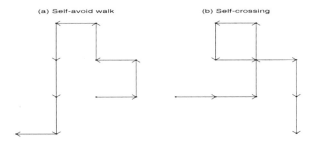

FIGURE 3.1. Two simple random walks on the square lattice. The first one is a self-avoiding walk, and the second one is not.

interest in this case is

$$\pi(\mathbf{x}) = \frac{1}{Z_N},$$

where the space of $\mathbf{x}$ is the set of all SAWs of length N and Z_N is the normalizing constant, which is just the total number of different SAWs with N atoms. Of interest to scientists is to understand certain descriptive statistics regarding such SAWs. For example, one may be interested in $E\|x_N - x_1\|^2$ (i.e., the mean squared extension of the chain).

The most naive way of simulating a SAW is to start a random walk at $(0, 0)$, and at each step i, the walker, for that he is not allowed to fall back to where it came from at step $i - 1$, chooses with equal probability one of the three allowed neighboring positions to go. If that position has already been visited earlier, the walker has to go back to $(0, 0)$ and start a new chain again. Otherwise, the walker keeps going until the presumed length N is reached. This simulation procedure is apparently inefficient: The rate (number of successes over the number of starts) of obtaining a successful SAW of given length N decreases exponentially as a function of N. For the two-dimensional lattice, this rate is roughly $\sigma_N = Z_N/(4 \times 3^{N-1})$. For $N = 20$, this rate is approximately $\sigma_{20} \approx 21.6\%$, and for $N = 48$, this number is as low as 0.79% (Lyklema and Kremer 1986)

3.1.2 Growing a polymer on the square lattice

To overcome the obvious attrition problem of the simple random walk method, Hammersley and Morton (1954) and Rosenbluth and Rosenbluth (1955) introduced essentially identical methods, one termed *"inversely restricted sampling"* and the other *"biased sampling."* Perhaps because of the names coined, the method is much more well known nowadays as the "Rosenbluth method" in the molecular simulation community. In order to be fair to both sets of authors, in this book we call their method the *growth method* since the approach can be intuitively seen as "growing" a polymer one monomer a time. This method is essentially a one-step-look-ahead strategy and is a special sequential importance sampler.

Without loss of generality, we assume that the simulated SAW is always started at position $(0,0)$ and pays its first visit to position $(1,0)$; that is, we let $x_1 = (0,0)$ and $x_2 = (1,0)$. Suppose at stage t, the random walker has made $t-1$ moves with no violation of the self-avoidance constraint and is located at position $x_t = (i,j)$. Then, in order to place x_{t+1}, the walker first examines all the neighbors of x_t [i.e., $(i \pm 1, j)$ and $(i, j \pm 1)$]. If all of the neighbors have been visited before, the walk is terminated and the whole trace is given a weight 0, otherwise the walker selects one of the available positions (not visited before) with equal probability and places his $(t+1)$st step. Mathematically, this scheme is equivalent to drawing the position of x_{t+1} conditional on the current configuration of $(x_1, \ldots, x_t)$, according to the probability distribution

$$P[x_{t+1} = (i', j') \mid x_1, \ldots, x_t] = \frac{1}{n_t},$$

where (i', j') is one of the unoccupied neighbors of x_t and n_t is the total number of such unoccupied neighbors. Clearly, this scheme is much more efficient than the simple random walk method. A self-avoiding walk with $N = 150$ produced by this method is shown in Figure 3.2

A Self-Avoiding Walk of Length N=150

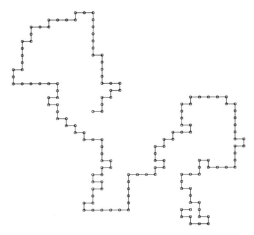

FIGURE 3.2. A SAW with 150 atoms on a 2-D lattice space, obtained by the growth method.

It is easy to check, however, that the SAWs produced by this "growth" method is not uniformly distributed (it tends to bias in favor of more "compact" configurations). For example, the probability of generating a chain as in Figure 3.3(a) by the growth method $\frac{1}{4} \times \frac{1}{3} \times \frac{1}{3} \times \frac{1}{2}$, whereas

58   3. Theory of Sequential Monte Carlo

the probability of chain in Figure 3.3(b) is $\frac{1}{4} \times \frac{1}{3} \times \frac{1}{3} \times \frac{1}{3}$. To correct for

FIGURE 3.3. Two different configurations for SAW with $N = 5$ nodes. (a) and (b) have different sampling probabilities according to the Rosenbluth method.

the bias introduced by this more efficient sampling scheme, Hammersley, Morton, and the Rosenbluths noticed that a successfully produced SAW in this way needs to be assigned a weight computed as

$$w(\mathbf{x}) = n_1 \times n_2 \times \cdots \times n_{N-1}.$$

Because the target distribution is $\pi(\mathbf{x}) \propto 1$ and the sampling distribution of $\mathbf{x}$ is $(n_1 \times \cdots \times n_{N-1})^{-1}$, the correctness of this weight assignment can be easily verified from the importance sampling principle.

Now, we would like to entertain the above procedure by the *sequential importance sampling* (SIS) framework introduced in Section 2.6.3. Clearly, the sampling distribution $g_t(x_t \mid \mathbf{x}_{t-1})$ in the growth method is equal to $1/n_{t-1}$ when $n_{t-1} > 0$ and x_t occupies one of x_{t-1}'s available neighbors, and is equal to 0 when $n_{t-1} = 0$. Here, n_{t-1} is the number of available neighbors of x_{t-1}. The sequence of auxiliary distributions, $\pi_t(\mathbf{x}_t)$, $t = 1, \ldots, N-1$, is just the sequence of *uniform distributions* of SAWs with t nodes (or $t-1$ steps); that is, $\pi_t(\mathbf{x}_t) = Z_t^{-1}$ for all SAWs with t nodes, where Z_t is the total number of such SAWs. It is easy to see that under π_t, the marginal distribution of the partial sample $\mathbf{x}_{t-1} = (x_1, \ldots, x_{t-1})$ is

$$\pi_t(\mathbf{x}_{t-1}) = \sum_{\text{all possible } x_t} \pi_t(\mathbf{x}_{t-1}, x_t) = \frac{n_{t-1}}{Z_t}.$$

Thus, the *conditional distribution* $\pi_t(x_t \mid \mathbf{x}_{t-1}) = 1/n_{k-1}$, which is exactly the same as our sampling distribution g_k. Thus, in this example, the sequence of auxiliary distributions, π_t, $t = 1, \ldots$, suggests to us a particular choice of g_t. This view of the growth method also suggests some possible improvements. For example, if we use another auxiliary sequence of distribution π_t^*, which is the *marginal distribution* of $\mathbf{x}_t$ under the uniform distribution of $(\mathbf{x}_t, x_{t+1})$:

$$\pi_t^*(\mathbf{x}_t) = \sum_{x_{t+1}} \frac{1}{Z_{t+1}} \propto n_t,$$

where n_t is the number of available neighbors of x_t, we can again let g_t be $\pi_t^*(x_t \mid \mathbf{x}_{t-1})$. Then, the implied SIS procedure is equivalent to a *two-step-look-ahead* approach.

As suggested in the SIS framework, the importance weight can be computed recursively as $w_{t+1} = w_t n_t$. Finally, we obtain the overall weight of the chain as

$$w_N = \prod_{t=2}^{N} \frac{1}{g_t(x_t \mid x_1, \ldots, x_{t-1})}.$$

Because our target distribution is $\pi(\mathbf{x}) = 1/Z_N$, the weight just computed satisfies the simple relationship

$$w_N = Z_N \pi(\mathbf{x})/p(\mathbf{x}).$$

Hence, $E_p(w_N) = Z_N$, which gives us a means to estimate the partition function (normalizing constant):

$$\hat{Z}_N = \bar{w}_N, \tag{3.2}$$

where $\bar{w}_N$ is the sample average of the weights obtained by repeating the biased sampling procedure multiple (m, say) times.

The Rosenbluths were initially interested in functional relationship between the mean squared extension, $\langle R_N^2 \rangle \equiv E_\pi(R_N^2)$, and the chain's length N. From the Monte Carlo simulations, they estimated a law for $E_\pi(R_N^2)$ as $E_\pi(R_N^2) \approx aN^\gamma$, with $\hat{\gamma} = 1.45$ and $\hat{a} = 0.917$. It was later shown that, asymptotically, $\gamma = 1.5$ for the square lattice (Nienhuis 1982). Another quantity of interest is the *partition function*, Z_N (i.e., the total number of possible configurations of SAWs with N nodes). By using the estimator in (3.2), the Rosenbluths guessed that $Z_N \approx c q_{\text{eff}}^N$, with $\hat{q}_{\text{eff}} \approx 2.66$ and $\hat{c} = 2.14$. It was later proven that the correct asymptotic law (as $N \to \infty$) for the partition function should be $Z_N \approx q_{\text{eff}}^N N^{\gamma-1}$, in which γ is equal to $43/32$ and q_{eff} is about 2.6385, for the two-dimensional square lattice (Kremer and Binder 1988, Nienhuis 1982).

3.1.3 Limitations of the growth method

A serious limitation of the growth method appears when one tries to simulate very large polymers, say with $N = 250$. First, the attrition problem persists and, second, Kong et al. (1994) argue that the weight process w_N after normalization is a martingale, thus having its variance stochastically increasing. So far, the most effective means to overcome this limitation is the *pruning* and *enrichment* approach used in structural simulations (Wall and Erpenbeck 1959, Grassberger 1997) and, equivalently, the *resampling* method invented by statisticians (Gordon et al. 1993, Liu and Chen 1995).

Liu, Chen and Wong (1998) introduced a rejection control method for improving the sequential sampling method (Section 2.6.4), of which we will describe a further twist — the *partial rejection control*, to improve simulation. In Section 3.4, we describe the general SIS methodology as an extension of the growth method of Hammersley, Morton, and the Rosenbluths and present, in a rather abstract form, a few strategies for improving the method. Then, in Chapter 4, we will illustrate in details how the general methodology can be applied to solve various interesting computational problems.

3.2 Sequential Imputation for Statistical Missing Data Problems

In a missing data problem (Section 1.9), we partition $\mathbf{y}$ as $(\mathbf{y}_{\mathrm{obs}}, \mathbf{y}_{\mathrm{mis}})$ where only $\mathbf{y}_{\mathrm{obs}}$ is observed and $\mathbf{y}_{\mathrm{mis}}$ is called the *missing data*. Furthermore, we assume that $\mathbf{y} \sim f(\mathbf{y} \mid \theta)$, where $f(\ )$ has a nice analytical form. Thus, both the complete-data likelihood $L(\theta \mid \mathbf{y})$ [which can be any function that is proportional to $f(\mathbf{y} \mid \theta)$] and the complete-data posterior distribution $p(\theta \mid \mathbf{y})$ are simple to compute. The central problem in this section is the computation of the *observed*-data likelihood, $L(\theta \mid \mathbf{y}_{\mathrm{obs}})$, and the *observed*-data posterior distribution, $p(\theta \mid \mathbf{y}_{\mathrm{obs}})$. This computation involves integrating out the missing data from $L(\theta \mid \mathbf{y}_{\mathrm{obs}}, \mathbf{y}_{\mathrm{mis}})$.

Suppose $\mathbf{y}_{\mathrm{obs}}$ and $\mathbf{y}_{\mathrm{mis}}$ can each be further decomposed into n corresponding components so that

$$\mathbf{y} = (y_1, \ldots, y_n) = (y_{\mathrm{obs},1}, y_{\mathrm{mis},1}, \ldots, y_{\mathrm{obs},n}, y_{\mathrm{mis},n}) \stackrel{\text{def}}{=} (\mathbf{y}_{\mathrm{obs}}, \mathbf{y}_{\mathrm{mis}}),$$

where $y_t = (y_{\mathrm{obs},t}, y_{\mathrm{mis},t})$ for $t = 1, \ldots, n$. In many applications, the y_t are independent and identically distributed given θ, but that is not a necessary assumption. For example, the unobserved state variables of a state-space model (or the hidden states of a hidden Markov model), which follow a Markov chain model, can be regarded as missing data in the sequential imputation framework. Sometimes, the missing pattern can be different for different t's. Indeed, an observation at t can be complete so that $y_t = y_{\mathrm{obs},t}$.

3.2.1 Likelihood computation

In order to estimate θ by the MLE method, the least (but not the last) we have to do is the evaluation of the likelihood function of the observed data, which can be derived by marginalizing the missing data from the complete-data likelihood function:

$$L(\theta \mid \mathbf{y}_{\mathrm{obs}}) = f(\mathbf{y}_{\mathrm{obs}} \mid \theta) \equiv \int f(\mathbf{y}_{\mathrm{obs}}, \mathbf{y}_{\mathrm{mis}} \mid \theta) d\mathbf{y}_{\mathrm{mis}}.$$

This integral, however, is often difficult to compute analytically. Suppose we can draw sample $\mathbf{y}_{\mathrm{mis}}^{(1)}, \ldots, \mathbf{y}_{\mathrm{mis}}^{(m)}$ from a trial distribution $g(\ )$. Then, we can estimate $L(\theta \mid \mathbf{y}_{\text{obs}})$ by

$$\hat{L}(\theta \mid \mathbf{y}_{\text{obs}}) = \frac{1}{m} \left\{ \frac{f(\mathbf{y}_{\text{obs}}, \mathbf{y}_{\text{mis}}^{(1)} \mid \theta)}{g(\mathbf{y}_{\text{mis}}^{(1)})} + \cdots + \frac{f(\mathbf{y}_{\text{obs}}, \mathbf{y}_{\text{mis}}^{(m)} \mid \theta)}{g(\mathbf{y}_{\text{mis}}^{(m)})} \right\} \quad (3.3)$$

as in a typical importance sampling (Section 2.5). Note that once the $\mathbf{y}_{\text{mis}}^{(j)}$ are produced, the formula (3.3) can be applied to estimate L for many different θ's. The question is, however, what would be a proper $g(\cdot)$? If only one θ, say θ_0, is in consideration, in the ideal situation, we would like to draw $\mathbf{y}_{\text{mis}}^{(j)}$ from $f(\mathbf{y}_{\text{mis}} \mid \mathbf{y}_{\text{obs}}, \theta_0)$, in which case the importance ratio is

$$\frac{f(\mathbf{y}_{\text{obs}}, \mathbf{y}_{\text{mis}} \mid \theta_0)}{f(\mathbf{y}_{\text{mis}} \mid \mathbf{y}_{\text{obs}}, \theta_0)} = f(\mathbf{y}_{\text{obs}} \mid \theta_0) \equiv L(\theta_0 \mid \mathbf{y}_{\text{obs}}),$$

implying that we can estimate L exactly with one random sample. However, sampling from $f(\mathbf{y}_{\text{mis}} \mid \mathbf{y}_{\text{obs}}, \theta_0)$ is perhaps more difficult than estimating $L(\theta_0 \mid \mathbf{y}_{\text{obs}})$ in most applications. A naive choice of $g(\ )$ may be $f(\mathbf{y}_{\text{mis}} \mid \theta_0)$ (which makes no use of the observed information). But this trial function often differs too much from the ideal one and will result in very variable importance weights. For an easy presentation, we use in the latter part of this section the following notations:

$$\mathbf{y}_{\text{obs},t} = (y_{\text{obs},1}, \ldots, y_{\text{obs},t}),$$
$$\mathbf{y}_{\text{mis},t} = (y_{\text{mis},1}, \ldots, y_{\text{mis},t}),$$
$$\mathbf{y}_t = (\mathbf{y}_{\text{obs},t}, \mathbf{y}_{\text{mis},t}).$$

Kong et al. (1994) propose the following *sequential imputation* method to produce Monte Carlo samples of $\mathbf{y}_{\text{mis}}$ and associate with them appropriate importance weights:

A. Draw $\mathbf{y}_{\text{mis}}$ (which is the same as $\mathbf{y}_{\text{mis},n}$ by our notations) from

$$g(\mathbf{y}_{\text{mis}}) = f(y_{\text{mis},1} \mid y_{\text{obs},1}, \theta) \times f(y_{\text{mis},2} \mid y_{\text{obs},1}, y_{\text{mis},1}, y_{\text{obs},2}, \theta)$$
$$\times \cdots \times f(y_{\text{mis},n} \mid \mathbf{y}_{\text{obs},n}, \mathbf{y}_{\text{mis},n-1}, \theta).$$

This step can be realized recursively by drawing $y_{\text{mis},t}$ from the predictive distribution

$$f(y_{\text{mis},t} \mid \mathbf{y}_{\text{obs},t}, \mathbf{y}_{\text{mis},t-1}).$$

B. Compute the weight of $\mathbf{y}_{\text{mis}}$ recursively as

$$w(\mathbf{y}_{\text{mis}}) = f(y_{\text{obs},1} \mid \theta) f(y_{\text{obs},2} \mid \mathbf{y}_1, \theta) \cdots f(y_{\text{obs},n} \mid \mathbf{y}_{n-1}, \theta).$$

Here, $f(\cdot \mid \cdot)$ and $f(\cdot)$ are the conditional and marginal distributions derived from the posited model.

By a careful examination, we see that the sampling distribution for $\mathbf{y}_{\text{mis}}$ satisfies

$$\int g(\mathbf{y}_{\text{mis}})w(\mathbf{y}_{\text{mis}})d\mathbf{y}_{\text{mis}} = f(\mathbf{y}_{\text{obs}} \mid \theta).$$

Thus, we can estimate the likelihood value $L(\theta \mid \mathbf{y}_{\text{obs}})$ by

$$\bar{w} = \frac{1}{m}\left\{w(\mathbf{y}_{\text{mis}}^{(1)}) + \cdots + w(\mathbf{y}_{\text{mis}}^{(m)})\right\}.$$

If one is also interested in the likelihood value at a different point θ' which is not too distant from θ, one can reweight the above Monte Carlo samples (instead of generating a new set) according to (3.3) so as to get a proper estimate.

In practice, the implementation of the sequential imputation method requires one to compute analytically the predictive distribution

$$f(y_{\text{obs},t} \mid \mathbf{y}_{\text{obs},t-1}, \mathbf{y}_{\text{mis},t-1}, \theta)$$

and to sample from $f(\mathbf{y}_{\text{mis},t} \mid \mathbf{y}_{\text{obs},t-1}, \mathbf{y}_{\text{mis},t-1}, y_{\text{obs},t})$. This requirement is rather mild because we have a complete freedom in choosing a decomposition $\mathbf{y} = (y_1, \ldots, y_n)$ to make the required computation feasible. Kong, Cox, Frigge and Irwin (1993) implemented such a procedure to compute likelihood function for a difficult linkage analysis problem. We have also applied this approach successfully to a wide variety of problems ranging from signal processing to bioinformatics (see Chapter 4).

If we relate the distribution $f(\mathbf{y}_{\text{mis},k} \mid \mathbf{y}_{\text{obs},k}, \theta)$ to $\pi_k(\mathbf{y}_{\text{mis},k})$ as in Section 2.6.3, we immediately see that sequential imputation is essentially an equivalent form of the sequential importance sampler described in Section 2.6.3.

3.2.2 Bayesian computation

Let us now take a Bayesian approach to the missing data problem just stated. Suppose a prior distribution $f(\theta)$ is imposed on the unknown parameter θ. In applications, this prior distribution can either be a summary of experts' opinions, or a summary of prior experimental results, or a rather diffused generic distribution [some guidance on the choice of prior distributions can be found in Gelman, Carlin, Stern and Rubin (1995)]. With this Bayesian model, the *joint* distribution of the parameter and the data can be written as

$$f(\theta, \mathbf{y}_{\text{mis}}, \mathbf{y}_{\text{obs}}) = f(\mathbf{y}_{\text{obs}}, \mathbf{y}_{\text{mis}} \mid \theta)f(\theta),$$

where $f(\theta)$ is called the prior distribution of the parameter. Therefore, if of interest is only the posterior distribution of θ, $f(\theta \mid \mathbf{y}_{\text{obs}})$, one would need

3.2 Sequential Imputation for Statistical Missing Data Problems

to carry out the computation

$$f(\theta \mid \mathbf{y}_{\text{obs}}) = \frac{\int f(\theta, \mathbf{y}_{\text{mis}}, \mathbf{y}_{\text{obs}}) d\mathbf{y}_{\text{mis}}}{\int f(\theta', \mathbf{y}_{\text{mis}}, \mathbf{y}_{\text{obs}}) d\theta' d\mathbf{y}_{\text{mis}}},$$

which can also be rewritten as

$$f(\theta \mid \mathbf{y}_{\text{obs}}) = \int f(\theta \mid \mathbf{y}_{\text{obs}}, \mathbf{y}_{\text{mis}}) f(\mathbf{y}_{\text{mis}} \mid \mathbf{y}_{\text{obs}}) d\mathbf{y}_{\text{mis}}.$$

In other words, if a random sample $\mathbf{y}_{\text{mis}}^{(1)}, \ldots, \mathbf{y}_{\text{mis}}^{(m)}$ can be generated from the distribution $f(\mathbf{y}_{\text{mis}} \mid \mathbf{y}_{\text{obs}})$, one can approximate the above posterior as

$$f(\theta \mid \mathbf{y}_{\text{obs}}) = \frac{1}{m} \sum_{j=1}^{m} f(\theta \mid \mathbf{y}_{\text{obs}}, \mathbf{y}_{\text{mis}}^{(j)})$$

(i.e., a mixture of complete-data posterior distributions). Since generating from $f(\mathbf{y}_{\text{mis}} \mid \mathbf{y}_{\text{obs}})$ is typically difficult, we can implement the *sequential imputation* procedure.

We start by drawing $y_{\text{mis},1}$ from $f(y_{\text{mis},1}|y_{\text{obs},1})$ and computing $w_1 = f(y_{\text{obs},1})$. Note that

$$f(y_{\text{mis},1} \mid y_{\text{obs},1}) \propto \int f(y_{\text{obs},1}, y_{\text{mis},1} \mid \theta) f(\theta) d\theta.$$

This integration can be done analytically for a class of statistical models (i.e., exponential families with conjugate priors). For $t = 2, \ldots, n$, the following are done sequentially:

A. Draw $y_{\text{mis},t}$ from the conditional distribution

$$f(y_{\text{mis},t}|\mathbf{y}_{\text{obs},t}, \mathbf{y}_{\text{mis},t-1}).$$

Notice that each $y_{\text{mis},t}$ is drawn conditional on the previously imputed missing components $\mathbf{y}_{\text{mis},t-1} = (y_{\text{mis},1}, \ldots, y_{\text{mis},t-1})$.

B. Compute the predictive probabilities $f(y_{\text{obs},t}|\mathbf{y}_{\text{obs},t-1}, \mathbf{y}_{mis,t-1})$ and

$$w_t = w_{t-1} f(y_{\text{obs},t}|\mathbf{y}_{\text{obs},t-1}, \mathbf{y}_{mis,t-1}). \tag{3.4}$$

Finally, we let

$$w(\mathbf{y}_{\text{mis}}) = w_n \equiv f(y_{\text{obs},1}) \prod_{t=2}^{n} f(y_{\text{obs},1} \mid \mathbf{y}_{\text{obs},t-1}, \mathbf{y}_{mis,t-1}).$$

Note that steps A and B are usually done simultaneously. Both steps A and B are required to be computationally simple, which is often the case if the *complete-data* predictive distributions $f(y_t|\mathbf{y}_{t-1})$ are simple. This is the key to the feasibility of sequential imputation. When integrating out θ analytically is not feasible, one can treat θ as an additional missing data, $y_{\text{mis},0}$, and use the same sequential imputation method.

Suppose steps A and B are done repeatedly and independently m times, which results in a *multiple imputation* of the missing data, $\mathbf{y}_{\text{mis}}^{(1)}, \ldots, \mathbf{y}_{\text{mis}}^{(m)}$, and their respective weights $w^{(1)}, \ldots, w^{(m)}$. Then, we can estimate the posterior distribution $f(\theta \mid \mathbf{y}_{\text{obs}})$ by the weighted mixture

$$\frac{1}{W} \sum_{j=1}^{m} w^{(j)} f(\theta \mid \mathbf{y}_{\text{obs}}, \mathbf{y}_{\text{mis}}^{(j)}), \qquad (3.5)$$

where $W = \sum w^{(j)}$. Again, the disadvantage of this method is that when the sample size n increases, the importance weights become very skewed. Kong et al. (1994) showed that the normalized weight w_n is a martingale sequence (with respect to n). Thus, the variance of w_n increases stochastically as n increases.

The similarity between the Bayesian missing data problem as the simulation of the SAW is transparent: mathematically, the ith missing component, $y_{\text{mis},i}$, plays the same role as the position of the ith monomer, x_i.

3.3 Nonlinear Filtering

The *state-space model* is a very popular dynamic system and has two major parts: (1) the observation equation, often written as $y_t \sim f_t(\cdot \mid x_t, \phi)$, and (2) the state equation, which can be represented by a Markov process as $x_t \sim q_t(\cdot \mid x_{t-1}, \theta)$. Putting the two together, we have

$$\begin{aligned}\text{(state equation):} &\quad x_t \sim q_t(\cdot \mid x_{t-1}, \theta), \\ \text{(observation equation):} &\quad y_t \sim f_t(\cdot \mid x_t, \phi).\end{aligned} \qquad (3.6)$$

The y_t are observations and the x_t are referred to as the (unobserved) state variables. This model is sometimes termed as the *hidden Markov model* (HMM) in many applications (most significantly in speech recognition and computational biology). Such a model can be represented graphically as in Figure 3.4. In practice, the x's can be the transmitted digital signals in wireless communication (Liu and Chen 1995), the actual words in speech recognition (Rabiner 1989), the target's position and velocity detected from a radar (Gordon et al. 1993, Gordon, Salmond and Ewing 1995, Avitzour 1995), the underlying volatility in an economical time series (Pitt and Shephard 1999), the structural types in the protein secondary structure prediction problem (Schmidler, Liu and Brutlag 2000), and many others.

A main challenge to researchers is to find efficient methods for on-line estimation and prediction (filtering) of the state variables (i.e., x_t) or other

FIGURE 3.4. A graphical representation of the state-space model, also called the hidden Markov model (HMM).

parameters when the signal y_t comes in sequentially. For the simplicity of our presentation, we assume in this section that the system's parameters (i.e., ϕ and θ) are all known and are omitted from the model description. Ideas on estimating model parameters together with the state variables can be found in Liu and Chen (1995), Gordon et al. (1995), and Liu and West (2000).

With known system parameters, the optimal (in terms of the mean squared errors) on-line estimate of x_t is then the *Bayes solution* (West and Harrison 1989)

$$\hat{x}_t = E(x_t \mid y_1, \ldots, y_t)$$
$$= \frac{\int \cdots \int x_t \prod_{s=1}^{t} [f(y_s \mid x_s) q_s(x_s \mid x_{s-1})] \, dx_1 \cdots dx_t}{\int \cdots \int \prod_{s=1}^{t} [f(y_s \mid x_s) q_s(x_s \mid x_{s-1})] \, dx_1 \cdots dx_t}.$$

Hence, of interest at any time t is the "current" posterior distribution of x_t, which can be theoretically derived recursively as

$$\pi_t(x_t) = P(x_t \mid y_1, \ldots, y_t) \qquad (3.7)$$
$$\propto \int q_t(x_t \mid x_{t-1}) f_t(y_t \mid x_t) \pi_{t-1}(x_{t-1}) dx_{t-1},$$

where $\pi_{t-1}(x_{t-1}) = p(x_{t-1} \mid y_1, \ldots, y_{t-1})$ is the "current" posterior distribution of x_{t-1}.

When both f_t and q_t are linear Gaussian conditional distributions in (3.6), the resulting model is called the *linear state-space model*, or the dynamic linear model, which has long been an active subject of research in automatic control, economics, engineering, and statistics (Harvey 1990, West and Harrison 1989). In this case, the posterior computation can be done analytically via recursion (3.8) because all the "current" posterior distributions are Gaussian. The resulting algorithm is the basis of the celebrated *Kalman filter* (Kalman 1960). When the x_t only takes on a finite, say K, possible values [e.g., binary signals in digital communication or nucleotides (with four different kinds) in DNA sequence analysis], such a state-space model is also referred to as the hidden Markov model (HMM). Its utility in many areas such as digital communication, speech recognition, protein sequence alignment, ion-channel analysis, etc. has been extensively studied.

Optimal on-line estimation of x_t can also be achieved via a recursive algorithm very similar to exact simulation and dynamic programming method described in Section 2.4.

Other than the two situations just described, exact computation of the optimal on-line estimation of x_t is generally impossible, in that the integrations needed to evaluate (3.8) and its normalizing constant quickly becomes infeasible as t increases.

A very general and simple method, named the *bootstrap filter* (and also called *particle filter*), for on-line estimation in a state-space model was proposed by Gordon et al. (1993), which makes use of the idea of sampling-importance-resampling (SIR) (Rubin 1987).

Suppose at time t that we have a random sample $\{x_t^{(1)}, \ldots, x_t^{(m)}\}$ of the state variable which follow approximately the current posterior distribution $\pi_t(x_t) = P(x_t \mid y_1, \ldots, y_t)$ (e.g., those spikes on the leftmost line in Figure 3.5). Gordon et al. (1993) suggested the following updating procedure when y_{t+1} is observed:

(a) Draw $x_{t+1}^{(*j)}$ from the state equation $q_t(x_{t+1} \mid x_t^{(j)})$, $j = 1, \ldots, m$.

(b) Weight each draw by $w^{(j)} \propto f_t(y_{t+1} \mid x_{t+1}^{(*j)})$.

(c) Resample from $\{x_{t+1}^{(*1)}, \ldots, x_{t+1}^{(*m)}\}$ with probability proportional to $w^{(j)}$ to produce a random sample $\{x_{t+1}^{(1)}, \ldots, x_{t+1}^{(m)}\}$ for time $t+1$.

It is easy to show that if the $x_t^{(m)}$ follow the "current" posterior distribution $\pi_t(x_t)$ and if m is large enough, then the new random sample $\{x_{t+1}^{(1)}, \ldots, x_{t+1}^{(m)}\}$ follows the updated posterior distribution $\pi_{t+1}(x_{t+1})$ approximately.

The connection between this problem and the statistical missing data problem analyzed in Section 3.2 is obvious: The state variable x_t can be treated as missing data and imputed sequentially. Under the SIS framework, we see that the "current" target distribution $\pi_t(\cdot)$ satisfies the recursion

$$\pi_t(\mathbf{x}_t) \propto f_t(y_t \mid x_t) q_t(x_t \mid x_{t-1}) \pi_{t-1}(\mathbf{x}_{t-1}). \tag{3.8}$$

A special feature of the state-space model is that the state variable x_t possess a Markovian structure. This feature makes it possible for one to consider only the marginal posterior distribution, $\pi_t(x_t)$, instead of the joint distribution $\pi_t(\mathbf{x}_t)$, and it is essential for the bootstrap filter to be applicable. In fact, if one shifts the step index t for the missing data $y_{\text{mis},t}$ forward by one unit (i.e., treating $y_{\text{mis},t-1}$ as $y_{\text{mis},t}$), then the sequential imputation method with stepwise resampling gives rise to the same algorithm. The advantage of the bootstrap filter is its simplicity, whereas the drawbacks of the method are two: (a) It did not use the current available information y_{t+1} in the sampling step; (b) its excessive use of resampling may decrease its efficiency.

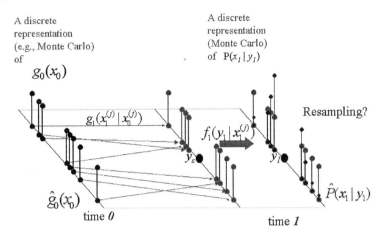

FIGURE 3.5. A graphical illustration of the bootstrap/particle filter. At time 0, one has a discrete representation of the current distribution. One propagates to time 1 by first sampling those $x_1^{(m)}$'s from the state equation (equivalent to sampling from its prior distribution) and then correcting this sampling by an importance reweighting and resampling.

Those models discussed in Sections 2.7, 3.1, and 3.2 and dealt with by the general SIS methodology do not share the special structure enjoyed by the state-space model (i.e., the x_t's are not Markovian). However, the generic SIS formulation forces one to think deeper into the general issues of efficiency improvement for importance sampling. For example, one may want to construct trial sampling distributions that are more efficient than the one used in the bootstrap filter and may think of applying techniques such as marginalization and rejection controls (Sections 3.4.6 and 2.6.2). Liu and Chen (1995, 1998) also discuss the role of resampling in a generic SIS framework.

3.4 A General Framework

The connection of the methods described in the previous three sections can be briefly summarized as follows. The growth method is a special application of the general SIS methodology described in Section 2.6.3. The sequential imputation method of Kong et al. (1994) is an equivalent form of the SIS under the statistical missing data setting. The bootstrap filter uses an SIS construction but has an additional resampling step inserted before each sequential sampling step. In this section, we show that the modifications to the growth method made by Wall and Erpenbeck (1959) and Grassberger (1997) is practically equivalent to an SIS with resampling (Liu and Chen 1995).

68   3. Theory of Sequential Monte Carlo

In summarizing the common features of the problems treated in the previous sections, we first define a *probabilistic dynamic system*. In the SIS framework, this system serves as a "guiding" (auxiliary) sequence of distributions.

Definition 3.4.1 *We call a sequence of probability distributions $\pi_t(\mathbf{x}_t)$, indexed by discrete time $t = 0, 1, 2, \ldots$, a probabilistic dynamic system (PDS). The state variable $\mathbf{x}_t$ can either increases its dimensionality as t increases [i.e., $\mathbf{x}_{t+1} = (\mathbf{x}_t, x_{t+1})$], or stay as it is (i.e., $\mathbf{x}_{t+1} = \mathbf{x}_t$).*

Here, $\pi(\ )$ always refers to the target distribution of the dynamic system and $p(\ )$ is a generic symbol for probability distributions.

For the SAW simulation problem in Section 3.1, the PDS is defined as $\pi_t(\mathbf{x}_t) = Z_t^{-1}$ on the set of all self-avoiding walks with t notes, where Z_t is the total number of such configurations. In words, π_t is the uniform distribution on the set of all SAWs of $t-1$ steps. In the literature of molecular simulation, one is often interested in more complex structures [e.g., protein structures, see Leach (1996)] in which the π_t is taken as the *Boltzmann distribution* of the form $\exp(-U_t/k_B T)/Z$. Here, U_t is the *interacting energy* generated by the monomers of the chain polymer (with t monomers), T is the absolute temperature, and k_B is the Boltzmann constant.

The state-space model example (Section 3.3) also fits very well into our PDS framework in that we can define $\pi_t(\mathbf{x}_t)$ as the posterior distribution $p(\mathbf{x}_t \mid \mathbf{y}_t, x_0)$. With this PDS, the estimation of x_t can be thought of as evaluating statistical (Monte Carlo) average of a random variable under distribution π_t. The methodology developed in this section will be generally called *sequential Monte Carlo*.

It is natural to think of the PDS as a sequence of posterior distributions conditional on the information "up to time t" for a system with random variables $x_1, x_2, \ldots$. In other words, the final probability distribution is "built up" sequentially as more and more structural details (i.e., information) are incorporated. In the traditional Bayesian analysis (Gelman, Carlin, Stern and Rubin 1995), the dimensionality of the random vector (i.e., $\mathbf{x}_t$) involved in the system does not vary when new information comes in; whereas in our PDS, the "configuration" space changes along with t. Once a proper PDS is employed, this PDS can further help us choose good sampling distributions $g_t(x_t \mid \mathbf{x}_{t-1})$.

It is helpful to recall how the growth method is used to simulate a SAW. At step t, a monomer is added to the existing partial chain $(\mathbf{x}_{t-1})$ according to a distribution that makes use of the information generated from a forward-looking step. This is equivalent to using the sequence of distributions, $\{\pi_t(\mathbf{x}_t), t = 1, 2, \ldots\}$, as the PDS, where π_t is the uniform distribution on all SAWs with t nodes, and letting $g_t(x_t \mid \mathbf{x}_{t-1}) = \pi_t(x_t \mid \mathbf{x}_{t-1})$. There is no reason why we cannot use information generated by more forward-looking steps. But as the one-step-forward-looking strategy requires us to check three neighboring positions of x_t, a k-step-forward-looking method generally requires us to check $\sim 3^k$ positions and this computation quickly becomes impractical. The next section discusses in detail some issues in choosing the sampling distribution.

3.4.1 The choice of the sampling distribution

The choice of the sampling distribution g_t is directly related to the efficiency of the SIS. In many problems, a good choice of g_t in light of the sequence of auxiliary distributions $\{\pi_t\}$ is

$$g_t(x_t \mid \mathbf{x}_{t-1}) = \pi_t(x_t \mid \mathbf{x}_{t-1}), \quad (3.9)$$

with the incremental weight

$$u_t = \frac{\pi_t(\mathbf{x}_{t-1})}{\pi_{t-1}(\mathbf{x}_{t-1})}. \quad (3.10)$$

Note that u_t in (3.10) does not depend on the value of x_t and this feature is important to several issues discussed later. The reason that drawing x_t from $\pi_t(x_t \mid \mathbf{x}_{t-1})$ is more desirable than from a more or less arbitrary function is clear from rewriting the incremental weight (3.10) as

$$u_t = \frac{\pi_t(\mathbf{x}_{t-1})}{\pi_{t-1}(\mathbf{x}_{t-1})} \frac{\pi_t(x_t \mid \mathbf{x}_{t-1})}{g_t(x_t \mid \mathbf{x}_{t-1})}.$$

Intuitively, the second ratio is needed to correct the discrepancy between $g_t(x_t \mid \mathbf{x}_{t-1})$ and $\pi_t(x_t \mid \mathbf{x}_{t-1})$ when they are different.

Other choices of g_t can also be desirable. For example, one may also consider a two-step-forward sampling distribution

$$g_t(x_t \mid \mathbf{x}_{t-1}) \propto \int \pi_{t+1}(x_{t+1}, \mathbf{x}_t) dx_{t+1}.$$

In the polymer simulation, this corresponds to a two-step look-ahead strategy (Section 3.1.2). In the content of polymer simulation, this idea was also proposed by Meirovitch (1982).

Another strategy is to "coarsen" the dynamic systems; that is, one can group, say, $(x_{bt+1}, \ldots, x_{b(t+1)})$ together as a new "mega"-state x'_t in the system. Then, the SIS method can be applied to draw a block of b x's at a time. This method is also proven useful in simulating long polymer chains (Wall, Rubin and Isaacson 1957).

3.4.2 Normalizing constant

In many scientific problems, computing the normalizing constant (i.e., the partition function) of an unnormalized probability distribution function is

70   3. Theory of Sequential Monte Carlo

of critical importance. Examples in physics and chemistry are abundant. Sometimes, even mathematicians and statisticians are interested in such problems, one of which is to count the total number Z of distinctive tables that contain only 0's and 1's and have the fixed row sums $r_1, \ldots, r_m$ and column sums $c_1, \ldots, c_n$. For example, there are twenty-seven 0-1 tables with the row sums 2, 2, 2, 3 and the column sums 3, 2, 3, 1. One such table is given in Table 3.1. This problem can be formulated as one of estimating

$$\begin{array}{cccc} 0 & 0 & 1 & 1 \\ 1 & 1 & 0 & 0 \\ 1 & 0 & 1 & 0 \\ 1 & 1 & 1 & 0 \end{array}$$

TABLE 3.1. A typical 0-1 table with given respective column sums: 3, 2, 3, 1 and row sums: 2, 2, 2, 3.

the normalizing constant Z in that the uniform distribution on space of all such tables is of the form $1/Z$. Problems such as evaluating likelihood (Section 3.2.1) and computing the Bayes factor (Kong et al. 1994, Meng and Wong 1996) also belong to this class.

The SIS approach can be an effective means for estimating the normalizing constant. Suppose the target probability distribution is $\pi(\mathbf{x})$. An auxiliary PDS is found to be $\Pi = \{\pi_t(\mathbf{x}_t), \ t = 1, \ldots, N\}$ (with $\pi_N \equiv \pi$), and each of its member's density function, π_t, is known up to a normalizing constant. In other words, we can write down the unnormalized distribution function, $q_t(\mathbf{x}_t) = Z_t \pi_t(\mathbf{x}_t)$, for everyone in the PDS. In statistics, Z_t often corresponds to the *Bayes factor* or the likelihood function evaluated at a given parameter value.

Suppose we implement an SIS scheme with the sequential sampling system $\{g_t(x_t \mid \mathbf{x}_{t-1}), \ t = 1, \ldots, N\}$. Then, the incremental importance weight u_t is computed as

$$u_t = \frac{q_t(\mathbf{x}_t)}{q_{t-1}(\mathbf{x}_{t-1}) g_t(x_t \mid \mathbf{x}_{t-1})} = \frac{Z_t \pi_t(\mathbf{x}_t)}{Z_{t-1} \pi_{t-1}(\mathbf{x}_{t-1}) g_t(x_t \mid \mathbf{x}_{t-1})}.$$

The final weight takes the form

$$w_N = \prod_{t=1}^N u_t = \frac{Z_N}{Z_1} \frac{\pi_N(\mathbf{x}_N)}{g_1(x_1) \cdots g_N(x_N \mid \mathbf{x}_{N-1})}. \tag{3.11}$$

Thus, the sample average of w_N gives us an unbiased estimate of $E(w_N) = Z_N/Z_1$, the ratio of two normalizing constants. In many cases, π_1 is a very simple system in which Z_1 can be easily obtained either analytically or numerically. Then, we can obtain a rather accurate estimate of Z_N. Based on this procedure, Chen (2001) designed an SIS algorithm to approximately count the total number of 0-1 tables with given marginal sums (more in Section 4.3).

In recent years, many techniques have been developed to improve the SIS-based methods for nonlinear filtering problems. But the estimation problem for the normalizing constants has been mostly neglected. We note, however, that expression (3.11) is broadly useful for estimating the normalizing constant with the following improvement methods.

3.4.3 Pruning, enrichment, and resampling

When the system grows, the variance of the importance weights w_t increases. Thus, after a certain number of steps, many of the weights become very small and a few very large. One of the earliest methods for improving the SIS procedure in polymer simulations, called the *enrichment* method, was proposed in Wall and Erpenbeck (1959). Their work, according to Kremer and Binder (1988), led to "one of the very first intriguing successes of Monte Carlo methods." Briefly, Wall and Erpenbeck suggested splitting those successfully simulated partial SAW chains into, say, r chains, each assigned $1/r$ of the original weight, and then continuing them in r independent ways. More precisely, once we have successfully simulated by the SIS method a chain up to step t, say, $\mathbf{x}_t$ with weight w_t, we can make r_t copies of it and assign each a weight w_t/r_t. Then, all the copies will be continued as if they were independently built up from scratch. An undesirable, and also unavoidable, feature of this enhancement is that the resulting full-length chains are no longer statistically independent.

Because choosing r_t can be a difficult task in general, Grassberger (1997) introduces some learning strategy for chain splitting and *pruning*. Instead of slitting each chain $\mathbf{x}_t$ indiscriminately, he suggested using a step-dependent upper cutoff value C_t and a lower cutoff value c_t. Then, if the weight w_t is greater than C_t, the chain $\mathbf{x}_t$ is split into r copies, each with weight w_t/r, whereas if $w_t \leq c_t$, one flips a fair coin to decide whether to keep it. If it is kept, its weight w_t is doubled to $2w_t$. He named the algorithm "PERM" for *Prune-Enriched Rosenbluth Method*. However, choosing the cutoff values c_t and C_t can also be tricky.

Independently, a resampling strategy (to be described in the next section) for improving SIS has been developed in the statistics and engineering community (Gordon et al. 1993, Liu and Chen 1995). Liu and Chen (1995) introduced a monitored resampling procedure for the general SIS and applied the method to a blind deconvolution problem. Their method produces the same enrichment and pruning effects on the SIS samples as that of Grassberger (1997) and is arguably more flexible than the PERM. In treating the state-space models, Gordon et al. (1993) also used a resampling strategy. In fact, a resampling step is enforced at every time step t of their *bootstrap filter*. The success of this method, however, relies heavily on the Markovian structure among the state variable $x_1, x_2, \ldots$; that is, given the realization of x_t, the next variable, x_{t+1}, is statistically independent of all the previous states, $\mathbf{x}_{t-1}$. Thus, resampling from set $\{\mathbf{x}_{t-1}^{(j)}, j=1,\ldots,m\}$ is equivalent to resampling from $\{x_{t-1}^{(j)}, j=1,\ldots,m\}$, the set of the "current state." Otherwise, frequent resampling will rapidly impoverish diversity of the partial samples produced earlier.

3.4.4 More about resampling

Suppose at step t we have a collection of m partial samples of length t, $\mathcal{S}_t = \{\mathbf{x}_t^{(j)}, j=1,\ldots,m\}$, which are properly weighted by the collection of weights $\mathcal{W}_t = \{w_t^{(j)}, j=1,\ldots m\}$ with respect to the PDS π_t. Each of these partial samples, $\mathbf{x}_t^{(j)}$, will also be called a *stream*. Instead of carrying the weight $w_t^{(j)}$ as the system evolves, it is also legitimate, and sometimes advantageous, to insert the following *resampling* step between SIS recursions (Liu and Chen 1998), and such a procedure is referred to as the *SIS with resampling*. The following two schemes are just two of many possible choices to achieve the resampling goal.

Simple Random Resampling

1. Sample a new set of streams (i.e., partial samples), denoted as $\mathcal{S}_t'$, from $\mathcal{S}_t$ according to the weights $w_t^{(j)}$.

2. Assign equal weights, W_t/m, to the streams in $\mathcal{S}_t'$, where $W_t = w_t^{(1)} + \cdots + w_t^{(m)}$.

Residual Resampling:

1'. Retain $k_j = [mw_t^{(*j)}]$ copies of $\mathbf{x}_t^{(j)}$, where $w_t^{(*j)} = w_t^{(j)}/W_t$, and $j = 1,\ldots,m$. Let $m_r = m - k_1 - \cdots - k_m$.

2'. Obtain m_r i.i.d. draws from $\mathcal{S}_t$ with probabilities proportional to $mw_t^{(*j)} - k_j, j=1,\ldots,m$.

3'. Reset all the weights to W_t/m.

Note that one can still use the average of the final weights to estimate the ratio Z_n/Z_1 of two normalizing constants (see Section 3.4.2) even after several resampling steps have been incurred. This is the primary reason why we set the weights of all the streams to their *current average* after each resampling step. It is easily shown that the residual sampling dominates the simple random sampling in having smaller Monte Carlo variance and favorable computation time, and it does not seem to have disadvantages in other aspects. Some new resampling method has recently been proposed by Doucet, Godsill and Andrieu (2000).

As one can clearly see here, the residual resampling scheme is very much similar to the enrichment and pruning (PERM) method of Grassberger (Wall and Erpenbeck 1959, Grassberger 1997). Both the PERM algorithm and the resampling scheme construct a proper set of importance samples which can be used to estimate any quantity of interest with respect to π_T (the target distribution). In particular, PERM will be proper as long as the cutoff values c_t and C_t are prescribed in advance; and resampling will cause no conceptual problem as long as the decision of resampling does not depend on the configurations. PERM has more flexibility in dealing with the weights, whereas the resampling method is more disciplined and requires less tuning. A prominent advantage of the resampling method over PERM is that the choice of cutoff values for splitting the streams is automatically and implicitly achieved by the cross-stream comparison of the weights. More precisely, in residual resampling, we split a stream into k more copies precisely because its weight is k times better than the average weight. This feature makes the resampling method very generally applicable and powerful. As we will show in the next section, resampling can be used, to a great advantage, in applications ranging from target tracking to population genetics and to biological sequence analysis.

Instead of using the w_t in resampling, we can also implement the following more general resampling strategy.

- For $j' = 1, \ldots, \tilde{m}$:

    - Draw $\tilde{\mathbf{x}}_t^{(j')}$ independently from the current sample $\{ \mathbf{x}_t^{(j)}, j = 1, \ldots, m \}$ according to the probability vector $(a^{(1)}, \ldots, a^{(m)})$; suppose we obtain that $\tilde{\mathbf{x}}_t^{(j')} = \mathbf{x}_t^{(j)}$.
    - A new weight $\tilde{w}_t^{(j')} = w_t^{(j)}/a^{(j)}$ is assigned to this sample.

- Return the new representation $\tilde{\mathcal{S}}_t = \{\tilde{\mathbf{x}}_t^{(j')}, j' = 1, \ldots, \tilde{m}\}$ and $\tilde{\mathcal{W}}_t = \{\tilde{w}_t^{(j')}, j' = 1, \ldots, \tilde{m}\}$.

The new set $\tilde{\mathcal{S}}_t$ thus formed is also properly weighted by $\tilde{\mathcal{W}}_t$ (approximately) with respect to π (Rubin 1987). Because the role of resampling is to prune away "bad" samples and to split good ones, we should choose $a^{(j)}$ as a monotone function of $w_t^{(j)}$. Having an additional flexibility in choosing the sampling weights $a^{(j)}$ is rather intriguing and can be potentially very useful. For example, the $a^{(j)}$ can be chosen to reflect certain "future trend" (Pitt and Shephard 1999) or be chosen to balance between the need of diversity (i.e., having multiple distinct samples) and the need of focus (i.e., giving more presence to those samples with large weights). A generic choice is

$$a^{(j)} = [w_t^{(j)}]^\alpha, \qquad (3.12)$$

where $0 < \alpha \leq 1$ can vary according to the coefficient of variation of the w_t. Professor W.H. Wong suggested choosing $\alpha = 1/2$, which seems to work well in several examples.

The *bootstrap filter* and its variations (Gordon et al. 1993, Kitagawa 1996, Pitt and Shephard 1999, Berzuini, Best, Gilks and Larizza 1997) can be seen as SIS with special choices of g_t and with resampling at every step. Since resampling at every step is neither necessary nor efficient, it is desirable to prescribe a schedule for the resampling step to take place. In the state-space models, frequent resampling does not produce much detrimental effect because of the Markov structure of the unobserved state variables. However, resampling too often gives very bad results in a general PDS, where no simple Markovian structure is present. These systems include the SAW model for polymer studies (Kremer and Binder 1988), the demographic tree model in population genetics (Stephens and Donnelly 2000, Chen and Liu 2000a), the model for wireless signal detection in flat-fading channels (Chen and Liu 2000a, Chen, Wang and Liu 2000), and many Bayesian missing data problems.

The resampling schedule (i.e., when to resample) can be either deterministic or dynamic. When the schedule is dynamic, some small bias may be introduced. However, our experience showed that this bias did not produce any adverse effect. With a *deterministic schedule*, we conduct resampling at time $t_0, 2t_0, \ldots$, where t_0 is given in advance and it may be tuned to suit for the particular problem of interest. In a *dynamic schedule*, a sequence of thresholds, $c_1, c_2, \ldots$, are given in advance. We monitor the coefficient of variation of the weights cv_t^2 and invoke the resampling step when event $cv_t^2 > c_t$ occurs. A typical sequence of c_t can be $c_t = a + bt^\alpha$. We summarize the resampling scheme in the following pseudo-code:

1. Check the weight distribution by performing one of the following two methods at time t:

    - Dynamic: go to step 2 if the coefficient of variation of the weight is modest [i.e., $cv_t^2(w) < c_t$]; otherwise go to step 3.
    - Deterministic: go to step 2 if $t \neq kt_0$ for integer k; otherwise go to step 3.

2. Invoke an SIS step. Set $t = t + 1$ and go to step 1.

3. Invoke a resampling step. Go to step 2.

It is not immediately clear why one needs resampling at a certain stage t. As much detailed theoretical discussion is given by Liu and Chen (1995), we only mention a few heuristics on the issue. First, if the weights $w_t^{(j)}$ are constant (or near constant) for all t (such a case occurs when one can draw from π_t directly), resampling only reduces the number of distinctive streams and introduces extra Monte Carlo variation. This suggests that one should not perform resampling when the coefficient of variation [as defined in (2.5)], cv_t^2, for the $w_t^{(j)}$ is small. As argued in Kong et al. (1994), the *effective sample size* is inversely proportional to $1+cv_t^2$. Second, it can be shown that as the system evolves, cv_t^2 *increases* stochastically (Kong et al. 1994). When the weights get very skewed at time t, carrying many streams with very small weights is apparently a waste. Resampling can provide chances for the good (i.e., "important") streams to amplify themselves and hence "rejuvenate" the sampler. The resampling step tends to result in a better group of "ancestors" so as to produce better "descendants" (i.e., Monte Carlo samples of the future states), although it does not improve inferences on the *current* state, $\mathbf{x}_t$.

3.4.5 Partial rejection control

As t increases, the distribution of w_t typically becomes more and more skewed, implying that the χ^2 distance between the sampling distribution of $\mathbf{x}_t$ and its current guiding distribution π_t increases. (This distance is equivalent to the coefficient of variation of w_t.) As a consequence, many streams carried by the SIS will have minimal impact on the final estimation. It is thus desirable to prune them away at an earlier stage. In the previous two sections, we discussed the ideas of PERM and resampling. In Section 2.6.4, we saw how the rejection control method can be used to achieve pruning without creating bias or correlations. However, the implementation of a full rejection control requires that we make up the lost streams by restarting from stage 1 [i.e., sampling from $g_1(\ )$ and proceeding forward] and passing through all the intermediate rejection steps (i.e., check-points). This procedure can be extremely time-consuming and is not very practical for a large probabilistic system.

Instead of employing the full rejection control, we can opt for the following more practical method, the *partial rejection control*:

1. At each check point t_k, start $\text{RC}(t_k)$ as described in Section 2.6.4 with the threshold value $c = c_k$. If stream $\mathbf{x}_{t_k}^{(j)}$ with weight $w_{t_k}^{(j)}$ passes this check point, one proceeds the same way as in a standard SIS with the old weight replaced by $w_{t_k}^{(*j)} = \max\{w_{t_k}^{(j)}, c_k\}$.

2. When rejected, go back to check point t_{k-1} to draw a stream $x_{t_{k-1}}^{(j)}$ from the pool $\mathcal{S}_{t_{k-1}}$, with probability proportional to $w_{t_{k-1}}^{(j)}$. Reset its weight as $\bar{w}_{t_{k-1}}$ and proceed with the SIS procedure. If the new stream formed in this way pass the check point t_k, then its weight is set as $w_{t_k}^{(*j)} = \max\{w_{t_k}^{(j)}, c_k\}$.

3. Reset all the weights as $w_{t_k}^{(j)} = \hat{p}_c w_{t_k}^{(*j)}$.

76   3. Theory of Sequential Monte Carlo

The partial rejection control method combines the three main ideas used in designing sequential Monte Carlo algorithms: weighting, rejection, and resampling. It is also related to the method of (Hurzeler and Kunsch 1998).

3.4.6 Marginalization, look-ahead, and delayed estimate

Two other general methods for improving the performance of an SIS algorithm are *marginalization* and *iterative updating*. The former is an analytical method and, when achievable, it can improve almost all Monte Carlo methods. Its general principle was described by Hammersley and Handscomb (1964) and Rubinstein (1981) as a dimension reduction technique. Using marginalization to improve a static importance sampling method has been discussed in Section 2.5.5. Its use with Markov chain Monte Carlo methods will be discussed in the next chapter. The latter approach is to make use of Metropolis or Gibbs sampling (heat-bath) algorithm to improve the underlying importance sampling distribution. MacEachern et al. (1999) provide some theory on why these operations help to improve the performance of the algorithm.

Another useful strategy in constructing the trial sampling distribution is to "look-ahead" for a few steps. This method has been successfully implemented for polymer simulations (Meirovitch 1982, Meirovitch 1985, Batoulis and Kremer 1988) and for nonlinear state-space models (Liu and Chen 1995). It is also called the "scanning future method" in statistical physics literature. More precisely, we can construct our sampling distribution as close to the future target as possible. One such construction is to let

$$g_t(x_t \mid \mathbf{x}_{t-1}) = \pi_{t+s}(x_t \mid \mathbf{x}_{t+1}).$$

Note that

$$\pi_{t+s}(x_t \mid \mathbf{x}_{t-1}) = \int \pi_{t+s}(x_{t+s}, \ldots, x_t \mid \mathbf{x}_{t-1}) dx_{t+1} \cdots dx_{t+s}.$$

If we regard each $\pi_t(\mathbf{x})$ as a "posterior distribution" conditional on the information up to time t, then the above approach is just trying to make use of as much future information as possible. If we had chosen s so that $t + s = n$, for example, then the resulting sampling distribution g_t from "forward-looking" is in fact the ideal conditional distribution for x_t under the target distribution (so that the product of g_t is equal to π_n). However, the forward-looking distribution can be difficult to deal with as s becomes large, which is the reason why we adopt the SIS approach in the first place. In Meirovitch (1985), he proposed the *mean-field scanning method*, in which one constructs g_t by making use of the future states information generated by a mean-field approximation.

3.5 Problems

1. Restate the growth method as a special form of SIS. Describe a proper PDS and the implied sampling distributions used for the growth method for simulating SAWs.

2. Describe a proper PDS for the nonlinear filtering problem with model (3.6) and the corresponding SIS strategy.

3. Design a SIS method for counting the number of 0-1 tables with given margins (Chen 2001).

4. Show that the weighting strategy in sequential imputation (Section 3.2) is proper and show that the average of the weight can be used to estimate the likelihood value.

5. Show that the (normalized) importance weight sequence in SIS is a martingale sequence and the variances increase stochastically (Kong et al. 1994).

6. Show that one can still use a similar method as in Section 3.4.2 to estimate the ratio of two normalizing constants, Z_n/Z_1, when a few resampling steps are incurred between time 1 and n.

7. Discuss why resampling can be useful in sequential Monte Carlo.

4
Sequential Monte Carlo in Action

The previous chapter outlines a general Monte Carlo framework based on the sequential buildup strategy. Several essential elements are (a) the choice of the trial densities, (b) the resampling method, (c) the marginalization strategy, and (d) the rejection control. This chapter will illustrate how these generic strategies are applied to various application problems.

4.1 Some Biological Problems

4.1.1 Molecular Simulation

Simulating molecular structures is one of the most important and challenging scientific problems. Starting with those simple SAW model for long-chain polymers (Section 3.1), chemists and structural biologists have developed numerous lattice-based models (and also more complicated models) for predicting native structures of important macromolecules, such as protein molecules. Here, we offer the reader a glimpse of this huge area.

The most well-known open problem in structural biology and biophysics is the the so-called *protein folding problem* in which one is required to predict the three-dimensional fold shape of a protein molecule based only on its sequence of amino acids. One simple model to imitate the real protein folding process is a 2-D or 3-D lattice-bead model. These models have identical structures as the SAW model described in Section 3.1, but they usually use a more complicated function for interactive energy between the "beads" on the lattice. In protein language, these beads correspond to amino acid residues, which are of 20 different types (i.e., a 20-letter alphabet).

One of the simplest models used by chemists (Unger and Moult 1993) has only two different kinds of beads, white and black, corresponding to hydrophilic and hydrophobic residues, respectively. An example of such a simple bead sequence of length 36 is

WWWBBWWBBWWWWWBBBBBBBWWBBWWWWBBWWBWW

It is of interest to find out its most "favorable fold" in a 2-D lattice space. To define what is meant by a favorable fold, we define an energy function for each configuration of this bead sequence:

$$U_n(\mathbf{x}_n) = - \sum_{|i-j|>1} c(x_i, x_j),$$

where $c(x_i, x_j) = 1$ if x_i and x_j are non-bonding neighbors and the identities of beads i and j are both black (hydrophobic), and $c(x_i, x_j) = 0$ otherwise. Clearly, this simple model favors the close packing of hydrophobic residues and, to certain extent, mimics some aspects of a real protein fold.[1]

With a target distribution

$$\pi_n(\mathbf{x}_n) \propto \exp\{-U_n(\mathbf{x}_n)/2\},$$

we applied the SIS, with various modifications, including resampling, rejection control, and one-step-look-ahead, to all the examples in Unger and Moult (1993). In light of a recent result of Bastolla, Frauenkron, Gerstner, Grassberger and Nadler (1998), it is not surprising that we were able to find the same or better minimum energy state than Unger and Moult (1993). In fact, Bastolla et al. (1998) applied the SIS method with pruning and enrichment modifications to the examples of Unger and Moult and many others and achieved some excellent results. Our results are comparable to theirs. Figure 4.1 (b) shows the best configuration we found for a 60-mer chain: it has 36 favorable contacts. In comparison, the best configuration found by Unger and Moult has only 34 favorable contacts. An additional experiment we did was to repeat the modified SIS simulation of this 60-mers (each took about 6 minutes) for 125 times, from which we obtained an estimate of the normalizing constant $\log(Z) = 89.65$ and the mean squared extension $E_\pi(R^2) \approx 157$.

To further investigate the effect of resampling and partial rejection controls, we applied the SIS to an earlier example of simulating SAWs in which

---

[1] Because most proteins are surrounded by water, these macromolecules tend to pack hydrophobic residues in the center of their fold and leave those hydrophilic ones on the surface to interact with water molecules.

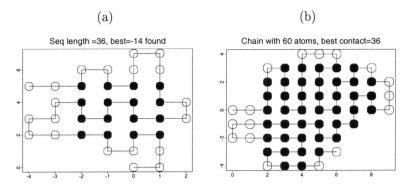

FIGURE 4.1. Configurations of two black-white bead chains with length 36 and 60, respectively. Configuration (a) has the minimum energy (verified by Unger and Moult), whereas (b) has a total favorable contacts of 36, the optimal one we have obtained. Its true minimum energy is unknown.

many accurate results have been obtained. The two-step-look-ahead strategy can increase simulation efficiency greatly, as shown by our example. We simulated a chain with $N = 100$ particles (99 links) and estimated the log-partition function, $\log Z_n \approx 96.39 \pm 0.037$. A numerical approximation formula gives

$$\log Z_n \approx (N-1) * \log(2.6385) + \left(\frac{43}{32} - 1\right) \log(N-1) - \log(4) = 96.24.$$

In all these simulations, either the original growth method (Rosenbluth and Rosenbluth 1955, Kong et al. 1994), which does not involve resampling and rejection control, or the bootstrap filtering method (Gordon et al. 1993), which resamples at every step, fails.

4.1.2 Inference in population genetics

In Section 2.7, we introduced a simple demographic model (also can be use for inferring phylogeny) for inferring relationships among different species based on comparisons of homologous DNA segments. The computational method described there is due to Griffiths and Tavare (1994) and can be seen as a special SIS method. The general method of resampling discussed in Section 3.4.4 can be applied to improve such computation (Chen and Liu 2000b).

As with many SIS applications, the trial distribution for simulating the evolutionary history $\mathcal{H}$ (Section 2.7) has the form

$$g(\mathcal{H}) = \prod_{t=1}^{k} g_t(H_{-t} \mid H_{-t+1}).$$

82   4. Sequential Monte Carlo in Action

We define the *current weight* (for $t \leq k$) for this trial density:
$$w_{-t} = \frac{p_\theta(H_{-t+1}|H_{-t}) \cdots p_\theta(H_0 \mid H_{-1})}{g_t(H_{-t}|H_{-t+1}) \cdots g_1(H_{-1} \mid H_0)} \equiv w_{-t+1} \frac{p_\theta(H_{-t+1}|H_{-t})}{g_t(H_{-t}|H_{-t+1})}.$$
The final weight is then $w = w_{-k}\, p_\theta(H_{-k})p_\theta(\text{stop} \mid H_0)$, where the last term is same as in (2.24).

In a parallel implementation of SIS, we first generate m samples from $q_\theta(H_{-1}|H_0)$, and then recursively generate $\{H_{-t}^{(1)}, \ldots, H_{-t}^{(m)}\}$, called the *current sample*, for $t = 2, 3, \ldots$, until coalescence in all m processes. Along with producing the current sample, we could also monitor the current weight and incur resampling steps at any time $-t$ when the coefficient of variation in $\{w_{-t}^{(1)}, \ldots, w_{-t}^{(m)}\}$ exceeds a threshold B. In resampling, one produces a new current sample by drawing with replacement from $\{H_{-t}^{(1)}, \ldots, H_{-t}^{(m)}\}$ according to probability $\propto \{w_{-t}^{(1)}, \ldots, w_{-t}^{(m)}\}$. The weight for each new sample after resampling is set as the *sample average* of the $w_{-t}^{(j)}$ so as to ensure that at the end we obtain a proper estimate of the likelihood function.

We note, however, that resampling among $\{H_{-t}^{(1)}, \ldots, H_{-t}^{(m)}\}$ is inefficient because these samples differ greatly in their coalescence speeds. Those $H_{-t}^{(j)}$ that have fast coalescence speeds (small population sizes) often have small current weights, but large final weights. Resampling among the $H_{-t}^{(m)}$ actually prunes away many "good" samples. To address this problem, we propose to conduct resampling at the same *coalescence time* instead of the same sequential sampling time. In other words, we wait until all the m processes reach the same population size, say, i, and then resample from $\{H_{-i_1}^{(1)}, \ldots, H_{-i_m}^{(m)}\}$, where $i_j = \min\{t : |H_{-t}^{(j)}| = i\}$. (Here, $|H_{-i}|$ denotes the population size of that generation.) Although early histories, $H_{-s}^{(j)}$, for $s < i_j$, are not needed in sequential sampling (due to a Markovian structure), we still need to keep all the early histories of those processes that survive the resampling.

The new resampling procedure can be implemented as follows. Suppose m parallel coalescence processes have been started as in Section 2.7. Then, for $i = n - 1, \ldots, 1$:

- for $j = 1, \ldots, m$, we run the jth process until its population size first reaches i; denote this time as $-i_j$;

- compute the coefficient of variation for $w_{-i_1}^{(1)}, \ldots, w_{-i_m}^{(m)}$; name this number CV_i;

- do resampling among $\{H_{-i_1}^{(1)}, \ldots, H_{-i_m}^{(m)}\}$ when CV_i is greater than a threshold, otherwise continue the usual sequential sampling;

- the weight for each new sample after resampling is set as the sample average of the $w_{-i_j}^{(j)}$.

At the end of this procedure, we produce a sample of histories $\mathcal{H}^{(*j)}$, $j = 1, \ldots, m$, and their associated weights $w^{(*j)}$, $j = 1, \ldots, m$. As in a standard SIS, we use the sample average of these weights,

$$\frac{1}{m} \left(w^{(*1)} + \cdots + w^{(*m)} \right),$$

to estimate the likelihood function $p_\theta(H_0)$. If we are interested in estimating $p_{\theta'}(H_0)$ for $\theta' \neq \theta$, we can use a modified weighted average

$$\hat{p}_{\theta'}(H_0) = \frac{1}{m} \left[w^{(*1)} \frac{p_{\theta'}(\mathcal{H}^{(*1)})}{p_\theta(\mathcal{H}^{(*1)})} + \cdots + w^{(*m)} \frac{p_{\theta'}(\mathcal{H}^{(*m)})}{p_\theta(\mathcal{H}^{(*m)})} \right].$$

Chen and Liu (2000b) applied this modified resampling step to the trial distribution described in Section 2.7. To compare with other approaches, the new method was implemented for a numerical example in Stephens and Donnelly (2000) (also treated in Section 2.7). With sample size $m = 10,000$ and bound $B = 4$, two resampling steps were incurred. We repeated this exercise five times and the results were rather similar. The extra computational cost was negligible. Figure 4.2(b) displays the likelihood curves estimated from five independent replications of our method. For comparison, Figure 4.2(a) shows the results (with $m = 10,000$) obtained by the plain SIS method discussed in Section 2.7. Figure 4.2(b) is almost indistinguishable from Figure 3(b) of Stephens and Donnelly (2000), which is produced by using a more efficient trial function. We note that the resampling modification described in this section can also be applied to their method.

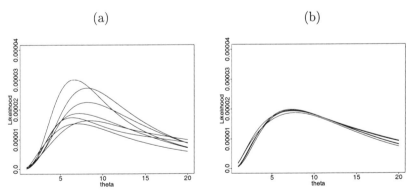

FIGURE 4.2. The estimated likelihood curve for a small dataset in Stephens and Donnelly (2000). (a) Seven independent runs of the plain SIS method of Griffiths and Tavare; (b) five independent runs of SIS with resampling.

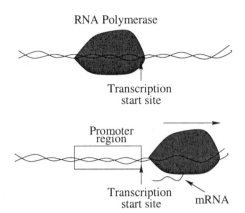

FIGURE 4.3. Cartoon illustration of gene transcription. Top figure: RNA polymerase binds to the promoter region and is about to start the transcription. Bottom figure: transcription is started when the RNA polymerase zips along the genome resulting in messenger RNA.

4.1.3 Finding motif patterns in DNA sequences

As we have briefly discussed in Section 1.5, all the hereditary information of a living thing is stored in its genome, which can be represented by linear sequences of letters from a four-letter alphabet (A, T, G, and C). Segments of the genome of 100 to 10,000 DNA bases long, called genes, code for proteins, which are responsible for almost all functions of life. An entire genome containing millions or billions of DNA bases and may encode tens of thousands of proteins. The process of transforming the information contained in a gene into its product (protein) is fairly complicated. First of all, the information of a gene has to be *transcribed* (copied) from the genome to messenger RNA (mRNA) by a molecular complex known as *RNA polymerase*. Then, these mRNAs are translated to proteins via *transfer RNAs* (tRNA). Free RNA polymerase molecules collide with the chromosomes at random positions, sliding along it but sticking only weakly to most DNA segments. When it meets a specific DNA sequence called the *promoter*, which signals the start of RNA synthesis, the polymerase binds tightly and starts the transcription. The promoter for a bacteria gene is located just "upstream" (at −10 and −35) of the transcriptional start site of the gene (Alberts, Bray, Lewis, Raff, Roberts and Watson 1994). Note that the transcriptional start site is different from the start of the gene (located further downstream).

It is an amazing fact that every single cell of an organism carries a complete copy of the full genome. However, diverse cells within an individual differ drastically (for example, cells in one's eyes and cells on one's skin have completely different forms and functions). The process by which cells acquire special characteristics, called differentiation, occurs because the expression of genes is regulated by various mechanisms. In single-cell organisms, genes are regulated so as to allow the cell to respond to environmental changes. A particular important form of gene regulation is achieved by proteins (often called *regulatory proteins*) bound to sites within or close to the promoter of a gene located in its upstream (5' end) non-coding regions. which then alters the rate of RNA polymerase binding to a transcriptional starting site. A protein bound to a site within the promoter region will prevent RNA polymerase from starting transcription. If the regulatory protein binds further upstream, it may enhance gene expression by attracting RNA polymerases. The locations on the genome where the regulatory proteins bind are called the regulatory *binding sites* and these sites are composed of one or more short DNA segments of 20 base pairs long (some are even shorter).

cole1	taatgtttgtgctggttttgtggcatcgggcgagaatagcgcgtggtgtgaaagactgttttttgatcgttttcacaaaaatggaagtccacagtcttgacag
ecoarabop	gacaaaaacgcgtaacaaaagtgtctataatcacggcagaaaagtccacattgattatttgcacggcgtcacactttgctatgccatagcattttatccataag
ecobglr1	acaaatcccaataacttaattattgggatttgttatatataactttataaattcctaaaattacacaaagttaataactgtgagcatggtcatattttatcaat
ecocrp	cacaaagcgaaagctatgctaaaacagtcaggatgctacagtaatacattgatgtactgcatgtatgcaaaggacgtcacattaccgtgcagtacagttgatagc
ecocya	acggtgctacacttgtatgtagcgcatctttctttacggtcaatcagcatggtgttaaattgatcacgttttagaccattttcgtcgtgaaactaaaaaaacc
ecodecop	agtgaattatttgaaccagatcgcattacagtgatgcaaacttgtaagtagatttccttaattgtgatgtgtatcgaagtgtgttgcggagtagatgttagaata
ecogale	gcgcataaaaaacggctaaattcttgtgtaaacgattccactaatttattccatgtcacactttcgcatctttgttatgctatggttatttcataccataagcc
ecoilvbpr	gctccggcggggttttttgttatctgcaattcagtacaaaacgtgatcaaccccctcaattttccctttgctgaaaaattttccattgtctcccctgtaaagctgt
ecolac	aacgcaattaatgtgagttagctcactcattaggcaccccaggctttacactttatgcttccggctcgtatgttgtgtggaattgtgagcggataacaatttcac
ecomale	acattaccgccaattctgtaacagagatcacacaaagcgacggtggggcgtaggggcaaggaggatggaaagaggttgccgtataaagaaactagagtccgttta
ecomalk	ggaggaggcgggaggatgagaacacggcttctgtgaactaaaccgaggtcatgtaaggaatttcgtgatgttgcttgcaaaaatcgtggcgattttatgtgcgca
ecomalt	gatcagcgtcgttttaggtgagttgttaataaagatttggaattgtgacacagtgcaaattcagacacataaaaaaacgtcatcgcttgcattagaaaggtttct
ecoompa	gctgacaaaaaagattaaacatacctatacaagacttttttttcatatgcctgacggagttcacacttgtaagttttcaactacgttgtagactttacatcgcc
ecotnaa	tttttaaacattaaaattcttacgtaatttataatctttaaaaaaagcatttaatattgctccccgaacgattgtgattcgattcacatttaaacaatttcaga
ecouxu1	cccatgagagtgaaattgttgtgatgtggttaacccaattagaattcgggattgacatgtcttaccaaaaggtagaacttatacgccatctcatccgatgcaagc
pbr-p4	ctggcttaactatgcggcatcagagcagattgtactgagagtgcaccatatgcggtgtgaaataccgcacagatgcgtaaggagaaaataccgcatcaggcgctc
trn9cat	ctgtgacggaagatcacttcgcagaataaataaatcctggtgtccctgttgataccgggaagccctgggccaacttttggcgaaaatgagacgttgatcggcacg
(tdc)	gatttttatactttaacttgttgatatttaaaggtatttaattgtaataacgatactctggaaagtattgaaagttaatttgtgagtggtcgcacatatcctgtt

TABLE 4.1. A dataset of 18 DNA segments, each of 105 base pairs long, taken from the upstream noncoding regions of 18 genes of *E. coli*. This dataset was created by Stormo and Hartzell (1989) who used a greedy sequential buildup strategy to find the binding sites.

The cyclic receptor protein (CRP) is a positive control factor necessary for the expression of catabolite repressible genes. Table 4.1 shows a set of DNA segments, each 105 bases long, cut from upstream non-coding regions of 18 genes of *E. coli*. It is known that there is at least one CRP-binding site in each of the 18 segments and the location of these binding sites have been experimentally determined (Stormo and Hartzell 1989). The width of

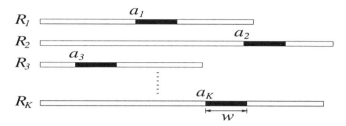

FIGURE 4.4. A schematic plot for finding subtle sites in multiple sequences. The blackened segment in each sequence represents the "common pattern" expected from these sequences; both its location and its content are unknown.

the binding motif is decided at $w = 20$. So this dataset allows one to test the ability of various new algorithms.

We can abstract the motif finding problem into a simpler form, as depicted by Figure 4.4: We are given K sequences, $\boldsymbol{R} = (R_1, \ldots, R_K)$, of letters from an alphabet of size d ($d = 20$ for proteins and 4 for DNAs), and we search within them for a "conserved" pattern of length w, as illustrated by the blackened region in the figure. In this model, we assume that every sequence R_k has exactly one binding site and their locations are called the *alignment variable* and denoted as $A = (a_1, \ldots, a_K)$.

Table 4.2 displays a typical alignment of a common pattern, corresponding to the CRP-binding sites, found from the CRP data in Table 4.1. Each sequence segment corresponds to a blackened part in Figure 4.4. The locations of these sites have been experimentally determined (Stormo and Hartzell 1989). Hence, the "true" value of the alignment variable A is $A_0 = (64, 58, \ldots, 81)$. Clearly, these patterns are not exactly conserved. Biologists also showed that there are multiple motif sites in several of the sequence segments. For example, the 20th positions of the first and second sequences (cole 1 and ecoarabop) are all real binding sites. Hence, the assumption that each sequence has exactly one motif segment is not very realistic. A few other models have been developed to account for this limitation (Liu, Neuwald and Lawrence 1995, Neuwald, Liu and Lawrence 1995).

A common approach in literature (Stormo and Hartzell 1989, Lawrence and Reilly 1990, Lawrence et al. 1993, Krogh, Brown, Mian, Sjolander and Haussler 1994, Liu 1994a, Liu, Neuwald and Lawrence 1995, Durbin, Eddy, Krogh and Mitchison 1998) is to assume that the specificity of the motif can be represented as an unknown matrix with w columns (a first-order model). Column j in the matrix represents the base-type preference for the jth position in the motif. For example, we see from Table 4.2 that the first position of the motif prefers T, and the second prefers both T and G, and so on. Thus, a pattern matrix can be thought of as a $4 \times w$ matrix in our example, in which each column shows the preference of nucleotide base

Gene's name	Motif Start	Binding Site
cole1	64	TTTGATCGTTTTCACAAAA
ecoarabop	58	TTTGCACGGCGTCACACTT
ecobglr1	79	TGTGAGCATGGTCATATTT
ecocrp	66	TGCAAAGGACGTCACATTA
ecocya	53	TGTTAAATTGATCACGTTT
ecodecop	10	TTTGAACCAGATCGCATTA
ecogale	45	TTTATTCCATGTCACACTT
ecoilvbpr	42	CGTGATCAACCCCTCAATT
ecolac	12	TGTGAGTTAGCTCACTCAT
ecomale	17	TGTAACAGAGATCACACAA
ecomalk	64	CGTGATGTTGCTTGCAAAA
ecomalt	44	TGTGACACAGTGCAAATTC
ecoompa	51	CCTGACGGAGTTCACACTT
ecotnaa	74	TGTGATTCGATTCACATTT
ecouxu1	20	TGTGATGTGGTTAACCCAA
pbr-p4	56	TGTGAAATACCGCACAGAT
(tdc)	81	TGTGAGTGGTCGCACATAT

TABLE 4.2. The alignment of a common motif in upstream regions of the 17 *E. coli* genes. These sites have been determined experimentally. The original data consisting of 18 segments were produced by Stormo and Hartzell (1989) and displayed in Table 4.1. One of the sequences is eliminated from the table because its binding site contains an extra insertion compared with others.

types, in terms of total counts of A, T, G, and C, for the corresponding position in the motif.

As stated in Stormo and Hartzell (1989), the problem of identifying the binding sites from a collection of unaligned sequences can be viewed as finding the best A that gives us the maximal "mutual similarity." This similarity among multiple DNA segments can be represented by the "information content"

$$I_A = \sum_{j=1}^{w} \sum_{b=A}^{T} f_{j,b} \log \frac{f_{j,b}}{p_b}, \tag{4.1}$$

where $f_{j,b}$ is the observed frequency of base b at the jth position of the site, and p_b is the fraction of base b in an appropriate background (e.g., the whole genome or the whole dataset in consideration). With the goal of optimizing (4.1), Stormo and Hartzell (1989) proposed the first effective method to search for conserved patterns in multiple sequences. Their method as outlined can be seen as a greedy sequential method:

Stormo-Hartzell Algorithm

1. Each of the w-*words* (k-long substring) of the first sequence, R_1, are considered as a possible motif pattern — they are indeed equally likely to be a binding site without further information. In the CRP example, one forms 86 "matrices" to represent possible motif patterns.

2. The next sequence on the list is added to the analysis. All the matrices formed previously (say, N of them) are paired with all possible w-words in the new sequence, and a "similarity score" is computed for each pair. There are a total of $N \times 86$ possible pairs for the CRP example.

3. The top N best scored pairs are kept, from which a new set of N "matrices" is formed. For each kept pair, the new matrix is formed by adding the site (w-word) found in the new sequence to the matrix it paired with.

4. Repeat the previous two steps until all sequences have been processed.

The foregoing algorithm can be understood from a sequential imputation viewpoint. Step 2 in the algorithm is, in fact, a predictive sampling step as in (1.4), $p(a_t \mid a_1, \ldots, a_{t-1}, R_1, \ldots, R_k)$. But instead of random sampling, they chose the most probable one according to this distribution. We now describe how the problem can be treated by a proper statistical model and how the computation can be completed by a sequential Monte Carlo method.

The matrix model can be more concisely stated as a *product multinomial* model (Liu, Neuwald and Lawrence 1995). Let $\boldsymbol{\theta}_0 = (\theta_{01}, \ldots, \theta_{0d})^T$ be the probability (column) vector describing the residue frequencies outside a motif ($d = 4$ in our example). For notational simplicity, we just use numbers $1, 2, \ldots, d$, instead of the actual names of the DNA base pairs or amino acid residues, to represent the letters in the alphabet. We let $\boldsymbol{\theta}_j$, $j = 1, \ldots, w$, represent the frequency of each base at the jth position of the motif. Then, $\Theta = [\boldsymbol{\theta}_1, \ldots, \boldsymbol{\theta}_w]$, called a *product-multinomial* model (Liu, Neuwald and Lawrence 1995), is equivalent to the pattern matrix we described previously by words. It is also called a *profile* matrix in literature. By treating the alignment variable $A = (a_1, \ldots, a_K)$ as missing data, we can now write down a simple statistical model:

$$
\begin{aligned}
p(\boldsymbol{R} \mid \boldsymbol{\theta}_0, \Theta, A) &= \boldsymbol{\theta}_0^{\mathbf{h}(\mathbf{R}_{\{A\}^c})} \prod_{j=1}^{w} \boldsymbol{\theta}_j^{\mathbf{h}(\mathbf{R}_{A+j-1})} \\
&= \boldsymbol{\theta}_0^{\mathbf{h}(\mathbf{R})} \prod_{j=1}^{w} \left(\frac{\boldsymbol{\theta}_j}{\boldsymbol{\theta}_0}\right)^{\mathbf{h}(\mathbf{R}_{A+j-1})}
\end{aligned}
\quad (4.2)
$$

In the expression, we use $A_{[-k]}$ to denote all of A but a_k, $A + l = \{a_1 + l, \ldots, a_K + l\}$ to denote the set of l-shifted positions of A, and $\{A\} = \{a_k + j - 1 : k = 1, \ldots, K, \ j = 1, \ldots, w\}$ to represent the set of residue indices occupied by the motif elements with alignment variable A. For any set C of indices, $\boldsymbol{R}_C$ represents the collection of the residues indexed by elements of C. For example, given any alignment variable A, we have $\boldsymbol{R}_{\{A\}} = \{r_{k, a_k+j-1} : \text{ for } j = 1, \ldots, w; \ k = 1, \ldots, K\}$. The counting function $h(\ )$, whose argument is a set of residues, counts how many of each letter types in a set of protein residues or DNA base pairs). For example, if $R = \{AATCCCTG\}$ is an oligonucleotide sequence, we obtain that $h(R) = (2, 2, 1, 3)$ for letter types A, T, G, and C. It is much simpler to state the complete-data model (4.2) in words: Given the alignment variable $\boldsymbol{A}$, the part of the dataset $\boldsymbol{R}$ outside the motif elements are like i.i.d. realizations from a multinomial model $\boldsymbol{\theta}_0$; and each position in a binding site follows a different multinomial model $\boldsymbol{\theta}_j$. Our task is to make inference on A, $\boldsymbol{\theta}_0$, and the motif matrix Θ. In order to take a Bayesian approach, we let the prior distribution for A be uniform on all allowable configurations; let the prior for $\boldsymbol{\theta}_0$ and $\boldsymbol{\theta}_j$ be Dirichlet$(N_0\alpha_{0,1}, \ldots, N_0\alpha_{0,4})$, where the α are base frequencies in the genome and the *pseudo-counts* N_0 were chosen as $\sqrt{K}$ in many of our examples. More details on this model can be found elsewhere (Liu 1994a, Liu, Neuwald and Lawrence 1995). A Gibbs sampling scheme for this problem will be shown in Section 6.5. Here, we will describe a SIS method to impute A.

Let $A_t = (a_1, \ldots, a_t)$ and let $\boldsymbol{R}_t = \{R_1, \ldots, R_t\}$ be the collection of the first t sequences. Suppose we have imputed multiple copies of the alignment variable, $A_{t-1}^{(1)}, \ldots, A_{t-1}^{(m)}$, with respective weights $w_{t-1}^{(1)}, \ldots, w_{t-1}^{(m)}$. Then, with the new sequence R_t, we can update the weight as

$$w_t^{(j)} = w_{t-1}^{(j)} p(R_t \mid A_{t-1}^{(j)}, \boldsymbol{R}_{t-1}), \qquad (4.3)$$

where the predictive probability can be computed:

$$p(R_t \mid A_{t-1}^{(j)}, \boldsymbol{R}_{t-1}) = \frac{1}{l_t - w + 1} \sum_{i=1}^{l_t - w + 1} p(R_t, a_t = i \mid A_{t-1}^{(j)}, \boldsymbol{R}_{t-1}).$$

Since $(A_{t-1}^{(j)}, \boldsymbol{R}_{t-1})$ determines an estimate of the pattern matrix Θ, the above predictive probability is just the likelihood of R_t under the jth estimated pattern matrix, given that it contains a binding site. Simultaneously with the weight updating, we can impute the motif locations for the new sequence; that is, $a_t^{(j)}$ is drawn from

$$P(a_t^{(j)} = i \mid A_{t-1}^{(j)}, \boldsymbol{R}_t) \propto p(R_t, a_t = i \mid A_{t-1}^{(j)}, \boldsymbol{R}_{t-1}). \qquad (4.4)$$

When the importance weight $w_t^{(j)}$ is too skewed, we *resample* among the alignment matrices with probability proportional to the $w_t^{(j)}$ (Section 3.4.4). One can also use a different set of weights to do resampling (Liu, Chen and Logvinenko 2000). For better efficiency, the resampling step should take place right after the weight updating step (4.3) and *before* the sampling step (4.4). One can see that the resampling and predict sampling steps play a similar role as Steps 2 and 3 of the Stormo-Hartzell algorithm.

4.2 Approximating Permanents

Let $A = (a_{ij})_{n \times n}$ be a 0-1 matrix (also called the restriction matrix), where each entry a_{ij} is either 0 or 1. Let Π be the set of all permutaions of $\{1, \ldots, n\}$. A useful representation of a permutation σ is

$$\begin{pmatrix} 1 & 2 & \cdots & n \\ \sigma(1) & \sigma(2) & \cdots & \sigma(n) \end{pmatrix},$$

where $\sigma(i)$ record the new position of label i after the permutation. For any $\sigma \in \Pi$, we define

$$a(\sigma) = \prod_{i=1}^{n} a_{i\sigma(i)}.$$

Following the notations in Diaconis et al. (2001), we let S_A be the set of all "permitted" permutations under A:

$$S_A = \{\sigma : a(\sigma) = 1\}.$$

The *permanent* of A is defined as

$$\operatorname{perm}(A) = \sum_{\sigma \in \Pi} \prod_{i=1}^{n} a_{i\sigma(i)} \equiv |S_A|, \tag{4.5}$$

There are many statistical applications that are related to permutations and the computation of the permanents (Diaconis et al. 2001). For example, for the test of independence for the astronomy data in Section 1.7, we want to approximate the tail probability of a test statistics under the uniform distribution of all "permitted" permutations. This task requires us to simulate from the uniform distribution on S_A. Sometimes one might also be interested in the total number of permitted permutations — the permanent perm(A). Approximating the permanent has been a great undertaking in recent years among computer scientists (Jerrum and Sinclair 1989).

A naive Monte Carlo method for approximating perm(A) is to generate random permutations $\sigma_1, \ldots, \sigma_N$ *uniformly* and to estimate

$$\operatorname{perm}(A) \approx n! \frac{\sum_{j=1}^{N} a(\sigma_j)}{N}.$$

However, this method becomes very inefficient when the number of zeros in A is relatively large. Clearly, if one of the entries in the product in $a_{1\sigma(1)}$, $\ldots$, $a_{n\sigma(n)}$ is zero, then $a(\sigma) = 0$. Thus, we want to design a Monte Carlo algorithm that samples only on those permutations whose $a_{i\sigma(i)}$ terms are all none-zero. This goal can be achieved by the following recursive strategy proposed in (Chen and Liu 2001).

Let r_i, $i = 1, \ldots, n$, and c_j, $j = 1, \ldots, n$, be the row sums and the column sums of A, respectively. For the first column, we draw an entry [i.e., $\sigma(1)$] among all those positions with $a_{i1} = 1$. However, we do not want to sample $\sigma(1)$ uniformly among all the c_1 positions. Note that for any permutation $\sigma \in \Pi$, if $\sigma(1) = s$, then none of the $\sigma(i)$, $i = 2, \ldots, n$ can be equal to s. Thus, we should prefer to use those nonzero entries that correspond to relatively small row sums (i.e., small r_s). More precisely, we sample $\sigma(1)$ from the distribution

$$P[\sigma(1) = s] \propto \frac{1}{r_s - 1}, \tag{4.6}$$

for s in the set $S_1 = \{s : a_{s1} = 1\}$. If one row sum, say, r_s equals one for some $s \in S_1$, then $q[\sigma(1) = s] = 1$. If we can find $s \neq s'$, both in S_1, such that $r_s = r_{s'} = 1$, then there is no allowable permutation; thus, perm$(A) = 0$.

After sampling the first column, or $\sigma(1)$ equivalently, we can update our "working matrix" by setting all the entries in the first column and the $\sigma(1)$th row to 0's and updating the corresponding row sums and column sums. Then, we proceed to the next column, applying the same sampling strategy. If this procedure can be carried out recursively to the last column, we obtain an "allowable" permutation σ whose sampling distribution is

$$P[\sigma = (s_1, s_2, \ldots s_n)] = \frac{(r_{(0)s_1} - 1)^{-1}}{D_{(0)}} \frac{(r_{(1)s_2} - 1)^{-1}}{D_{(1)}} \cdots \frac{(r_{(n-1)s_n} - 1)^{-1}}{D_{(n-1)}},$$

where $r_{(k)s_{k+1}}$ is the kth modification of the s_{k+1}th row sum (after resetting the first k columns)and the denominator is

$$D_{(k)} = \sum_{i:\, a^{(k)}_{i(k+1)}=1} (r_{(k)i} - 1)^{-1},$$

where $a^{(k)}_{ij}$ is the (i,j)th entry of the kth modified restriction matrix (i.e., after resetting the first k columns and the rows $s_1, \ldots, s_k$ to 0).

The weight of each generated permutation σ is updated recursively as

$$w_{k+1}(s_1, \ldots, s_{k+1}) = \begin{cases} w_k(r_{(k)s_{k+1}} - 1)D_{(k)} & \text{if } D_{(k)} < \infty \\ w_k & \text{otherwise,} \end{cases} \tag{4.7}$$

If at some stage it is impossible to proceed to produce a valid permutation, we assign a weight 0 to this unsuccessful trial. After performing m such trials of sequential sampling, we obtained m weighted permutations, $(\sigma^{(j)}, w^{(j)})$, for $j = 1, \ldots, m$, including unsuccessful ones (corresponding to a weight of 0), we can estimate the permanent of A as

$$\widehat{\text{perm}}(A) = \frac{w^{(1)} + \cdots + w^{(m)}}{m}.$$

Other statistical tasks such as the hypothesis testing can also be accommodated.

This simple method was shown very efficient in a number of simulation examples we have tested on. We started with an 8 × 8 matrix of which we know the true answer. The cv^2 of our procedure, which approximates perm(A) accurately, was about 0.3. We then simulated a random matrix of 50×50. The cv^2 of our SIS procedure for this case only increased moderately to 0.5. It took about half a second to generate one permutation on a Sun Ultra 60 workstation. When we increased the matrix size to 100 × 100, the cv^2 was still impressively small, only about 0.7.

It should be noted that the methods of Kuznetsov (1996) and Beichl and Sullivan (1999) can all be viewed as SIS samplers. The computational and conceptual complexity of our approach is similar to that of Kuznetsov (1996). But our method tends to be more efficient. Compared with Beichl and Sullivan, our approach is much simpler, yet more efficient.

4.3 Counting 0-1 Tables with Fixed Margins

The problem of counting 0-1 tables with fixed margins was introduced in Section 3.4.2. It is a long-standing problem in the fields of applied mathematics and computer science and has been subject to active research for many years. In a recent work, Chakraborty, Chen, Diaconis, Holmes and Liu (2001) developed a number of techniques for solving the problem. These include a MCMC method, an exact counting method, and an SIS method. Here, we focus on their SIS approach.

Briefly, the SIS method begins by filling in columns (or rows) of the $m \times n$ table from left to right sequentially. After the first $t-1$ columns are filled, the row sums are updated and the tth column is filled in by sampling c_t of its m possible positions to put in 1's. These c_t positions are sampled according to a distribution that is related to the vector of the updated row sums. For example, the probability that positions $i_1, \ldots, i_{c_t}$ of the tth column are sampled can be proportional to

$$\left(\frac{r_{i_1}}{n - r_{i_1}} \times \cdots \times \frac{r_{i_{c_t}}}{n - r_{i_{c_t}}} \right)^{1+\delta}. \tag{4.8}$$

More choices of the sampling distribution and other mathematical details can be found in Chen (2001). This sequential sampling method gives us a rather accurate estimate for fairly large tables.

To test the method, they examined Darwin's finch data (Sanderson 2000), which, in the form of an "occurrence matrix," recorded the presence and absence for each of the 13 species of Galápagos finch in 17 islands of an east archipelago. This data is shown as in Table 4.3.

0	0	1	1	1	1	1	1	1	1	0	1	1	1	1	1	1	14
1	1	1	1	1	1	1	1	1	1	0	1	0	1	1	0	0	13
1	1	1	1	1	1	1	1	1	1	1	1	0	1	1	0	0	14
0	0	1	1	1	0	0	1	0	1	0	1	1	0	1	1	1	10
1	1	1	0	1	1	1	1	1	1	0	1	0	1	1	0	0	12
0	0	0	0	0	0	0	0	0	0	1	0	1	0	0	0	0	2
0	0	1	1	1	1	1	1	1	0	0	1	0	1	1	0	0	10
0	0	0	0	0	0	0	0	0	0	0	1	0	0	0	0	0	1
0	0	1	1	1	1	1	1	1	1	0	1	0	0	1	0	0	10
0	0	1	1	1	1	1	1	1	1	0	1	0	1	1	0	0	11
0	0	1	1	1	0	1	1	0	1	0	0	0	0	0	0	0	6
0	0	1	1	0	0	0	0	0	0	0	0	0	0	0	0	0	2
3	3	10	9	9	7	8	9	7	8	2	9	3	6	8	2	2	

TABLE 4.3. The finch dataset records the occurrences of 13 species(rows) of Galápagos finch in 17 islands (columns) of an east Pacific archipelago. Ecologists are interested in testing if the occurrence pattern is a random draw from the uniform distribution of all such patterns.

The problem of couting the number of 0-1 tables that have the same marginal sums as the finch data was first raised by Susan Holmes in Stanford. Based on 1000 sequentially simulated tables, the SIS method estimated that the total count of the 0-1 tables with the given margins is $(6.72 \pm .02) \times 10^{16}$. The computation took 12 minutes on a Pentium 400 machine. With more computing time, 10^8 tables were generated and the total number of the tables was estimated as 6.715×10^{16}. The coefficient of variation of the importance weights (defined as the sample variance divided by the square of the sample mean) was around 0.7. The "truth" they obtained by using an exact-counting algorithm is $6.71... \times 10^{16}$.

To push further for the algorithm, we simulated a 50×50 random table and recorded its marginal sums. The row sums are 1, 1, 1, 1, 1, 1, 1, 1, 1, 1, 1, 1, 2, 2, 2, 2, 2, 2, 3, 3, 3, 3, 3, 3, 4, 4, 4, 4, 5, 5, 6, 6, 7, 8, 8, 9, 10, 11, 12, 12, 12, 12, 12, 13, 14, 14, 15, 16, 18, 19, respectively, and the column sums are 1, 1, 1, 1, 2, 2, 2, 2, 2, 2, 2, 2, 3, 3, 3, 3, 3, 3, 4, 4, 4, 5, 5, 5, 5, 5, 6, 6, 6, 6, 6, 7, 7, 8, 8, 8, 8, 8, 9, 9, 9, 10, 10, 11, 12, 13, 13, 13, 14, 14, 14, respectively. The column sums were ordered from the largest to the smallest and the columns were sampled sequentially according to (4.8) with $\delta = 0$. This gave us a cv^2 of around 0.2. Based on 1,000 samples, which took about 5 minutes to generate, we estimated that the total number of 0-1 tables with these marginal sums is $(8.9 \pm 0.1) \times 10^{242}$. Based on 10,000 samples, the estimate was improved to 8.78×10^{242} with standard error 0.05×10^{242}. This example shows that the SIS method is still very efficient even for large tables.

4.4 Bayesian Missing Data Problems

4.4.1 Murray's data

Let us revisit the Gaussian missing data problem described in Section 2.5.7. The previous approach was based on the plain importance sampling in which the trial density was chosen as the posterior distribution based on the first four complete observations. Here, we consider an SIS approach.

Recall that the covariance matrix is assigned the Jeffreys non-informative prior (2.13) The posterior distribution of Σ given complete data is

$$p(\mathcal{Z}|\text{complete data}) \propto \mathcal{Z}^{\frac{3}{2}-1} \exp\left\{-\frac{1}{2}\operatorname{tr}[\mathcal{Z} \cdot S]\right\},$$

where $S = (s_{ij})_{2\times 2}$ is the uncorrected sum of squares matrix and $\mathcal{Z} = \Sigma^{-1}$.

Let the complete data be $y_1, \ldots, y_{12}$, where $y_t = (y_{t,1}, y_{t,2})$ for $t = 1, \ldots, 12$. Thus, the $y_{t,2}$ are missing for observations 5 to 8 and the $y_{t,1}$ are missing for observations 9 to 12. The predictive distribution of y_{t+1} given $\mathbf{y}_t = (y_1, \ldots, y_t)$ is

$$y_{t+1}|\mathbf{y}_t \sim t_2\left(0, \frac{S_t}{t-1}, t-1\right),$$

where $S_t = (s_{ij})$, $s_{ij} = \sum_{s=1}^{t} y_{s,i} y_{s,j}$, and t_2 is the bivariate t-distribution. It follows that, conditional on $\mathbf{y}_t = (y_1, \ldots, y_t)$, the marginal and conditional distributions are

$$y_{t+1,1} \mid \mathbf{y}_t \sim t_1\left(0, \frac{s_{11}}{t-1}, t-1\right),$$

$$y_{t+1,2} \mid \mathbf{y}_t, y_{t+1,1} \sim t_1\left[\frac{s_{12}}{s_{11}} y_{t+1,1}, \frac{|S_t|}{ts_{11}}\left(1 + \frac{y_{t+1,1}^2}{s_{11}}\right), t\right],$$

respectively. Similar results can be obtained for conditional distributions $[y_{t+1,2} \mid \mathbf{y}_t]$ and $[y_{t+1,1} \mid \mathbf{y}_t, y_{t+1,2}]$. Based on these distributional results, Steps A and B of the sequential imputation (Section 3.2.2) can both be easily implemented. Figure 4.5 gives the exact posterior distribution of ρ and the approximated posterior distribution based on sequential imputation. The approximation is a weighted mixture of $m = 1000$ complete-data posterior distributions.

The complete-data posterior distribution of ρ is still difficult to compute since it has the form

$$p(\rho \mid \mathbf{y}) \propto (1-\rho^2)^{\frac{\nu-2}{2}} \int_0^\infty \omega^{-1}\left(\omega + \frac{1}{\omega} - 2\rho r\right)^{-(\nu+1)} d\omega, \qquad (4.9)$$

4.4 Bayesian Missing Data Problems

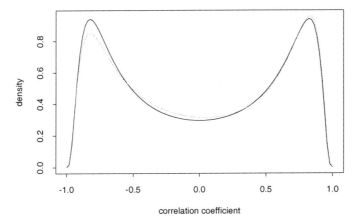

FIGURE 4.5. An approximation to the posterior distribution of the sample correlation coefficient ρ for Murray's data. Solid line: true density; dashed line: approximation.

where **y** is the completed data after the missing part being imputed, r is the sample correlation coefficient, and $\nu = n - 2 = 10$. To avoid numerical integration in (4.9), we can use the algorithm of Odell and Feiveson (1966) to generate observations from the inverse Wishart distribution, which gives rise to a sample from $p(\rho \mid \mathbf{y})$. Note that we do not have to draw from an inverse Wishart distribution *during* sequential imputation.

4.4.2 Nonparametric Bayes analysis of binomial data

Kong et al. (1994) and Liu (1996b) considered the following *hierarchical* binomial model:

$$y_t \sim \text{Binomial}(l_t, \zeta_t)$$
$$\zeta_t \stackrel{\text{i.i.d.}}{\sim} F, \quad 1 \leq t \leq n.$$

Our interests are in drawing inference about both F and the ζ_t's based on the observed data y_t. As a concrete example, consider a dataset (Figure 4.6) of n thumbtacks randomly drawn from a certain population. Here, ζ_t can be interpreted as the inherent probability for the tth tack to point up when being flicked. The observed data for the tth tack is y_t, the number of times the tack landed point up out of a total of l_t flicks. The data in Figure 4.6 was generated by Beckett and Diaconis (1994), of whom each flicked 16 different thumbtacks on 10 different surfaces. For each person-tack-surface combination (a total of $2 \times 16 \times 10$ scenarios), the experiment was repeated 9 times. For simplicity, we treat the data as though they came from 320 different tacks and each being flicked 9 times independently. A histogram of the data (the y_i) is shown in Figure 4.6.

96   4. Sequential Monte Carlo in Action

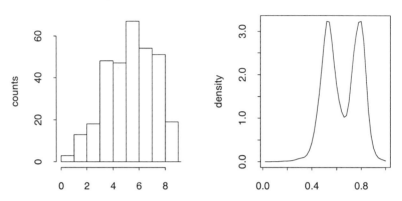

FIGURE 4.6. Left: The histogram of the tack data produced by Beckett and Diaconis (1994). Right: The predictive distribution of ζ for a future tack (which is also the posterior mean of F).

Different from the method outlined in Section 1.8, we take a *nonparametric* Bayes approach by assigning a *Dirichlet process* prior, $\mathcal{D}(\alpha)$, to the infinite-dimensional parameter F. The hyperparameter α is a finite positive measure on interval $[0, 1]$. We define the norm $\|\alpha\|$ as the total measure of the support of F. Note that $\mathcal{D}(\alpha)$ is a probability measure on $\mathcal{P}$, the set of all probabilities on the Borel sets of $[0, 1]$. Readers not familiar with this special class of Dirichlet process distribution and nonparametric Bayes inference are referred to Ferguson (1974). In the tacks example, we may be interested in estimating ζ_t for tack t, given its performance in 9 independent flips, or in obtaining the posterior mean of the unknown distribution F [i.e., $E(F \mid \zeta_1, \ldots, \zeta_n)$].

For simplicity, suppose α is the uniform measure on $[0, 1]$ (hence, the norm $\|\alpha\| = 1$). Then, a key result (Ferguson 1974) is that conditional on $\zeta_1, \ldots, \zeta_{t-1}$, the posterior distribution of F is again a Dirichlet process, $\mathcal{D}(\alpha_{t-1})$ with $\alpha_{t-1} = \alpha + \sum_{i=1}^{t-1} \delta_{\zeta_i}$, where δ_{ζ_i} is a delta measure with mass 1 located at ζ_s. Thus, the predictive distribution for the next thumbtack, ζ_t, is

$$\zeta_t \mid \zeta_1, \ldots, \zeta_{t-1} \sim \frac{1}{t}\left(\alpha + \sum_{i=1}^{t-1} \delta_{\zeta_i}\right), \qquad (4.10)$$

which should be interpreted as a probabilistic mixture of α and delta measures concentrated at the ζ_i's (Antoniak 1974). Note that the posterior mean of F, $E(F \mid \zeta_1, \ldots, \zeta_t)$, is the same as (4.10). It is easy to see from

(4.10) that

$$[\zeta_t|\zeta_1,\ldots,\zeta_{t-1},y_t] = \frac{1}{Z}\Bigg[B(y_t+1,l_t-y_t+1)\text{Beta}(y_t+1,l_t-y_t+1)$$
$$+ \sum_{i=1}^{t-1}\zeta_i^{y_t}(1-\zeta_i)^{l_t-y_t}\delta_{\zeta_i}\Bigg] \quad (4.11)$$

where

$$B(y_t+1,l_t-y_t+1) = \int_0^1 \zeta^{y_t}(1-\zeta)^{l_t-y_t}d\zeta = \frac{\Gamma(y_t+1)\Gamma(l_t-y_t+1)}{\Gamma(l_t+2)}$$

is the Beta function, Beta$(\cdot,\cdot)$ is the standard Beta distribution, and

$$Z = B(y_t+1,l_t-y_t+1) + \sum_{i=1}^{t-1}\zeta_i^{y_t}(1-\zeta_i)^{l_t-y_t}$$

is the normalizing constant. Note that (4.11) is a mixture of a Beta distribution and discrete point masses. From (4.10), we also get

$$p(y_t|\zeta_1,\ldots,\zeta_{t-1}) = \frac{1}{t}B(y_t+1,l_t-y_t+1) + \frac{1}{t}\sum_{i=1}^{t-1}\zeta_i^{y_t}(1-\zeta_i)^{l_t-y_t},$$

which is the term needed for updating the importance sampling weights. Hence, both Steps A and B of sequential imputation can be easily implemented (Liu 1996b).

Note that a direct application of the data augmentation algorithm (Section 6.4) is difficult because sampling F from

$$[F \mid \zeta_1,\ldots,\zeta_n] = \mathcal{D}(\alpha + \delta_{\zeta_1} + \cdots + \delta_{\zeta_n})$$

is infeasible. [Some approximations exist; see Doss (1994)]. Escobar (1994) described a "collapsed" Gibbs sampling algorithm (Section 6.7) in which the sampling of F is not needed. His method also takes advantage of the simplicity of the predictive distributions (4.10) and (4.11). For a related problem where the $\zeta_t's$ are ordered, Gelfand and Kuo (1991) used a similar idea to avoid sampling the infinite-dimensional parameter F.

The sequential imputation procedure derived from using (4.10) and (4.11) was applied to the thumbtack data with α being uniform and $\|\alpha\| = 1$. Figure 4.6 displays the posterior mean of the unknown density function F, $E(F \mid \mathbf{y})$. Liu (1996b) studied the sensitivity of this result to the prior assumption and presented a method to estimate the norm of α.

A more sophisticated "second-generation" SIS procedure for nonparametric Bayes problems was proposed by MacEachern et al. (1999). They noticed that when sampling ζ_t from (4.11), either a new ζ is produced from the Beta distribution or a previous ζ is drawn. As a consequence, the ζ_t forms natural clusters because of the Dirichlet process assumption on F. The plain SIS procedure tends to fix the location of these clusters at a relatively early stage, which is inefficient because an early cluster location may be invalidated by the subsequent new observations. The new SIS method introduces the *cluster* indicator variable I_t, which tells us whether ζ_t should be in one of the clusters formed by $\zeta_1, \ldots, \zeta_{t-1}$ or form a new cluster. Conditional on the indicators $I_1, \ldots, I_n$, one can integrate out all the ζ_t. The new sequential importance sampler is then constructed on the space of $(I_1, \ldots, I_n)$. When applied to the tack data under the same prior setting, the new method yields a substantial improvement over the plain SIS described earlier in this section (a cv^2 of 11 for the new method versus 43 for the old one).

4.5 Problems in Signal Processing

4.5.1 Target tracking in clutter and mixture Kalman filter

Tracking a target in clutter is of interest to engineers and computer scientists. The problem has received much attention recently since the proposal of the bootstrap filter (Avitzour 1995, Gordon et al. 1995). Here, we use the simple one-dimensional tracking problem in Avitzour (1995) to show how the SIS methods can be applied in this area.

Avitzour (1995) modeled the tracking problem as a state-space model with the state variable $x_t = (x_{t,1}, x_{t,2})$, where $x_{t,1}$ is the location of the target on a straight line and $x_{t,2}$ is the target velocity. The x_t evolve in the following way:

$$x_{t,1} = x_{t-1,1} + x_{t-1,2} + \frac{1}{2} w_t,$$
$$x_{t,2} = x_{t-1,2} + w_t,$$

where the noise term w_t are i.i.d. and follow distribution $N(0, q^2)$. If we could identify the object (without confusing it with others) at all times, then our observation would have been the object's location $x_{t,1}$ plus a small random noise. In other words, we observe

$$z_t = x_{t,1} + v_t,$$

where the v_t are i.i.d. Gaussian noises with distribution $N(0, r^2)$. If all the z_t are directly observable, then the tracking problem can be solved satisfactorily by using a Kalman filter (Bar-Shalom and Fortmann 1988).

When confusing objects are present in the detection window, however, our observation at time t becomes y_t, which is a vector of length m_t, where m_t is the total number of observed objects in the window and each component of y_t represents an object's position. Among these m_t measured locations, at most one corresponds to the true target we are interested in tracking (i.e., is equal to z_t). The occurrence of the confusing objects is assumed to follow a Poisson process with rate α. We further assume that there is only a probability $p_d < 1$ for the observation window to actually include the target's location (z_t) in y_t. Therefore, if the range of the detection window is Δ, the distribution of m_t is Bernoulli(p_d)+Poisson($\lambda\Delta$) and the false signals are uniformly distributed in the detection region. By introducing an indicator variable I_t,

$$I_t = \begin{cases} 0 & \text{if the target object is not in the detection range} \\ k & \text{if the } k\text{th object corresponds to the target,} \end{cases}$$

we can formulate this problem as a state-space model; that is, when $I_t = 0$,

$$p(y_t \mid x_t, I_t = 0) = \Delta^{-m_t} \frac{(\lambda\Delta)^{m_t}}{m_t!} e^{-\lambda\Delta} = \frac{\lambda^{m_t}}{m_t!} e^{-\lambda\Delta};$$

and when $I_t = k$,

$$p(y_t \mid x_t, I_t = k) = \frac{\lambda^{m_t - 1}}{(m_t - 1)!} e^{-\lambda\Delta} \frac{1}{\sqrt{2\pi}r} \exp\left\{-\frac{(y_{t,k} - x_t)^2}{2r^2}\right\}.$$

Since *a priori* $P(I_t = 0) = 1 - p_d$ and $P(I_t = k) = p_d/m_t$, we have

$$f_t(y_t, I_t \mid x_t) \propto \begin{cases} (1 - p_d)\lambda & \text{if } I_t = 0; \\ p_d(2\pi r^2)^{-1/2} \exp\left[-\frac{(y_{t,k} - x_t)^2}{2r^2}\right] & \text{otherwise.} \end{cases}$$

Since I_t is not observable, we need to sum out the I_t to obtain the *observation distribution* $f_t(y_t \mid x_t)$.

Using $q(x_t \mid x_{t-1})$ to denote the state evolution relationship, we can obtain a sequence of auxiliary distributions, $\pi_t(x_1, \ldots, x_t)$, as in (3.8):

$$\pi_t(\mathbf{x}_t) \propto f_t(y_t \mid x_t) q(x_t \mid x_{t-1}) \pi_{t-1}(\mathbf{x}_{t-1})$$

and

$$\pi_t(x_t) \propto \int f_t(y_t \mid x_t) q(x_t \mid x_{t-1}) \pi_{t-1}(x_{t-1}) dx_{t-1}.$$

The second equation holds because of the Markovian structure among the x_t. With this formulation, the current position of the target can be estimated as $E_{\pi_t}(x_t)$.

Conditional on the current value of x_{t-1}, we can easily simulate x_t from the state equation $q(\ )$. Given a sampled value of x_t, the computation of its weight, $f_t(y_t \mid x_t)$, is also easy. Thus, the *bootstrap filter* (Section 3.3) can be easily applied to this problem. However, if we pay a little more attention to the model's special structure, we can come up with a much more efficient algorithm.

Let $\Lambda_t = (I_1, \ldots, I_t)$ be called a *trajectory* up to time t. We note that an important feature of our tracking model is that if we know the values of the trajectory Λ_t of the target, the tracking system becomes linear and Gaussian and the computation of the Bayes estimator, $E(x_t \mid y_1, \ldots, y_t)$, can be achieved *exactly* by a standard Kalman filter. Therefore, conditional on Λ_t, we can integrate out $\mathbf{x}_t$ exactly. This feature enables us to design a SIS system only on the reduced space of Λ_t (Liu and Chen 1998, Chen and Liu 2000a). This approach is a vivid demonstration of the power of *marginalization* technique (Section 3.4.6). Since this new algorithm takes the form of mixing over a number of Kalman filters, we call the method *mixture Kalman filter* (MKF). Chen and Liu (2000a) give more details. Briefly, a general MKF algorithm can be stated as follows.

Suppose at time $t-1$, we have sampled m trajectories $\{\Lambda_{t-1}^{(1)}, \ldots, \Lambda_{t-1}^{(m)}\}$ with weights $w_{t-1}^{(1)}, \ldots, w_{t-1}^{(m)}$. For each trajectory $\Lambda_{t-1}^{(j)}$, we can compute via a Kalman filter the mean vector $\mu_{t-1}^{(j)}$ and the covariance matrix $\Sigma_{t-1}^{(j)}$ for the target. These are the sufficient statistics because the system is linear and Gaussian given Λ_{t-1}. We denote $KF_{t-1}^{(j)} = (\mu_{t-1}^{(j)}, \Sigma_{t-1}^{(j)})$. At time t, we run the following MKF updates: For $j = 1, \ldots, m$,

- generate $I_t^{(j)}$ from a trial distribution $g(I_t \mid \Lambda_{t-1}^{(j)}, KF_{t-1}^{(j)}, y_t)$;

- conditional on each $\{KF_{t-1}^{(j)}, y_t, I_t^{(j)}\}$, obtain $KF_{t+1}^{(j)}$ by a one-step Kalman filter (Chen and Liu 2000a);

- update the new weight as $w_t^{(j)} = w_{t-1}^{(j)} \times u_t^{(j)}$, where

$$u_t^{(j)} = \frac{p(\Lambda_{t-1}^{(j)}, I_t^{(j)} \mid \mathbf{y}_t)}{p(\Lambda_{t-1}^{(j)} \mid \mathbf{y}_{t-1}) g(I_t^{(j)} \mid \Lambda_{t-1}^{(j)}, KF_{t-1}^{(j)}, y_t)};$$

- if the coefficient of variation of the w_t exceeds a threshold value, we resample a new set of KF_t from $\{KF_t^{(1)}, \ldots, KF_t^{(m)}\}$ with probability proportional to the weights $w_t^{(j)}$.

Smith and Winter (1978) proposed a deterministic filtering method, called *split-track filter* (STF), which has a similar flavor to the MKF we just outlined. In STF, one always keeps m trajectories of the latent indicators. At a future time step, it evaluates the likelihoods of all possible propagations from the m trajectories kept at the previous step, then finds and keeps the m new trajectories with the highest likelihood values. In contrast, our MKF selects these trajectories randomly, according to the weights (which is the predictive likelihood value), and uses the associated weights to measure how good each trajectory is. The important step of resampling is naturally built into MKF, which can overcome some weaknesses of STF. More sophisticated sampling and estimation methods can also be incorporated.

4.5 Problems in Signal Processing

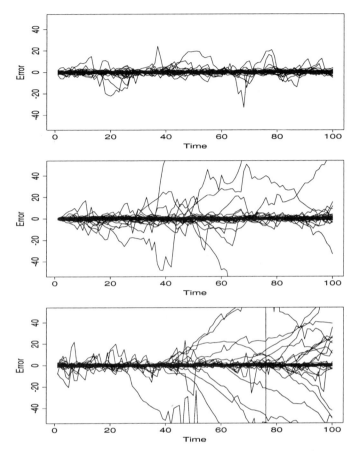

FIGURE 4.7. The tracking errors of 50 runs of the MKF (top), a sequential importance sampler (middle), and the split-track filter (bottom) for a simulated target moving system.

Figure 4.7 shows the plots of tracking errors (estimated location − true location) of 50 simulated runs, with $r^2 = 1.0$, $q^2 = 0.1$, $p_d = 0.9$, and $\lambda = 0.1$. These parameter combination is slightly different from that of Avitzour (1995), with smaller clutter density but larger state equation variance. With their configuration, the results are similar, but the differences between different procedures are smaller. Five hundred streams ($m=500$) were used, with resampling done at every step. The top part of Figure 4.7 resulted from using the MKF method, and the middle part shows the result from using a slightly improved bootstrap filter (Avitzour 1995). We also implemented the split-track filter for this problem, which, at each step, saves the 500 trajectories with the highest likelihood values (bottom part).

4.5.2 Digital signal extraction in fading channels

Many mobile communication channels can be modeled as Rayleigh flat-fading channels, which have the following form:

$$\text{State equations:} \begin{cases} \mathbf{x}_t = F\mathbf{x}_{t-1} + W w_t \\ \alpha_t = G\mathbf{x}_t \\ s_t \sim p(\cdot \mid s_{t-1}), \end{cases}$$

$$\text{Observation equation:} \quad y_t = \alpha_t s_t + V v_t,$$

where s_t are the input digital signals (symbols), y_t are the received complex signals, and α_t are the unobserved (changing) fading coefficients. Both w_t and v_t are complex Gaussian with identity covariance matrices. It is important to note that this model has some similarity to the tracking problem; that is, given the input signals s_t, the system is linear in $\mathbf{x}_t$ and y_t. Therefore, we can design a MKF which focuses solely on the s_t (with $\mathbf{x}_t$ integrated out). The algorithmic detail is similar to the MKF updating steps described in the previous subsection and will be omitted here. Readers interested in more details and related applications are referred to Chen and Liu (2000a) and Chen, Wang and Liu (2000).

Consider a special example in which input signals are binary (i.e., $s_t = -1$ or $+1$). The fading coefficient takes complex values, with independent real and imaginary parts following the same state equation. Simulation were done with the following configurations:

$$F = \begin{pmatrix} 0 & 1 & 0 & 0 \\ 0 & 0 & 1 & 0 \\ 0 & 0 & 0 & 1 \\ 0 & 0.94 & -2.88 & 2.94 \end{pmatrix} ; \quad G' = 10^{-4} \begin{pmatrix} 0.04 \\ 0.11 \\ 0.11 \\ 0.04 \end{pmatrix} ; \quad W = \begin{pmatrix} 0 \\ 0 \\ 0 \\ 1 \end{pmatrix} ;$$

and $V = r$. That is, both of the real and the imaginary parts of α_t follow an ARMA(3,3) process

$$\alpha_t - 0.94\alpha_{t-1} + 2.88\alpha_{t-2} - 2.94\alpha_{t-3}$$
$$= 0.04e_t + 0.11e_{t-1} + 0.11e_{t-2} + 0.04e_{t-3}$$

where $e_t \sim N(0, 0.01^2)$. In the communication literature, this is called a (low-pass) Butterworth filter of order 3 with cutoff frequency 0.01. It is normalized to have a stationary variance 1.

We are interested in estimating the differential code $d_t = s_t s_{t-1}$. Figure 4.8 shows the bit error rate of different signal-to-noise ratios (SNR), using a form of the MKF, the differential detection $\hat{d}_t = \text{sgn}[\text{real}(y_t y_{t-1}^*)]$

and a lower bound. The lower bound is obtained using the true fading coefficients α_t and $\hat{d}_t = \text{sgn}[\text{real}(\alpha_t^* y_t y_{t-1}^* \alpha_{t-1})]$. The Monte Carlo sample size m was 100 for the MKF. We also include the result of a delayed estimation, in which s_t is estimated using the samples $s_t^{(j)}$ generated by MKF and the weight $w_{t+1}^{(j)}$ at time $t+1$ (Liu and Chen 1998). This delayed estimation is able to utilize the substantial information contained in the future information y_{t+1}, hence more accurate, due to the strong memory in the fading channel.

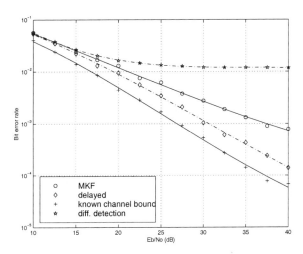

FIGURE 4.8. The bit error rate of extracting differential binary signals from a fading channel using MKF and differential detection. A lower bound that assumes the exact knowledge of the fading coefficients is also shown.

We can see that the simple differential detection works very well in low-SNR cases and no significant improvement can be expected. However, it has an apparent bit error rate floor for high-SNR cases. The MKF managed to break that floor, by using the structure of the fading coefficients.

4.6 Problems

1. Prove that the method described in Section 4.2 is valid. In particular, show that the weight updating strategy is correct. That is, the average of the weights indeed provide us an unbiased estimate on perm(A).

2. Describe how to use the SIS method of Section 4.2 to conduct the hypothesis test of independence for the astronomy data in Section 1.7.

3. Test to see whether you can find a better sampling distribution than (4.6) for approximating permanents. One possible choice is $P(\sigma(1) =$

$s) \propto (r_s - 1)^{-\alpha}$ for some $\alpha > 0$. But it might be possible to use a more complex function of all the r_j.

4. Study the "scanning future" method (Meirovitch 1982, Meirovitch 1985). Investigate its potential for the nonlinear filtering problem and the molecular structural optimization problem.

5. Investigate further the difference between the method of "split-track filter" and the mixture Kalman filter.

5
Metropolis Algorithm and Beyond

We have discussed in the previous chapters the important role of Monte Carlo methods in evaluating integrals and simulating stochastic systems. The most critical step in developing an efficient Monte Carlo algorithm is the simulation (sampling) from an appropriate probability distribution $\pi(\mathbf{x})$. When directly generating independent samples from $\pi(\mathbf{x})$ is not possible, we have to either opt for an *importance sampling* strategy, in which random samples are generated from a trial distribution *different* from (but close to) the target one and then weighted according to the importance ratio; or produce statistically *dependent* samples based on the idea of *Markov chain Monte Carlo* sampling. The importance sampling approach and its extensions have been discussed in Chapters 2–4. In this chapter, we introduce the cornerstone of all Markov chain-based Monte Carlo methods: the algorithm proposed in a very short paper (four pages) by Nicholas Metropolis, Arianna Rosenbluth, Marshall Rosenbluth, Augusta Teller, and Edward Teller in 1953.

Let $\pi(\mathbf{x}) = Z^{-1} \exp\{-h(\mathbf{x})\}$ be the target distribution under investigation (presumably all probability distribution functions can be written in this form), where the normalizing constant, or the *partition function*, Z, is often unknown to us. In principle, $Z = \int \exp\{-h(\mathbf{x})\}d\mathbf{x}$ is "knowable," but evaluating Z is no easier (and often harder) than the original problem of simulating from π. Motivated by computational problems in statistical physics, Metropolis et al. (1953) introduced the fundamental idea of evolving a Markov process to achieve the sampling of π. This idea, later known as the *Metropolis algorithm*, is of great simplicity and power — its variations and extensions have now been widely adopted by researchers in many different scientific fields, including biology, chemistry, computer sciences, economics, engineering, material sciences, physics, statistics, and others.

The Metropolis algorithm can be used to generate random samples from virtually any target distribution $\pi(\mathbf{x})$ known up to a normalizing constant, regardless of its analytical complexity and its dimensionality. Although this claim is true in theory, a potential problem with these Markov-chain-based Monte Carlo methods is that the resulting samples are often highly correlated. Therefore, the estimates resulting from these samples tend to have greater (often much greater) variances than those resulting from independent samples. Various attempts have been made in many different fields (e.g., physics, chemistry, structural biology, and statistics) to overcome these limitations. Interesting research topics include the design of Markov-chain-based Monte Carlo algorithms that can generate less correlated samples, the finding of more efficient ways to use generated Monte Carlo samples, the assessment of statistical errors of the estimates, etc. Detailed discussions regarding these topics will be given in the later chapters.

5.1 The Metropolis Algorithm

The basic idea of the Metropolis algorithm is to simulate a Markov chain in the state space of $\mathbf{x}$ so that the limiting/stationary/equilibrium[1] distribution of this chain is the target distribution π. Note that in traditional Markov chain analysis, one is often given a *transition rule*[2] and is interested in knowing what the stationary distribution is (see Section 12.1 for an introduction to Markov chains), whereas in Markov chain Monte Carlo simulations, one *knows* the equilibrium distribution and is interested in prescribing an efficient transition rule so as to reach this equilibrium.

Starting with any configuration $\mathbf{x}^{(0)}$, the Metropolis algorithm proceeds by iterating the following two steps.

M1: Propose a random "unbiased perturbation" of the current state $\mathbf{x}^{(t)}$ so as to generate a new configuration $\mathbf{x}'$. Mathematically, $\mathbf{x}'$ can be seen as being generated from a *symmetric* probability transition function[3] (often called the *proposal function* or *trial proposal*) $T(\mathbf{x}^{(t)}, \mathbf{x}')$ [i.e., $T(\mathbf{x}, \mathbf{x}') = T(\mathbf{x}', \mathbf{x})$]; calculate the change $\Delta h = h(\mathbf{x}') - h(\mathbf{x}^{(t)})$.

---

[1] There are subtle differences among these three concepts: limiting, stationary, or equilibrium distributions. But for most practical examples, they are the same thing. See the Appendix and Karlin and Taylor (1998) for more details.

[2] A *transition rule* is a probabilistic law, or more precisely, a conditional distribution, that dictates the chances of moving from one point in the state space to another.

[3] A function $T(\mathbf{x}, \mathbf{y})$ is called a probability transition function if it is non-negative and satisfies $\sum_{\text{all } \mathbf{y}} T(\mathbf{x}, \mathbf{y}) = 1$, for all $\mathbf{x}$.

M2: Generate a random number $U \sim \text{Uniform}[0,1]$. Let $\mathbf{x}^{(t+1)} = \mathbf{x}'$ if
$$U \leq \pi(\mathbf{x}')/\pi(\mathbf{x}^{(t)}) \equiv \exp(-\Delta h),$$
and let $\mathbf{x}^{(t+1)} = \mathbf{x}^{(t)}$ otherwise.

A more casual but perhaps better known description of the Metropolis algorithm is as follows: At each iteration, (a) a small but random perturbation of the current configuration is made, (b) the "gain" in an objective function [i.e., $-h(\mathbf{x})$] resulting from this perturbation is computed, (c) a random number U is generated independently, and (d) the new configuration is accepted if $\log(U)$ is smaller than or equal to the "gain" and is rejected otherwise. Heuristically, the Metropolis algorithm is constructed based on a "trial-and-error" strategy.

Metropolis et al. (1953) restricted their choices of the "perturbation rule" to the symmetric ones. According to this perturbation rule, the chance of obtaining $\mathbf{x}'$ from perturbing $\mathbf{x}$ is always equal to that of obtaining $\mathbf{x}$ from perturbing $\mathbf{x}'$. Intuitively, this means that there is no "trend bias" at the proposal stage. Mathematically, this symmetry requirement can be expressed as
$$T(\mathbf{x}, \mathbf{x}') = T(\mathbf{x}', \mathbf{x}).$$

The Metropolis scheme has been extensively used in statistical physics over the past five decades and is the cornerstone of all Markov chain Monte Carlo (MCMC) techniques recently adopted and further developed in the statistics community.

As an illustration, we consider the simulation of a simple hard-shell ball model for gas. In this model, the positions of K nonoverlapping hard-shell balls, with equal diameters, are required to be uniformly distributed in the box $[0, A] \times [0, B]$. Let $(X, Y) = \{(x_i, y_i), i = 1, \ldots, K\}$ denote the positions of these balls. The target distribution of interest, $\pi(X, Y)$, is then *uniform* for all allowable configurations (i.e., nonoverlapping and within the box). The Metropolis algorithm for this simulation can be implemented as follows: (a) Pick a ball at random, say, the ball at position (x_i, y_i); (b) propose to move this ball to a new position $(x_i', y_i') = (x_i + \delta_1, y_i + \delta_2)$, where $\delta_j \sim N(0, \sigma_0^2)$; and (c) accept the proposed position (x_i', y_i') if it does not violate the constraints, otherwise stay put. With $K = 6$, $d = 0.8$, $A = B = 3.5$, and starting positions of the balls at regular grids, we adjusted σ_0^2 to 0.5, which gave us an acceptance rate of about 30%. Figure 5.1 shows two snapshots of this simulation: The first one was taken after 1000 iterations, and the second one taken after 2000 iterations.

Another example is the simulation of the Ising model. As described in Section 1.3, the Ising model takes a probabilistic form
$$\pi(\mathbf{x}) \propto \exp\{-U(\mathbf{x})/\beta T\},$$
where $\mathbf{x} = (x_s, s \in \mathcal{L})$ and $x_s = \pm 1$, $\mathcal{L}$ is a lattice space; and $U(\mathbf{x}) = -J \sum_{s \sim s'} \delta_{x_s = x_{s'}}$. To simulate from this model, one needs to prescribe a

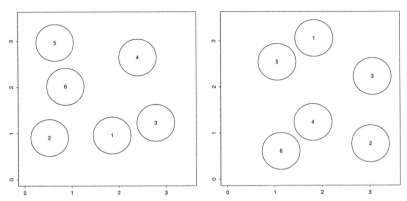

FIGURE 5.1. The simulation of a hard-shell ball model by the Metropolis algorithm. Left: after 1000 iterations; right: after 2000 iterations.

way to "perturb" the current configuration. A convenient proposal transition function is as follows: Pick a site, say, σ, at random, and negate its current value x_σ to $-x_\sigma$. Thus, the proposed new configuration $\mathbf{x}'$ differs from the initial one $\mathbf{x}$ only by a single site.

To be more concrete, consider the simulation of a one-dimensional Ising model in which $\mathbf{x} = (x_1, \ldots, x_d)$ and $U(\mathbf{x}) = -J \sum_{s=1}^{d-1} x_s x_{s+1}$. By letting $J = \mu \beta T$, we write the target distribution as

$$\pi(\mathbf{x}) = \frac{1}{Z} \exp\left\{ \mu \sum_{s=1}^{d-1} x_s x_{s+1} \right\}. \tag{5.1}$$

Suppose the current configuration is $\mathbf{x}^{(t)} = (x_1^{(t)}, \ldots, x_d^{(t)})$; then the next state $\mathbf{x}^{(t+1)}$ is produced by the Metropolis rule as follows:

- Choose a site, say, the jth site at random and set its current spin $x_j^{(t)}$ to the opposite. Thus, the newly proposed configuration is $\mathbf{x}' = (x_1^{(t)}, \ldots, -x_j^{(t)}, \ldots, x_d^{(t)})$.

- Compute the Metropolis ratio. In this case, it is easy to check that the random flipping process is completely symmetric; hence, $T(\mathbf{x}^{(t)}, \mathbf{x}') = T(\mathbf{x}', \mathbf{x}^{(t)})$, and, when $j \neq 1$ or d,

$$r = \pi(\mathbf{x}')/\pi(\mathbf{x}^{(t)}) = \exp\left\{ -2\mu x_j^{(t)} \left(x_{j-1}^{(t)} + x_{j+1}^{(t)} \right) \right\}.$$

- Simulate an independent uniform random variable U. Let $\mathbf{x}^{(t+1)} = \mathbf{x}'$ if $U \leq r$ and let $\mathbf{x}^{(t+1)} = \mathbf{x}^{(t)}$ otherwise.

Since this distribution has many components, it is difficult to have an overall sense of how the chain moves in the space. In Figure 5.2(a), we plot the traces of the total magnetization, defined as $M^{(t)} = \sum_{i=1}^{d} x_i^{(t)}$, for the first 2000 steps of a simulation. In this example, we took $\mu = 1$, $d = 50$, and the "all-up" starting configuration [i.e., $\mathbf{x}^{(0)} = (1, \ldots, 1)$].

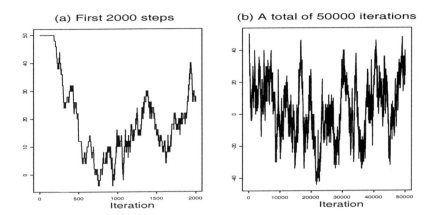

FIGURE 5.2. Simulation of the 1-D Ising model. The trace plots of (a) the first 2000 steps and (b) the total of 50,000 steps.

For this example, we can, in fact, compute the normalizing constant Z analytically and conduct an *exact* simulation (Section 2.4). Note that

$$\sum_{x_1} g(\mathbf{x}) = (e^{\mu x_2} + e^{-\mu x_2}) \exp\left\{\mu \sum_{i=2}^{d-1} x_i x_{i+1}\right\}$$

$$= (e^{-\mu} + e^{\mu}) \exp\left\{\mu \sum_{i=2}^{d-1} x_i x_{i+1}\right\}.$$

We can recursively sum out x_2, x_3, etc., and obtain that

$$\sum_{x_1,\ldots,x_d} g(\mathbf{x}) = \sum_{x_d} \left[\cdots \sum_{x_2} \left\{\sum_{x_1} g(\mathbf{x})\right\} \cdots\right] = 2\left(e^{-\mu} + e^{\mu}\right)^{d-1},$$

for $d \geq 2$. Thus, the marginal distribution of x_d is $x_d = 1$ or -1 with equal probability (which is actually obvious without doing any computation). Conditional on x_d, the distribution of x_{d-1} is

$$\Pr(x_{d-1} = x_d) = e^{\mu}/(e^{\mu} + e^{-\mu}),$$

and $\Pr(x_{d-1} = -x_d) = e^{-\mu}/(e^{\mu} + e^{-\mu})$. Thus, we can recursively simulate $\mathbf{x}$ backward (Section 2.4).

We implemented both the exact simulation method and the Metropolis algorithm for the 1-D Ising model (5.1) with $\mu = 1$ and $\mu = 2$, respectively.

110   5. Metropolis Algorithm and Beyond

Figures 5.3(a) and 5.3(c) show the histograms of the total magnetization variable M from 20,000 exact samples, for model (5.1) with $\mu = 1$ and $\mu = 2$, respectively. Figures 5.3 (b) and 5.3(d) show the corresponding histograms from 20,000 Monte Carlo samples generated from 1,000,000 Metropolis sampling steps. The chosen samples were 1 in every 50 lags [i.e., $\mathbf{x}^{(50)}, \mathbf{x}^{(100)}, \ldots, \mathbf{x}^{(50k)}, \ldots$].

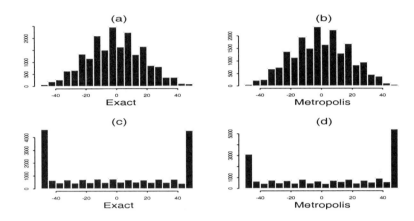

FIGURE 5.3. Histograms of the total magnetization M. (a) and (c): using 20,000 exact samples from the 1-D Ising models with $\mu = 1$ and $\mu = 2$; (b) and (d): using 20,000 samples chosen from 1 million Metropolis steps for each model.

From these simple graphs, we can make several interesting observations: (i) The computational effort for each Metropolis step is roughly 1/50th of that of exact simulation, so 1 million Metropolis steps took roughly the same CPU time as the production of 20,000 exact samples; (ii) when $\mu = 1$, the samples produced by the Metropolis algorithm were almost as good as those produced by independent sampling (at least to our eyes); (iii) when $\mu = 2$, the independent sampling showed an obvious advantage. These observations, in fact, reflect some important features of and deeper issues about the Metropolis algorithm: Each step of the Metropolis algorithm is usually very simple, but the Monte Carlo samples produced by a Metropolis sampler may become very "sticky"(e.g., getting stuck in local modes) in distributions with "low temperature" (i.e., high-energy barrier). In fact, the "stickiness" is already observable when $\mu = 1$ from Figure 5.2 (a): If one starts from an "all-up" configuration, it took about 1000 steps to get to the "ball park" of the interesting region and it took about 2000 steps to complete an "up-down" cycle. Quantitative measurement of this "stickiness" is often expressed as *autocorrelations*. A lag-k autocorrelation for a time series $M^{(1)}, M^{(2)}, \ldots$, is defined as

$$\rho_k = \text{corr}(M^{(1)}, M^{(k+1)}),$$

under their stationary distribution. Higher autocorrelations imply that the produced samples are stickier. Figure 5.4 shows the autocorrelation plots (i.e., a plot for ρ_k versus k) for those Monte Carlo samples in Figures 5.3(b) and 5.3(d), respectively.

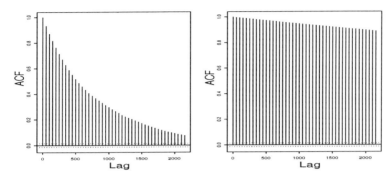

FIGURE 5.4. Autocorrelation plots of the time series $M^{(n)}$ produced by the Metropolis sampler for the 1-D Ising model with (a) $\mu = 1$ and (b) $\mu = 2$.

One observes that the 2000-lag autocorrelation of the $M^{(n)}$ series produced by the Metropolis sampler for $\mu = 2$ is still as high as about 0.9, implying that roughly one independent sample is as good as 20,000 Metropolis steps. More discussions on efficiency analysis of Markov chain Monte Carlo methods are discussed in Section 5.8 and Chapter 12.

5.2 Mathematical Formulation and Hastings's Generalization

The Metropolis algorithm prescribes a *transition rule* for a Markov chain. It uses a symmetric proposal function $T(\mathbf{x}, \mathbf{y})$ to suggest a possible move and then employs an acceptance-rejection rule to "thin it down." Hastings (1970) later extended the algorithm to the case when T is not necessarily symmetric. In Hastings' generalization, the only serious restriction on the proposal function is that $T(\mathbf{x}, \mathbf{y}) > 0$ if and only if $T(\mathbf{y}, \mathbf{x}) > 0$. With this transition function, one can implement the following iteration:

Metropolis-Hastings Algorithm. Given current state $\mathbf{x}^{(t)}$:

- Draw $\mathbf{y}$ from the proposal distribution $T(\mathbf{x}^{(t)}, \mathbf{y})$.
- Draw $U \sim \text{Uniform}[0,1]$ and update

$$\mathbf{x}^{(t+1)} = \begin{cases} \mathbf{y}, & \text{if } U \leq r(\mathbf{x}^{(t)}, \mathbf{y}) \\ \mathbf{x}^{(t)} & \text{otherwise.} \end{cases} \quad (5.2)$$

where Metropolis et al. (1953) and Hastings (1970) suggested using

$$r(\mathbf{x},\mathbf{y}) = \min\left\{1, \frac{\pi(\mathbf{y})T(\mathbf{y},\mathbf{x})}{\pi(\mathbf{x})T(\mathbf{x},\mathbf{y})}\right\}.$$

Clearly, this algorithm is identical to the original Metropolis algorithm when $T(\mathbf{x},\mathbf{y}) = T(\mathbf{y},\mathbf{x})$.

Barker (1965) suggested another acceptance function:

$$r_B(\mathbf{x},\mathbf{y}) = \frac{\pi(\mathbf{y})T(\mathbf{y},\mathbf{x})}{\pi(\mathbf{y})T(\mathbf{y},\mathbf{x}) + \pi(\mathbf{x})T(\mathbf{x},\mathbf{y})}.$$

A more general formula for $r(\mathbf{x},\mathbf{y})$ is given by Charles Stein (personal communication):

$$r(\mathbf{x},\mathbf{y}) = \frac{\delta(\mathbf{x},\mathbf{y})}{\pi(\mathbf{x})T(\mathbf{x},\mathbf{y})}, \tag{5.3}$$

where $\delta(\mathbf{x},\mathbf{y})$ is any *symmetric function* in $\mathbf{x}$ and $\mathbf{y}$ that makes $r(\mathbf{x},\mathbf{y}) \leq 1$ for all $\mathbf{x},\mathbf{y}$. The intuition behind the ratio $T(\mathbf{y},\mathbf{x})/T(\mathbf{x},\mathbf{y})$ is that it compensates the "flow bias" of the proposal function.

If a rejection function of the form (5.3) is used, then for any $\mathbf{y} \neq \mathbf{x}$, the actual transition probability from $\mathbf{x}$ to $\mathbf{y}$ is

$$A(\mathbf{x},\mathbf{y}) = T(\mathbf{x},\mathbf{y})r(\mathbf{x},\mathbf{y}) = T(\mathbf{x},\mathbf{y})\frac{\delta(\mathbf{x},\mathbf{y})}{\pi(\mathbf{x})T(\mathbf{x},\mathbf{y})} = \pi(\mathbf{x})^{-1}\delta(\mathbf{x},\mathbf{y}). \tag{5.4}$$

Because $\delta(\mathbf{x},\mathbf{y}) = \delta(\mathbf{y},\mathbf{x})$, we have that $\pi(\mathbf{x})A(\mathbf{x},\mathbf{y}) = \pi(\mathbf{y})A(\mathbf{y},\mathbf{x})$. This implies that the Markov chain induced by the Metropolis-Hastings rule is *reversible* and has π as its invariant distribution (the next section).

For discrete state spaces, Peskun (1973) showed that the optimal choice of $r(\mathbf{x},\mathbf{y})$ in terms of statistical efficiency is the one in the original Metropolis algorithm (see Section 13.3.1 for more details). But the issue is less clear in terms of convergence rate of the induced Markov chain (Frigessi, Distefano, Hwang and Sheu 1993, Liu 1996c) As we will show in the next section, a main criterion used in the design of a Markov transition rule such as the Metropolis-Hastings algorithm is to ensure that the target distribution $\pi(\mathbf{x})$ is the invariant distribution of this chain.

5.3 Why Does the Metropolis Algorithm Work?

We first verify that the Metropolis-Hastings algorithm prescribes a transition rule with respect to which the target distribution $\pi(\mathbf{x})$ is invariant. Let $A(\mathbf{x},\mathbf{y})$ be the *actual* transition function of the algorithm. It differs from the proposal function $T(\mathbf{x},\mathbf{y})$ because an acceptance-rejection step is involved. We are required to show that

$$\int \pi(\mathbf{x})A(\mathbf{x},\mathbf{y})d\mathbf{x} = \pi(\mathbf{y}). \tag{5.5}$$

5.3 Why Does the Metropolis Algorithm Work?

Fortunately, there is an easier-to-check, but more restrictive, condition than (5.5), the *detailed balance*, which can be stated as

$$\pi(\mathbf{x})A(\mathbf{x},\mathbf{y}) = \pi(\mathbf{y})A(\mathbf{y},\mathbf{x}). \tag{5.6}$$

Clearly, if the detailed balance (5.6) holds, we have

$$\int \pi(\mathbf{x})A(\mathbf{x},\mathbf{y})d\mathbf{x} = \int \pi(\mathbf{y})A(\mathbf{y},\mathbf{x})d\mathbf{x} = \pi(\mathbf{y})\int A(\mathbf{y},\mathbf{x})d\mathbf{x} = \pi(\mathbf{y}).$$

Thus, the detailed balance *ensures* invariance. The converse is not true. In Markov chain literature, chains that satisfy the detailed balance condition are called *reversible*.

We can write down $A(\mathbf{x},\mathbf{y})$ explicitly for the Metropolis algorithm: For any $\mathbf{x} \neq \mathbf{y}$, the probability that we actually make the move from $\mathbf{x}$ to $\mathbf{y}$ is equal to the proposal probability, $T(\mathbf{x},\mathbf{y})$, multiplied by the acceptance probability; that is,

$$A(\mathbf{x},\mathbf{y}) = T(\mathbf{x},\mathbf{y})\min\left\{1, \frac{\pi(\mathbf{y})T(\mathbf{y},\mathbf{x})}{\pi(\mathbf{x})T(\mathbf{x},\mathbf{y})}\right\}, \tag{5.7}$$

for $\mathbf{x} \neq \mathbf{y}$. Hence,

$$\pi(\mathbf{x})A(\mathbf{x},\mathbf{y}) = \pi(\mathbf{x})T(\mathbf{x},\mathbf{y})\min\left\{1, \frac{\pi(\mathbf{y})T(\mathbf{y},\mathbf{x})}{\pi(\mathbf{x})T(\mathbf{x},\mathbf{y})}\right\}$$
$$= \min\{\pi(\mathbf{x})T(\mathbf{x},\mathbf{y}), \pi(\mathbf{y})T(\mathbf{y},\mathbf{x})\},$$

which is a *symmetric function* in $\mathbf{x}$ and $\mathbf{y}$. Thus, the detailed balance condition is satisfied.

More generally, as long as the transition function $A(\mathbf{x},\mathbf{y})$ is of the form

$$A(\mathbf{x},\mathbf{y}) = \pi(\mathbf{y})\delta(\mathbf{x},\mathbf{y}),$$

where $\delta(\mathbf{x},\mathbf{y})$ is a symmetric function in $\mathbf{x},\mathbf{y}$, one can easily verify that the detailed balance condition has to be satisfied. The difficulty in the construction of A, however, is that the symmetric function $\delta(\mathbf{x},\mathbf{y})$ has to be chosen properly so that the integral $\int A(\mathbf{x},\mathbf{y})d\mathbf{y} = 1$. One can easily check that the Metropolis transition can be written as (for $\mathbf{x} \neq \mathbf{y}$)

$$A(\mathbf{x},\mathbf{y}) = \pi(\mathbf{y})\min\left\{\frac{T(\mathbf{x},\mathbf{y})}{\pi(\mathbf{y})}, \frac{T(\mathbf{y},\mathbf{x})}{\pi(\mathbf{x})}\right\}.$$

The acceptance rule proposed by Barker (1965) corresponds to a transition (for $\mathbf{x} \neq \mathbf{y}$)

$$A(\mathbf{x},\mathbf{y}) = \pi(\mathbf{y})\frac{T(\mathbf{x},\mathbf{y})T(\mathbf{y},\mathbf{x})}{\pi(\mathbf{y})T(\mathbf{y},\mathbf{x}) + \pi(\mathbf{x})T(\mathbf{x},\mathbf{y})}.$$

It was not clear, then, which acceptance rule is better. Peskun (1973) later showed that the Metropolis rule generally works better in terms of statistical efficiency.

114   5. Metropolis Algorithm and Beyond

By the standard Markov chain theory, if the chain is irreducible[4], aperiodic[5] [this is almost surely true for the Metropolis algorithm (Tierney 1994)], and possesses an invariant distribution, then the chain will become stationary at its invariant distribution, π. Therefore, if we run this chain long enough (say, after a burn-in period of n_0 steps), the samples $\mathbf{x}_{n_0+1}, \mathbf{x}_{n_0+2}, \ldots$ produced by the chain can be regarded as approximately following the target distribution π. One then realizes the task of drawing random (but correlated) samples from a given distribution.

5.4 Some Special Algorithms

To illustrate how the Metropolis-Hastings rule is practiced, we describe a few special algorithms that have appeared frequently in the literature.

5.4.1 Random-walk Metropolis

Suppose the target distribution $\pi(\mathbf{x})$ is defined on the d-dimensional Euclidean space $\mathbb{R}^d$. A natural "perturbation" of the current configuration $\mathbf{x}^{(t)}$ is the addition of a random "error;" that is, the next candidate position is proposed as $\mathbf{x}' = \mathbf{x}^{(t)} + \epsilon_t$, where $\epsilon_t \sim g_\sigma(\cdot)$ is independent and identically distributed for different t. Here, σ represents the "range" of the proposal exploration and is controlled by the user. In problems where we do not have much information on the shape of the target distribution, we often end up letting $g_\sigma(\cdot)$ be a spherically symmetric distribution. Typical choices include the spherical Gaussian distribution $N(0, \sigma^2 I)$ or the uniform distribution in a ball of radius σ. Clearly, if one does not exercise the Metropolis rejection rule to this proposal, the resulting walk will drift away to infinity and never come back (when $d \geq 3$). It is thus worthwhile to point out that in a Metropolis algorithm, the proposal chain is not required to have any good "global properties" other than being irreducible (see footnote 4 on page 114). But we do require some local properties for $T(\mathbf{x}, \mathbf{y})$ [e.g., $T(\mathbf{x}, \mathbf{y}) > 0$ whenever $T(\mathbf{y}, \mathbf{x}) > 0$].

Given the current state $\mathbf{x}^{(t)}$, the random-walk Metropolis algorithm iterates the following steps:

- Draw $\epsilon \sim g_\sigma$ and set $\mathbf{x}' = \mathbf{x}^{(t)} + \epsilon$, where g_σ is a spherically symmetric distribution and σ can be controlled by the user.

---

[4] A Markov chain is said to be *irreducible* if the chain has nonzero probability (density) to move from one position in the state space to any other position in a finite number of steps.

[5] A Markov chain is said *aperiodic* if the maximum common divider of the number of steps it takes for the chain to come back to the starting point (any) is equal to one.

- Simulate $u \sim \text{Uniform}[0,1]$ and update

$$x^{(t+1)} = \begin{cases} y & \text{if } u \leq \dfrac{\pi(x')}{\pi(x^{(t)})} \\ x^{(t)} & \text{otherwise.} \end{cases}$$

In the example of simulating six hard-shell balls in a box, we used the random-walk method for each individual ball. In an interesting study, Gelman, Roberts and Gilks (1995) suggested that a rule of thumb in choosing σ in a random-walk Metropolis is to maintain a 25% to 35% acceptance rate. This rule is supported by a theoretical analysis for a Gaussian target density (more details in Section 5.8).

5.4.2 Metropolized independence sampler

A very special choice of the proposal transition function $T(x, y)$ is an *independent* trial density $g(y)$; that is, the proposed move y is generated from $g(\cdot)$ independent of the previous state $x^{(t)}$. This method, as first suggested in Hastings (1970), appears to be an alternative to the rejection sampling and importance sampling. Its convergence properties was studied in Liu (1996a), where all the eigenvalues and eigenfunctions of the actual transition function are derived (see Section 13.4).

The MIS Scheme: given the current state $x^{(t)}$,

- Draw $y \sim g(y)$.

- Simulate $u \sim \text{Uniform}[0,1]$ and let

$$x^{(t+1)} = \begin{cases} y & \text{if } u \leq \min\left\{1, \dfrac{w(y)}{w(x^{(t)})}\right\} \\ x^{(t)} & \text{otherwise,} \end{cases}$$

where $w(x) = \pi(x)/g(x)$ is the usual *importance sampling weight*.

As with the rejection method, the efficiency of MIS depends on how close the trial density $g(y)$ is to the target $\pi(y)$. To ensure robust performance, it is advisable to let $g(\cdot)$ be a relatively long-tailed distribution. Gelman and Rubin (1992) and Tierney (1994) suggested that one can insert a couple of MIS steps into Gibbs iteration when correctly sampling from a conditional distribution is difficult. The idea is useful in many Bayesian computations in which each conditional density can be approximated reasonably well by a Gaussian distribution. To accommodate irregular tail behaviors, it is essential to use a long-tailed t-distribution as $g(x)$.

5.4.3 Configurational bias Monte Carlo

The configurational bias Monte Carlo (CBMC) algorithm can be viewed as an SIS-based Metropolized independence sampler. Suppose the argument of the target distribution, $\mathbf{x}$, can be decomposed as $\mathbf{x} = (x_1, \ldots, x_d)$. As in a sequential importance sampler (Section 2.6.3 and Chapters 3 and 4), we assume that there is a sequence of *auxiliary distributions*

$$\pi_1(x_1), \pi_2(x_1, x_2), \ldots, \pi_{d-1}(\mathbf{x}_{d-1}), \pi(\mathbf{x})$$

that can help us construct the trial sampling distribution.

Let the trial sampling distribution of $\mathbf{x}$ be

$$g(\mathbf{x}) = g_1(x_1) g_2(x_2 \mid x_1) \cdots g_d(x_d \mid \mathbf{x}_{d-1}).$$

To implement a CBMC algorithm (Siepmann and Frenkel 1992), we first draw $\mathbf{x}^{(0)}$ from $g(\mathbf{x})$ via a SIS strategy and compute its importance weight $w^{(0)}$ (up to a normalizing constant). Suppose that currently we have $\mathbf{x}^{(t)}$ with weight $w(\mathbf{x}^{(t)})$; then, at the next iteration, we do the following:

- Independently generate a trial configuration $\mathbf{y}$ from $g(\ )$ (using the sequential approach); compute its importance weight

$$w(\mathbf{y}) = \pi(\mathbf{y})/g(\mathbf{y}),$$

which can often be derived recursively as

$$w(\mathbf{y}) = \frac{\pi_1(y_1)}{g_1(y_1)} \frac{\pi_2(y_1, y_2)}{g_2(y_2 \mid y_1)\pi_1(y_1)} \cdots \frac{\pi_d(y_1, \ldots, y_d)}{g_d(y_d \mid \mathbf{y}_{d-1})\pi_{d-1}(\mathbf{y}_{d-1})}.$$

(This recursion is key to the SIS approach.)

- Accept $\mathbf{y}$; that is, let $\mathbf{x}^{(t+1)} = \mathbf{y}$, with probability $\min\left\{1, \frac{w(\mathbf{y})}{w(\mathbf{x}^{(t)})}\right\}$; and let $\mathbf{x}^{(t+1)} = \mathbf{x}^{(t)}$ otherwise.

This procedure is exactly the same as the *Metropolized independence sampler* described in the previous subsection, except that the trial density is built up sequentially and the importance weight computed recursively.

A useful modification of the foregoing CBMC procedure is to incorporate a stage-wise rejection decision. Suppose all the previous $k-1$ steps in the sequential simulation of the trial configuration $\mathbf{x}'$ have been accepted. Then, at the kth stage of SIS, we accept $\mathbf{x}'_k = (x'_1, \ldots, x'_k)$ with probability

$$p_k = \min\left\{1, \frac{\pi_k(\mathbf{x}'_k)\pi_{k-1}(\mathbf{x}_{k-1})g_k(x_k \mid \mathbf{x}_{k-1})}{\pi_k(\mathbf{x}_k)\pi_{k-1}(\mathbf{x}'_{k-1})g_k(x'_k \mid \mathbf{x}'_{k-1})}\right\} = \min\left\{1, \frac{u_k(\mathbf{x}'_k)}{u_k(\mathbf{x}_k)}\right\},$$

where u_k is the same as that in (3.10). In other words, the acceptance probability is equal to the ratio of the incremental importance weights between the trial and the current configurations. When rejected, we go back to the first stage to rebuild the whole configuration. It should be noted that one does not need to perform the acceptance-rejection decision at every stage and she/he has a complete control on when to conduct conduct the acceptance-rejection step.

This multistage method has been shown effective in simulating superfluid Helium 4 and other quantum mechanical systems (Ceperley 1995). Compared to the CBMC, this multistage sampler can force an early stop so as to save computing power. However, the chance of the final acceptance of a complete configuration in the multistage approach should be smaller than that in the CBMC because

$$\min(1, a_1) \times \cdots \times \min(1, a_d) \leq \min(1, \, a_1 \times \cdots \times a_d).$$

To show that the multistage modification of CBMC is proper, we can write down its actual transition function and prove that it satisfies the detailed balance. Here, we provide only a proof for the case when $d = 2$ [i.e. $\mathbf{x} = (x_1, x_2)$]. The general proof is left to the reader. Suppose the auxiliary distributions when $d = 2$ are $\pi_1(x_1)$ and $\pi_2(\mathbf{x}) \equiv \pi(\mathbf{x})$. Suppose the current state is $\mathbf{x}$. Then, the probability of accepting a new configuration $\mathbf{x}' = (x_1', x_2') \neq \mathbf{x}$ is

$$\begin{aligned}
P(\mathbf{x} \to \mathbf{x}') &= g_1(x_1')g_2(x_2'|x_1') \min\left\{1, \frac{\pi_1(x_1')g_1(x_1)}{\pi_1(x_1)g_1(x_1')}\right\} \\
&\quad \times \min\left\{1, \frac{\pi(x_1', x_2')\pi_1(x_1)g_2(x_2 \mid x_1)}{\pi(x_1, x_2)\pi_1(x_1')g_1(x_2' \mid x_1')}\right\} \\
&= \pi(\mathbf{x}') \min\{g_1(x_1')\pi_1(x_1), \pi_1(x_1')g_1(x_1)\} \\
&\quad \times \min\left\{\frac{\pi_1(x_1')g_1(x_2'|x_1')}{\pi(\mathbf{x}')}, \frac{\pi_1(x_1)g_1(x_2|x_1)}{\pi(\mathbf{x})}\right\}.
\end{aligned}$$

Hence, this transition function is indeed of the form $\pi(\mathbf{x}')\delta(\mathbf{x}, \mathbf{x}')$, where δ is a symmetric function. The detailed balance condition is thus satisfied (see the argument in Section 5.3).

5.5 Multipoint Metropolis Methods

In principle, the Metropolis sampling method discussed in Section 5.2 can be applied to almost any target distribution. In practice, however, it is not infrequent to discover that finding a good proposal transition kernel is rather difficult. Although the important generalization of Hastings (1970) enables one to use asymmetric proposal functions, a simple random-walk-type proposal is still most frequently seen in practice simply because there is no obviously advantageous alternative available. It is then often the case that a small step-size in the proposal transition (for the algorithms similar to those described in Section 5.4.1) will result in exceedingly slow movement of the corresponding Markov chain, whereas a large step-size will result in very low acceptance rate. In both cases, the mixing rate of the algorithm would be very slow.

Here, we describe a generalization of the Metropolis-Hastings's transition rule. This new rule (Frenkel and Smit 1996, Liu et al. 2000, Qin and Liu 2001) enables a MCMC sampler to make large step-size jumps without lowering the acceptance rate.

5.5.1 Multiple independent proposals

Suppose $T(\mathbf{x}, \mathbf{y})$ is an arbitrary proposal transition function and $\delta(\mathbf{x}, \mathbf{y})$ is an arbitrary symmetric and non-negative function. A modest requirement is that $T(\mathbf{x}, \mathbf{y}) > 0$ if and only if $T(\mathbf{y}, \mathbf{x}) > 0$. Define

$$w(\mathbf{x}, \mathbf{y}) = \pi(\mathbf{x}) T(\mathbf{x}, \mathbf{y}) \lambda(\mathbf{x}, \mathbf{y}), \tag{5.8}$$

where $\lambda(\mathbf{x}, \mathbf{y})$ is a non-negative symmetric function in $\mathbf{x}$ and $\mathbf{y}$ that can be chosen by the user. The only requirement is that $\lambda(\mathbf{x}, \mathbf{y}) > 0$ whenever $T(\mathbf{x}, \mathbf{y}) > 0$. We present a few choices of $\lambda(\mathbf{x}, \mathbf{y})$ in the latter part of this section. Suppose the *current state* is $\mathbf{x}^{(t)} = \mathbf{x}$; then, a MTM transition is defined as follows:

Multiple-Try Metropolis (MTM)

- Draw k independent trial proposals, $\mathbf{y}_1, \ldots, \mathbf{y}_k$, from $T(\mathbf{x}, \cdot)$. Compute $w(\mathbf{y}_j, \mathbf{x})$ as in (5.8) for $j = 1, \ldots, k$.

- Select $\mathbf{y}$ among the trial set $\{\mathbf{y}_1, \ldots, \mathbf{y}_k\}$ with probability proportional to $w(\mathbf{y}_j, \mathbf{x})$, $j = 1, \ldots, k$. Then, produce a "reference set" by drawing $\mathbf{x}_1^*, \ldots, \mathbf{x}_{k-1}^*$ from the distribution $T(\mathbf{y}, \cdot)$. Let $\mathbf{x}_k^* = \mathbf{x}$.

- Accept $\mathbf{y}$ with probability

$$r_g = \min\left\{1, \frac{w(\mathbf{y}_1, \mathbf{x}) + \cdots + w(\mathbf{y}_k, \mathbf{x})}{w(\mathbf{x}_1^*, \mathbf{y}) + \cdots + w(\mathbf{x}_k^*, \mathbf{y})}\right\} \tag{5.9}$$

and reject it with probability $1 - r_g$. The quantity r_g is called the *generalized M-H ratio*.

When $T(\mathbf{x}, \mathbf{y})$ is symmetric, for example, one can choose $\lambda(\mathbf{x}, \mathbf{y}) = T^{-1}(\mathbf{x}, \mathbf{y})$. Then, $w(\mathbf{x}, \mathbf{y}) = \pi(\mathbf{x})$. In this case, the MTM algorithm is simplified as the following algorithm, known as *orientational bias* Monte Carlo (OBMC) in the field of molecular simulation.

OBMC Algorithm

5.5 Multipoint Metropolis Methods

- Draw k trials $\mathbf{y}_1, \ldots, \mathbf{y}_k$ from a *symmetric* proposal function $T(\mathbf{x}, \mathbf{y})$.

- Select $\mathbf{Y} = \mathbf{y}_l$ among the $\mathbf{y}$'s with probability proportional to $\pi(\mathbf{y}_j)$, $j = 1, \ldots, k$; then, draw the reference points $\mathbf{x}'_1, \ldots, \mathbf{x}'_{k-1}$ from the distribution $T(\mathbf{y}_l, \mathbf{x}')$. Let $\mathbf{x}'_k = \mathbf{x}$.

- Accept $\mathbf{y}_l$ with probability

$$\min\left\{1, \frac{\pi(\mathbf{y}_1) + \cdots + \pi(\mathbf{y}_k)}{\pi(\mathbf{x}'_1) + \cdots + \pi(\mathbf{x}'_k)}\right\}$$

and reject with the remaining probability.

The proof of the correctness of this method is straightforward (Liu et al. 2000). Roughly speaking, one can directly check the detailed balance by writing down what the algorithmic instructions mean mathematically. To illustrate the idea, we prove the case for $k = 2$.

Proof: Let $A(\mathbf{x}, \mathbf{y})$ be the actual transition probability for moving from $\mathbf{x}$ to $\mathbf{y}$ in a MTM sampler. Suppose $\mathbf{x} \neq \mathbf{y}$ and let I indicate which of $\mathbf{y}_j$ has been selected. Since $w(\mathbf{y}, \mathbf{x}) = \pi(\mathbf{y}) T(\mathbf{y}, \mathbf{x}) \lambda(\mathbf{y}, \mathbf{x})$ and the $\mathbf{y}_j$ are exchangeable, we have

$$\begin{aligned}
\pi(\mathbf{x}) A(\mathbf{x}, \mathbf{y}) &= 2\, \pi(\mathbf{x}) P[(Y_1 = \mathbf{y}) \cap (I = 1) \mid \mathbf{x}] \quad \text{(symmetry)} \\
&= 2\, \pi(\mathbf{x}) \int T(\mathbf{x}, \mathbf{y}) T(\mathbf{x}, \mathbf{y}_2) \frac{w(\mathbf{y}, \mathbf{x})}{w(\mathbf{y}, \mathbf{x}) + w(\mathbf{y}_2, \mathbf{x})} \\
&\quad \times \min\left\{1, \frac{w(\mathbf{y}, \mathbf{x}) + w(\mathbf{y}_2, \mathbf{x})}{w(\mathbf{x}, \mathbf{y}) + w(\mathbf{x}^*_2, \mathbf{y})}\right\} T(\mathbf{y}, \mathbf{x}^*_2) d\mathbf{y}_2 d\mathbf{x}^*_2 \\
&= 2\, \frac{w(\mathbf{x}, \mathbf{y}) w(\mathbf{y}, \mathbf{x})}{\lambda(\mathbf{y}, \mathbf{x})} \int T(\mathbf{x}, \mathbf{y}_2) T(\mathbf{y}, \mathbf{x}^*_2) \\
&\quad \times \min\left\{\frac{1}{w(\mathbf{y}, \mathbf{x}) + w(\mathbf{y}_2, \mathbf{x})}, \frac{1}{w(\mathbf{x}, \mathbf{y}) + w(\mathbf{x}^*_2, \mathbf{y})}\right\} d\mathbf{y}_2 d\mathbf{x}^*_2.
\end{aligned}$$

The final expression is symmetric in $\mathbf{x}$ and $\mathbf{y}$ because $\lambda(\mathbf{x}, \mathbf{y}) = \lambda(\mathbf{y}, \mathbf{x})$. Thus, we proved that $\pi(\mathbf{x}) A(\mathbf{x}, \mathbf{y}) = \pi(\mathbf{y}) A(\mathbf{y}, \mathbf{x})$, which is the detailed balance condition. ◇

Another interesting application is to combine the MTM approach with the Metropolized independence sampler (Section 5.4.2). Because the trial samples are generated independently, one does not need to generate another "reference set." More precisely, suppose the current state is $\mathbf{x}^{(t)} = \mathbf{x}$ in our MCMC iteration; then, the next state can be generated by the following *multiple-trial Metropolized independence sampler* (MTMIS):

MTMIS:

- Generate a trial set of i.i.d. samples by drawing $\mathbf{y}_j \sim p(\mathbf{y})$, $j = 1, \ldots, k$, independently, where $p(\ )$ is a trial distribution chosen by the user. Compute $w(\mathbf{y}_j) = \pi(\mathbf{y}_j)/p(\mathbf{y}_j)$ and $W = \sum_{j=1}^{k} w(\mathbf{y}_j)$.

- Draw $\mathbf{y}$ from the trial set $\{\mathbf{y}_1, \ldots, \mathbf{y}_k\}$ with probability proportional to $w(\mathbf{y}_j)$.

- Let $\mathbf{x}^{(t+1)} = \mathbf{y}$ with probability

$$\min\left\{1, \frac{W}{W - w(\mathbf{y}) + w(\mathbf{x})}\right\}$$

and let $\mathbf{x}^{(t+1)} = \mathbf{x}$ with the remaining probability.

The idea of using MTM to make large-step moves along certain favorable directions is a useful heuristic and can be applied broadly. We will show later (Chapter 11) how the MTM can be applied to improve the algorithm's performance in more complicated settings.

5.5.2 Correlated multipoint proposals

Based on the work of OBMC and MTM, Qin and Liu (2001) provide a more general scheme, termed as the *multipoint* method, which allows one to choose from multiple *correlated* proposals at each iteration. Its application in hybrid Monte Carlo has shown promising results. Suppose the current state is $\mathbf{x}^{(t)} = \mathbf{x}$. We generate k trial proposals as follows: Let $\mathbf{y}_1 \sim P_1(\cdot \mid \mathbf{x})$ and let

$$\mathbf{y}_j \sim P_j(\cdot \mid \mathbf{x}, \mathbf{y}_1, \ldots, \mathbf{y}_{j-1}), \quad j = 2, \ldots, k.$$

For brevity, we also let $\mathbf{y}_{[1:j]} = (\mathbf{y}_1, \ldots, \mathbf{y}_j)$, $\mathbf{y}_{[j:1]} = (\mathbf{y}_j, \ldots, \mathbf{y}_1)$ and let

$$P_j(\mathbf{y}_{[1:j]} \mid \mathbf{x}) = P_1(\mathbf{y}_1 \mid \mathbf{x}) \cdots P_j(\mathbf{y}_j \mid \mathbf{x}, \mathbf{y}_{[1:j-1]}).$$

A weight function is defined as

$$w_j(\mathbf{x}, \mathbf{y}_{[1:j]}) = \pi(\mathbf{x}) P_j(\mathbf{y}_{[1:j]} \mid \mathbf{x}) \lambda_j(\mathbf{x}, \mathbf{y}_{[1:j]}), \tag{5.10}$$

where $\lambda_j(\ )$ is a *sequentially symmetric* function; that is,

$$\lambda_j(a, b, \ldots, z) = \lambda_j(z, \ldots, b, a).$$

The general algorithm is as follows:

Multipoint Method:

- Sample $\mathbf{y}$ from the trial set $\{\mathbf{y}_1, \ldots, \mathbf{y}_k\}$ with probability proportional to $w(\mathbf{y}_{[l:1]}, \mathbf{x})$; suppose $\mathbf{y}_j$ is chosen.

- Create a reference set by letting $\mathbf{x}_l^* = \mathbf{y}_{j-l}$ for $l = 1, \ldots, j-1$, $\mathbf{x}_j^* = \mathbf{x}$, and drawing

$$\mathbf{x}_m^* \sim P_m(\cdot \mid \mathbf{y}, \mathbf{x}_{[1:m-1]}^*),$$

for $m = j+1, \ldots, k$.

- Let $\mathbf{x}^{(t+1)} = \mathbf{y}$ with probability

$$r_{mp} = \min\left\{1, \frac{\sum_{l=1}^k w(\mathbf{y}_{[l:1]}, \mathbf{x})}{\sum_{l=1}^k w(\mathbf{x}_{[l:1]}^*, \mathbf{y})}\right\},$$

and let $\mathbf{x}^{(t+1)} = \mathbf{x}$ with the remaining probability.

The simplest choice of $\lambda_j(\ )$ for (5.10) is the constant function. But we may, in some cases, want to give larger weights to larger j's since these points are "farther" away from the initial point $\mathbf{x}$. When P_j is constructed by composing a symmetric transition kernel j times, the resulting function is *sequentially symmetric*. Thus, we can choose λ_j as v_j/P_j, where v_j is a constant, so that $w_j(\mathbf{y}_{[j:1]}, \mathbf{x}) = v_j \pi(\mathbf{y}_j)$. The resulting algorithm is very similar to OBMC.

When the state space is $\mathbb{R}^d$, we can create a random-grid Monte Carlo algorithm similar to the random-ray Monte Carlo method (Liu et al. 2000) to be described in Section 6.3.3. At each iteration, we do the following steps.

Random-Grid Method:

- Randomly generate a direction $\mathbf{e}$ and a grid size r.

- Construct the candidate set as

$$\mathbf{y}_l = \mathbf{x} + l \cdot r \cdot \tilde{\mathbf{e}}, \quad \text{for } l = 1, \ldots, k.$$

- Draw $\mathbf{y} = \mathbf{y}_j$ from $\{\mathbf{y}_1, \ldots, \mathbf{y}_k\}$ with probability proportional to $u_j \pi(\mathbf{y}_j)$, where u_j is a constant chosen by the user (e.g., $u_j = \sqrt{j}$).

- Construct the reference set by letting $\mathbf{x}_l^* = \mathbf{y} - l \cdot r \cdot \mathbf{e}$, for $l = 1, \ldots k$. Therefore, $\mathbf{x}_l^* = \mathbf{y}_{j-l}$ for $l < j$ and $\mathbf{x}_l^* = \mathbf{x} - (l-j) \cdot r \cdot \tilde{\mathbf{e}}$ for $l \geq j$.

- Accept the candidate $\mathbf{y}$ with probability

$$p = \min\left\{1, \sum_{l=1}^k \pi(\mathbf{y}_l) / \sum_{l=1}^k \pi(\mathbf{x}_l^*)\right\},$$

and reject otherwise.

5.6 Reversible Jumping Rule

In applications such as image analysis (Grenander and Miller 1994) and Bayesian model selections (Green 1995), one often needs to design a sampler that jumps between different dimensional spaces. In principle, one still can follow the Metropolis-Hastings's rule to guide for the design of such a sampler. The only technical complication is in ensuring the reversibility of proposals for jumping between two different dimensional spaces.

Suppose $\mathcal{X}$ is the state space of interest and $\mathcal{Y}$ is a subspace of $\mathcal{X}$ with a lower dimensionality. For example, $\mathcal{Y}$ can be a manifold defined as $\mathcal{Y} = \{\mathbf{x} : f(\mathbf{x}) = 0\}$ for some differentiable function f. Furthermore, we suppose that the target distribution $\pi(\mathbf{x})$, known up to a normalizing constant, lives on these two spaces *simultaneously*. This distribution can be represented as

$$\pi(\mathbf{x}) \propto q_0(\mathbf{x})|_{\mathbf{x}\in\mathcal{Y}} + q_1(\mathbf{x}), \tag{5.11}$$

where q_0 and q_1 are two unnormalized probability density functions defined on their respective spaces [i.e., $q_i(\mathbf{x}) = c_i \pi_i(\mathbf{x})$ with c_i unknown]. Therefore, if we draw a random sample from $\mathcal{X}$ according to $\pi_1(\mathbf{x})$, the chance that it lies in $\mathcal{Y}$ is zero! To make things worse, we assume that the ratio of the two constants, $\gamma = c_0/c_1$, is generally unknown to us. If we can design a Monte Carlo algorithm to sample from $\pi(\mathbf{x})$, the ratio γ can be estimated by the ratio of the number of samples lying in $\mathcal{Y}$ over that lying in $\mathcal{X}$.

The above setting is of particular interest in Bayesian hypothesis testing problems. Suppose we have a probability model $f(\mathbf{y} \mid \boldsymbol{\theta})$ where $\boldsymbol{\theta} = (\theta_0, \theta_1)$ and we are interested in testing H_0: $\theta_0 = \theta_1$ versus H_1: $\theta_0 \neq \theta_1$ in light of an observation $\mathbf{y}$. We let model M_1 correspond to the unrestricted parameter space $\boldsymbol{\theta}$ (two dimensional) and let model M_0 correspond to the subspace defined by $\theta_0 - \theta_1 = 0$. It is sometimes natural assume that the two models are equally likely *a priori*; then, the posterior distribution of $\boldsymbol{\theta}$ under the "mixture" of two plausible models is

$$\pi(\boldsymbol{\theta}) \propto f(\mathbf{y} \mid \boldsymbol{\theta}) f_0(\boldsymbol{\theta})|_{\theta_0=\theta_1} + f(\mathbf{y} \mid \boldsymbol{\theta}) f_0(\boldsymbol{\theta})|_{\theta_0 \neq \theta_1}.$$

Statisticians are often interested in estimating the ratio of the two normalizing constants,

$$\frac{c_0}{c_1} \equiv \frac{\int_{\theta_0=\theta_1} f(\mathbf{y} \mid \boldsymbol{\theta}) f_0(\boldsymbol{\theta}) d\boldsymbol{\theta}}{\int f(\mathbf{y} \mid \boldsymbol{\theta}) f_1(\boldsymbol{\theta}) d\boldsymbol{\theta}},$$

which reflects the posterior odds ratio of model M_0 versus model M_1. As we mentioned in the previous paragraph, this ratio can be estimated if we can design a Monte Carlo scheme to sample π.

In order to design a Monte Carlo Markov chain that lives on both the general state space and the restricted space, we need to have two *different* proposals, one for $\mathcal{Y} \to \mathcal{X}$ and another for $\mathcal{X} \to \mathcal{Y}$. Since $\mathcal{Y}$ is of lower dimensional, any transition from $\mathcal{Y}$ to $\mathcal{X}$ must have a degenerate density with respect to the dominant measure on $\mathcal{X}$, implying that no proposal for moves from $\mathcal{X} \to \mathcal{Y}$ can be properly "reversed" by a proposal from $\mathcal{Y}$ to $\mathcal{X}$. To overcome this difficulty, we must have a "matching space" $\mathcal{Z}$, so that $\mathcal{Y} \times \mathcal{Z}$ has the same dimension as $\mathcal{X}$ and a matching proposal $g(\mathbf{z} \mid \mathbf{y})$. With the matched space, one can come up with two non-degenerate proposals and follow the Metropolis-Hastings's rule to design jumps.

Green (1995) presented a formal treatment of this type of moves (involving change-of-variables and Jacobians) and named them *reversible jumps*. Here, we study only a sufficiently instructive special case: $\mathcal{X} = \mathcal{Y} \times \mathcal{Z}$. Therefore, each $\mathbf{x}$ can be written as $\mathbf{x} = (\mathbf{y}, \mathbf{z})$, and "subspace" $\mathcal{Y}$ in fact corresponds to $\mathcal{Y} \times \{\mathbf{z}_0\}$ for some $\mathbf{z}_0 \in \mathcal{Z}$. In order to jump from $\mathcal{Y}$ to $\mathcal{X}$, we may first propose $\mathbf{y} \to \mathbf{y}'$ by a proposal transition $T_1(\mathbf{y}, \mathbf{y}')$, match $\mathbf{y}'$ with a $\mathbf{z}'$ drawn from $g(\cdot \mid \mathbf{y}')$, and then let $\mathbf{x}' = (\mathbf{y}', \mathbf{z}')$. This can be viewed as an *expansion* transition. A *contraction* transition is needed to propose from $\mathbf{x} \in \mathcal{X}$ back into $\mathcal{Y}$. This can be achieved by first dropping the $\mathbf{z}$ component in $\mathbf{x}$, and then proposing $\mathbf{y}'$ from $T_2(\mathbf{y}, \mathbf{y}')$. According the Metropolis-Hastings rule, the expansion proposal $\mathbf{y} \to \mathbf{x}'$ is accepted with probability

$$\alpha = \min\left\{1, \frac{q_1(\mathbf{y}', \mathbf{z}')T_2(\mathbf{y}', \mathbf{y})}{q_0(\mathbf{y})T_1(\mathbf{y}, \mathbf{y}')g(\mathbf{z}' \mid \mathbf{y}')}\right\},$$

where q_0 and q_1 are as defined in (5.11). The contraction proposal $\mathbf{x} \to \mathbf{y}'$ (where $\mathbf{x} = (\mathbf{y}, \mathbf{z})$) is accepted with probability

$$\beta = \min\left\{1, \frac{q_0(\mathbf{y}')T_1(\mathbf{y}', \mathbf{y})g(\mathbf{z} \mid \mathbf{y})}{q_1(\mathbf{y}, \mathbf{z})T_2(\mathbf{y}, \mathbf{y}')}\right\}.$$

The expansion proposal in the foregoing procedure can be seen as first "proposing" and then "lifting" (from a lower dimensional space to the higher one). Similarly, we can conduct "lifting" first and "proposing" afterward. More precisely, in order to accomplish the proposal $\mathbf{y} \to \mathbf{x}'$, we can first draw $\mathbf{z} \sim g(\cdot \mid \mathbf{y})$ and then draw $\mathbf{x}'$ from $S_1[(\mathbf{y}, \mathbf{z}), \cdot]$. The contraction move is achieved by first proposing $\mathbf{x} \to \mathbf{x}' = (\mathbf{y}', \mathbf{z}')$ according to $S_2(\mathbf{x}, \mathbf{x}')$ and then dropping $\mathbf{z}'$. Thus, the acceptance probabilities are respectively

$$\alpha' = \min\left\{1, \frac{q_1(\mathbf{x}')S_2[\mathbf{x}', (\mathbf{y}, \mathbf{z})]}{q_0(\mathbf{y})g(\mathbf{z} \mid \mathbf{y})S_1[(\mathbf{y}, \mathbf{z}), \mathbf{x}']}\right\},$$

$$\beta' = \min\left\{1, \frac{q_0(\mathbf{y}')g(\mathbf{z}' \mid \mathbf{y}')S_1[(\mathbf{y}', \mathbf{z}'), \mathbf{x}]}{q_1(\mathbf{x})S_2[\mathbf{x}, (\mathbf{y}', \mathbf{z}')]}\right\}.$$

Note that both S_1 and S_2 are proposals in the higher-dimensional space $\mathcal{X}$, whereas both T_1 and T_2 are proposals in the lower-dimensional subspace $\mathcal{Y}$. Thus, without having other justifications, we prefer lifting *after* proposing than the other way around because it is often easier to propose lower-dimensional moves. It is conceivable, however, that lifting *before* proposing can sometimes help the chain escape from a local energy trap of the lower-dimensional space. Similar ideas have been employed in the clustering method and simulated tempering (Chapters 7 and 10).

A useful strategy in improving Monte Carlo sampling efficiency is to introduce a number of related probabilistic systems with different levels of "difficulties" (in terms of Monte Carlo sampling) and then simulate them together. These auxiliary systems are often made by varying a "temperature" parameter in the original target distribution for the sake of easy manipulation (i.e., simulated tempering and parallel tempering; see Chapter 10). However, it may be more efficient to consider a system consisting of spaces with different dimensions (Liu and Sabatti 1998). To sample from this augmented system, one needs to use the reversible jumping rule. It should be noted that the basic principle behind the reversible jumps is similar to that behind the sequential importance sampling and the CBMC (Sections 2.6.3 and 5.4.3) because the move from $\mathbf{y}$ to $\mathbf{x}$ can be seen as a one-step SIS update.

5.7 Dynamic Weighting

Wong and Liang (1997) introduced the use of a dynamic weighting variable for controlling Markov chain simulation. By using this scheme, they were able to obtain better results for many optimization problems, such as the traveling salesman problem and neural network training, and high-dimensional integration problems, such as the Ising model simulation.

To start a dynamic weighting scheme, we first augment the sample space $\mathcal{X}$ to $\mathcal{X} \times \mathbb{R}^+$ so as to include a weight variable. Similar to the Metropolis algorithm, we also need a proposal function $T(\mathbf{x}, \mathbf{y})$ on the space $\mathcal{X}$. Suppose at iteration t, we have $(\mathbf{x}^{(t)}, w^{(t)}) = (\mathbf{x}, w)$. Then an R-type move is defined as follows:

- Draw $\mathbf{y}$ from $T(\mathbf{x}, \mathbf{y})$ and compute the Metropolis-Hastings ratio

$$r(\mathbf{x}, \mathbf{y}) = \frac{\pi(\mathbf{y})T(\mathbf{y}, \mathbf{x})}{\pi(\mathbf{x})T(\mathbf{x}, \mathbf{y})}.$$

- Choose $\theta = \theta(w, \mathbf{x}) > 0$, and draw U from Uniform(0,1). Then let

$$(\mathbf{x}^{(t+1)}, w^{(t+1)}) = \begin{cases} (\mathbf{y}, wr(\mathbf{x}, \mathbf{y}) + \theta) & \text{if } U \leq \dfrac{wr(\mathbf{x}, \mathbf{y})}{wr(\mathbf{x}, \mathbf{y}) + \theta} \\ \left(\mathbf{x}^{(t)}, \dfrac{w(wr(\mathbf{x}, \mathbf{y}) + \theta)}{\theta}\right) & \text{otherwise.} \end{cases}$$

(5.12)

It is easy to check that the R-type move does not have π as its equilibrium distribution. Wong and Liang (1997) propose to use *invariance with respect to importance weighting* (IWIW) for justifying the above scheme; that is, if the joint distribution of $(\mathbf{x}, w)$ is $f(\mathbf{x}, w)$ and $\mathbf{x}$ is said *correctly weighted* by w with respect to π if $\sum_w w f(\mathbf{x}, w) \propto \pi(\mathbf{x})$. A transition rule is said to satisfy IWIW if it *maintains* the *correctly weightedness* for the joint distribution of $(\mathbf{x}, w)$. Clearly, the R-type move satisfies IWIW.

The purpose of introducing importance weights into the dynamic Monte Carlo process is to provide a means for the system to make large transitions not allowable by the standard Metropolis transition rules. The weight variable is updated in a way that allows for an adjustment of the bias induced by such non-Metropolis moves. Although this algorithm has been applied successfully in many difficult optimization and simulation problems [see Section 10.6 and Liang (1997)], theoretical properties of this algorithm are still rather subtle. A first theory is recently given by Liu, Liang and Wong (2001) and some of which, together with another type of dynamic weighting scheme, the *Q-type* move, will be presented in Section 13.6. An important application of the dynamic weighting method is to be combined with a *simulated tempering* algorithm, and this aspect will be discussed in more detail in Section 10.6.

5.8 Output Analysis and Algorithm Efficiency

In analyzing outputs from a Markov chain Monte Carlo algorithm (this applies to all the later chapters), one of the major concerns is its *statistical efficiency* in estimating the expectation of interest. Let us suppose that the Markov chain is irreducible and aperiodic (see footnotes 4 and 5 on page 114) and converges to its unique stationary distribution, $\pi(\mathbf{x})$. At the heart of every MCMC computation is the estimation of $E_\pi h(\mathbf{x})$ for a certain $h(\ )$ of interest. Thus, what we really care about at the end is how accurate we can estimate this quantity. Suppose we have drawn samples $\mathbf{x}^{(1)}, \ldots, \mathbf{x}^{(m)}$ via a MCMC sampler with $\pi(\mathbf{x})$ as its equilibrium distribution. Let us further assume that we have run the process long enough (in the previous day, say) and have thrown away the initial n_0 iterations needed for the equilibration of the chain (i.e., we assume that $\mathbf{x}^{(0)} \sim \pi$). Then,

$$m \mathrm{var} \left\{ \frac{h(\mathbf{x}^{(1)}) + \cdots + h(\mathbf{x}^{(m)})}{m} \right\} = \sigma^2 \left[1 + 2 \sum_{j=1}^{m-1} \left(1 - \frac{j}{m}\right) \rho_j \right]$$

$$\approx \sigma^2 \left[1 + 2 \sum_{j=1}^{\infty} \rho_j \right] \quad (5.13)$$

where $\sigma^2 = \mathrm{var}[h(\mathbf{x})]$ and $\rho_j = \mathrm{corr}\{h(\mathbf{x}^{(1)}), h(\mathbf{x}^{(j+1)})\}$. In physics literature (Goodman and Sokal 1989), one defines the *integrated autocorrelation time* of $h(\mathbf{x})$ as

$$\tau_{\text{int}}(h) = \frac{1}{2} + \sum_{j=1}^{\infty} \rho_j.$$

Then we have

$$m\text{var}(\hat{h}) = 2\tau_{\text{int}}(h)\sigma^2.$$

This variance is, in effect, equal to that of an estimator with $m/[2\tau_{\text{int}}(h)]$ independent random samples. Thus, $m/[2\tau_{\text{int}}(h)]$ is often called the *effective sample size*.

If is often observed that ρ_j decays exponentially. Therefore, we can model the autocorrelation curve as

$$|\rho_j| \sim \exp\left\{-\frac{j}{\tau_{\exp}(h)}\right\},$$

which gives rise to the expression

$$\tau_{\exp}(h) = \limsup_{j \to \infty} \frac{j}{-\log|\rho_j|},$$

where $\tau_{\exp}(h)$ is called the *exponential autocorrelation time*. When $\tau_{\exp}(h)$ is large, we can see that

$$\tau_{\text{int}}(h) \approx \sum_{j=0}^{\infty} e^{-j/\tau_{\exp}(h)} - \frac{1}{2} = \frac{1}{1 - e^{-1/\tau_{\exp}(h)}} - \frac{1}{2} \approx \tau_{\exp}(h).$$

The "relaxation time" of the system is defined as

$$\tau_{\exp} = \sup_{h \in L^2(\pi)} \tau_{\exp}(h).$$

The concepts of the autocorrelation and relaxation time are also closely related to the convergence rate of the algorithm (i.e., the second largest eigenvalue of the Markov chain transition matrix). More precisely, if we let h be an eigenfunction that corresponds to an eigenvalue λ of the transition matrix, then we have $\rho_j(h) = \lambda^j$. Hence,

$$\tau_{\text{int}}(h) = \frac{1+\lambda}{2(1-\lambda)}, \quad \tau_{\exp}(h) = -\frac{1}{\log|\lambda|},$$

and the relaxation time is

$$\tau_{\exp} = -\frac{1}{\log|\lambda_2|},$$

where λ_2 is the second largest eigenvalue in modular of the transition matrix. Thus, $\tau_{\exp}$ reflects the convergence speed of an MCMC sampler, whereas τ_{int} is most relevant when the statistical efficiency of the algorithm is of interest.

It is commonly agreed that finding an ideal proposal chain is an art. In fact, the Metropolis algorithm aided with Hastings's (1970) generalization is so general that one always tends to feel unsatisfactory in settling down on any specific proposal chain. It is important, therefore, to analyze autocorrelation curves of an algorithm in order to obtain its behavioral characteristics. Peskun (1973) suggests that two Markov chains, with transition functions P_1 and P_2, respectively, and the same equilibrium distribution, should be compared based on a statistical criterion — the asymptotic variance of the corresponding estimator (5.13). When the state space is finite, he found an explicit asymptotic formula for (5.13), from which he concluded that for a given proposal transition, the Metropolis acceptance-rejection rule is "optimal" in the sense of having the smallest asymptotic variance for the resulting estimates. Consequently, Barker's proposal is less desirable in general. An interesting twist of Peskun's result will be described more details in Section 5.4.2.

Another interesting result is given by Gelman, Roberts and Gilks (1995) for a continuous state space. Suppose the target distribution is $N(0, 1)$ and a random-walk Metropolis algorithm (Section 5.4.1) is used for its simulation. The proposal transition is of the form $\mathbf{x}' = \mathbf{x}^{(t)} + \epsilon$, where $\epsilon \sim N(0, \sigma^2)$. Gelman, Roberts and Gilks (1995) show that the optimal choice for σ that gives the smallest autocorrelation is $\sigma = 2.38$. A slightly larger σ does not affect the efficiency much, but a smaller one has a significant adverse effect. These assertions can be verified by direct simulation. This optimal σ corresponds to an acceptance rate of 44%, suggesting that this is a useful reference number to be watched when tuning a proposal distribution. Roberts, Gelman and Gilks (1997) recommended calibrating the acceptance rate to about 25% for a high-dimensional model and to about 50% for models of dimensions 1 or 2.

5.9 Problems

1. Implement a Metropolis algorithm to sample from a Poisson (λ) distribution. Test it for $\lambda=3$, 5, and 10. A suggestion for the proposal function: the simple random walk on a line.

2. Let $\pi(\mathbf{x})$ be $N(0, 1)$. Implement the random-walk Metropolis algorithm to sample from π, using $N(\cdot, \sigma^2)$ as the proposal function. Plot the autocorrelation curve for the corresponding chain. Argue why $\sigma = 2.38$ is a good choice.

3. Prove that the multistage modification of the CBMC method is proper for all d (Section 5.4.3).

4. Show that the detailed balance condition $\pi(\mathbf{x})A(\mathbf{x},\mathbf{y}) = \pi(\mathbf{y})A(\mathbf{y},\mathbf{x})$ guarantees that π is the invariant distribution of $A(\mathbf{x},\mathbf{y})$.

5. Suppose the Metropolized independence sampler (MIS) is applied to sample from $\pi(\mathbf{x})$, where $\mathbf{x}$ is defined on a finite state space and the trial distribution is $g(\mathbf{x})$.

    (a) Write down the actual transition matrix A for the MIS.
    (b) Show that the second largest eigenvalue of A is $\min_{\mathbf{x}}\{\pi(\mathbf{x})/g(\mathbf{x})\}$.
    (c) Find its corresponding eigenvector.

6. Show that the random-grid method is proper [i.e., it leaves the target distribution $\pi(\mathbf{x})$ invariant]. Implement the random-grid method to sample from a multidimensional Gaussian distribution. Study empirically how the choices of k and the grid-size distribution affect algorithmic efficiency.

7. Show that the reversible jump algorithm as described in Section 5.6 leaves π invariant.

8. Show that the R-type dynamic weighting rule satisfies the IWIW property.

9. Show that the Q-type dynamic weighting rule does not satisfy the IWIW property.

10. Implement both the random-ray and the random-grid methods to replace the griddy-Gibbs method in Example 6.1 of Ritter and Tanner (1992).

11. Prove that the MTMIS algorithm gives rise to a reversible Markov chain whose equilibrium distribution is π.

6
The Gibbs Sampler

The proposal transition $T(\mathbf{x}, \mathbf{y})$ in a Metropolis sampler is often an arbitrary choice out of convenience. In many applications, the proposal is chosen to be a locally uniform move. In fact, the use of symmetric and locally uniform proposals is so prevailing that these are often referred to as "unbiased proposals" in the literature. If not subjecting to the Metropolis-Hastings rejection rule, this type of move would have led to a form of simple random walk in the configuration space $\mathcal{X}$. Although such a proposal is simple both conceptually and operationally, the performance of the resulting Metropolis algorithm is often inefficient because the proposal is too "noisy." In contrast, the conditional sampling techniques to be discussed in this and the next chapters enable a MCMC sampler to follow the local dynamics of the target distribution. A distinctive feature of these MCMC algorithms is that at each iteration, they use conditional distributions (i.e., those distributions resulting from constraining the target distribution π on certain subspaces) to construct Markov chain moves. As a consequence, no rejection is incurred at any of its sampling steps. The multipoint methods described in Section 5.5 are similar in spirit but are computationally more expensive than conditional sampling.

6.1 Gibbs Sampling Algorithms

The Gibbs sampler (Geman and Geman 1984) is a special MCMC scheme. Its most prominent feature is that the underlying Markov chain is constructed by composing a sequence of conditional distributions along a set of directions (often along the coordinate axis).

Suppose we can decompose the random variable into d components [i.e., $\mathbf{x} = (x_1, \ldots, x_d)$]. In the Gibbs sampler, one randomly or systematically chooses a coordinate, say x_1, and then updates it with a new sample x_1' drawn from the conditional distribution $\pi(\cdot \mid \mathbf{x}_{[-1]})$, where $\mathbf{x}_{[-A]}$ refers to $\{x_j, j \in A^c\}$ for any subset A of the coordinate indices. Algorithmically, we can describe two types of Gibbs sampling strategy.

Random-Scan Gibbs Sampler. Let $\mathbf{x}^{(t)} = (x_1^{(t)}, \ldots x_d^{(t)})$ for iteration t. Then, at iteration $t+1$, we conduct the following steps:

- Randomly select a coordinate i from $\{1, \ldots, d\}$ according to a given probability vector $(\alpha_1, \ldots, \alpha_d)$ [e.g., $(1/d, \ldots, 1/d)$].

- Draw $x_i^{(t+1)}$ from the conditional distribution $\pi(\cdot \mid x_{[-i]}^{(t)})$ and leave the remaining components unchanged; that is, let

$$\mathbf{x}_{[-i]}^{(t+1)} = \mathbf{x}_{[-i]}^{(t)}.$$

Systematic-Scan Gibbs Sampler. Let $\mathbf{x}^{(t)} = (x_1^{(t)}, \ldots x_d^{(t)})$. At the $t+1$ iteration:

- We draw $x_i^{(t+1)}$ from the conditional distribution

$$\pi(x_i \mid x_1^{(t+1)}, \ldots, x_{i-1}^{(t+1)}, x_{i+1}^{(t)}, \ldots, x_d^{(t)})$$

For $i = 1, \ldots, d$.

It is easy to check that *every* single conditional update step in both the random-scan and the systematic-scan Gibbs samplers leaves π invariant. To see this point, suppose $\mathbf{x}^{(t)} \sim \pi$. Then, $\mathbf{x}_{[-i]}^{(t)}$ follows its marginal distribution under π. Thus,

$$\pi(x_i^{(t+1)} \mid \mathbf{x}_{[-i]}^{(t)}) \times \pi(\mathbf{x}_{[-i]}^{(t)}) = \pi(\mathbf{x}_{[-i]}^{(t)}, x_i^{(t+1)}),$$

which means that after one conditional update, the new configuration still follows distribution π.

Under regularity conditions, one can show that a Gibbs sampler chain converges geometrically and its convergence rate is related to how the variables correlate with each other (Liu 1991, Liu, Wong and Kong 1995, Schervish and Carlin 1992). Based on a finding that the Gibbs sampler's convergence rate is controlled by the *maximal correlation* (Section 12.6.3) between the states of two consecutive Gibbs iterations, Liu, Wong and Kong (1994) and Liu (1994a) argued that grouping (some researchers also call it *blocking*) highly correlated components together (i.e., update them jointly) in the Gibbs sampler can greatly improve its efficiency. Some researchers have also shown that random scan can outperform systematic scan in terms of convergence speed (Roberts and Sahu 1997).

A simple restatement of the conditional updates used in the Gibbs sampler can be potentially useful: Each Gibbs update can be seen as a random *relocation* (we used the word "perturbation" in a Metropolis sampler) of the current state **x** along a chosen direction. For example, if the first coordinate direction is chosen, then this "relocation" can be represented as

$$(x_1, \ldots, x_d) \to (x_1 + \gamma, x_2, \ldots, x_d),$$

where γ is a random draw from an appropriate distribution. It is not difficult to show that if γ is drawn from $p(\gamma) \propto \pi(x_1 + \gamma, \mathbf{x}_{[-1]})$, then the move leaves π invariant. This view is critical in generalizing the Gibbs updates to more versatile conditional moves [e.g., Markov chain updates under a transformation group setting (Liu and Wu 1999)], which are useful for designing more efficient MCMC samplers. See Chapter 8 for more discussions.

The Gibbs sampler's popularity in statistics community stems from its extensive use of *conditional distributions* in each iteration. The data augmentation of Tanner and Wong (1987) first links the Gibbs sampling structure with statistical missing data problems and the EM algorithm (see Section A.4 of the appendix for a detailed description of the algorithm). The generality and the basic theory behind Gibbs sampler were noted by Li (1988). Gelfand and Smith (1990) further demonstrated that the conditional distributions needed in Gibbs iterations are commonly available in many Bayesian and likelihood computations.

6.2 Illustrative Examples

Consider the simulation from a bivariate Gaussian distribution. Let $\mathbf{x} = (x_1, x_2)$ and let the target distribution be

$$N\left\{\begin{pmatrix} 0 \\ 0 \end{pmatrix}, \begin{pmatrix} 1 & \rho \\ \rho & 1 \end{pmatrix}\right\}.$$

The Markov chain $\mathbf{x}^{(t)} = (x_1^{(t)}, x_2^{(t)})$ corresponding to a systematic-scan Gibbs sampler is generated as

$$x_2^{(t+1)} \mid x_1^{(t+1)} \sim N\{\rho x_1^{(t+1)}, (1-\rho^2)\},$$
$$x_1^{(t+1)} \mid x_2^{(t)} \sim N\{\rho x_2^{(t)}, (1-\rho^2)\}.$$

It is seen from simple computation that

$$\begin{pmatrix} x_1^{(t)} \\ x_2^{(t)} \end{pmatrix} \sim N\left\{\begin{pmatrix} \rho^{2t-1} x_2^{(0)} \\ \rho^{2t} x_2^{(0)} \end{pmatrix}, \begin{pmatrix} 1-\rho^{4t-2} & \rho-\rho^{4t-1} \\ \rho-\rho^{4t-1} & 1-\rho^{4t} \end{pmatrix}\right\}.$$

132   6. The Gibbs Sampler

Thus, as $t \to 0$, the joint distribution of $(x_1^{(t)}, x_2^{(t)})$ converges to the target distribution. Furthermore, the rate of convergence is equal to the maximal correlation between $x_i^{(t)}$ and $x_i^{(t+1)}$, which is ρ^2. A more general analysis along this line is given in Chapter 12.

Another simple example is given by Casella and George (1992), in which the target distribution is

$$\pi(x,y) \propto \binom{n}{x} y^{x+\alpha-1}(1-y)^{n-x+\beta-1}$$

for $x = 0, 1, \ldots, n$ and $0 \leq y \leq 1$. It is easy to see that the two necessary conditional distributions are

$$x \mid y \sim \text{Binom}(n, y)$$
$$y \mid x \sim \text{Beta}(x + \alpha, n - x + \beta)$$

A Gibbs sampler iterates between the above two conditional sampling steps. Figure 6.1 shows some simulation results for $n = 20$ and $\alpha = \beta = 0.5$.

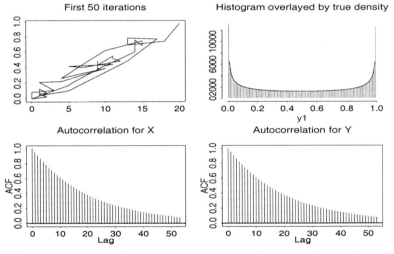

FIGURE 6.1. Using the Gibbs sampler to simulate from Beta-Binomial distribution. (a) Trace plot of first 50 iterations; (b) estimating the density of y using the Monte Carlo samples (from 200,000 iterations); (c) and (d), autocorrelation plots for x and y.

It is noted that the autocorrelation plots of x and y are almost identical — this is not an accident. It has been shown (Liu et al. 1994) that in a two-component Gibbs sampler, the two components share a common convergence rate and can be thought of as "interleaving chains" (see Chapter 12 for detailed analysis).

6.3 Some Special Samplers

6.3.1 Slice sampler

Suppose $\pi(\mathbf{x})$ is a density function of interest and $\mathbf{x} \in \mathbb{R}^d$. Then, drawing $\mathbf{x} \sim \pi(\mathbf{x})$ is equivalent to generating $\mathbf{z} = (z_1, \ldots, z_{d+1})$ so that it is uniformly distributed in the region S under the surface of π; that is,

$$S = \{\mathbf{z} \in \mathbb{R}^{d+1} : z_{d+1} \leq \pi(z_1, \ldots, z_d)\}.$$

However, generating uniformly distributed random variables in an arbitrary region is equally as difficult as the original Monte Carlo simulation problem. One can apply the following Gibbs iteration to achieve the sampling:

- Draw $\mathbf{y}^{(t+1)} \sim \text{Uniform}[0, \pi(\mathbf{x}^{(t)})]$.
- Draw $\mathbf{x}^{(t+1)}$ uniformly from region $S^{(t+1)} = \{\mathbf{x} : \pi(\mathbf{x}) \geq \mathbf{y}^{(t+1)}\}$.

However, region $S^{(t+1)}$ in the iteration is still difficult to deal with. When π can be written as the product of k functions [i.e., $\pi(\mathbf{x}) = f_1(\mathbf{x}) \times \cdots \times f_k(\mathbf{x})$], Edwards and Sokal (1988) introduced k auxiliary variables $y_1, \ldots, y_k$ and described a Gibbs sampler for sampling $(\mathbf{x}, y_1, \ldots, y_k)$ uniformly over the region $0 < y_i < f_i(\mathbf{x})$, $i = 1, \ldots, k$:

- Draw $y_i^{(t+1)} \sim \text{Uniform}[0, f_i(\mathbf{x}^{(t)})]$, $i = 1, \ldots, k$.
- Draw $\mathbf{x}^{(t+1)}$ uniformly from the region

$$S^{(t+1)} = \bigcap_{i=1}^{k} \{\mathbf{x} : f_i(\mathbf{x}) \geq y_i^{(t+1)}\}.$$

This method is also related to the clustering algorithms for Ising model simulations pioneered by Swendsen and Wang (1987) (Chapter 7). Damien, Wakefield and Walker (1999) showed that in many cases, one can find a decomposition of π so that the intersection set $S^{(t+1)}$ is easy to compute, which leads to an easily implemented sampler. Applications of this approach to image analysis have been discussed by Besag and Green (1993) and Higdon (1998). However, the convergence rate of the slice sampler may generally be rather slow because of the presence of many auxiliary variables.

6.3.2 Metropolized Gibbs sampler

When the state space of interest is discrete, Liu (1996c) suggested an "over-relaxation" strategy to improve the ordinary Gibbs sampler. Let $\mathbf{x} = (x_1, \ldots, x_d)$, where x_i takes m_i possible values, and let $\pi(\mathbf{x})$ be the distribution of interest. In the random-scan Gibbs sampler described in Section 6.1, a coordinate i is first chosen at random and the current value

x_i is replaced by a value y_i drawn from the corresponding full-conditional distribution. Here, we consider a modification of this procedure in which a value y_i, different from x_i, is drawn with probability

$$\frac{\pi(y_i \mid \mathbf{x}_{[-i]})}{1 - \pi(x_i \mid \mathbf{x}_{[-i]})},$$

then y_i replaces x_i with the Metropolis-Hastings acceptance probability,

$$\min\left\{1, \frac{1 - \pi(x_i \mid \mathbf{x}_{[-i]})}{1 - \pi(y_i \mid \mathbf{x}_{[-i]})}\right\};$$

else x_i is retained. Liu (1996c) proves that the modified Gibbs sampler for discrete random variables as defined earlier is statistically more efficient than the random-scan Gibbs sampler (see Section 13.3.1).

When $m_i = 2$, the Gibbs sampler is essentially the method of Barker (1965), whereas the modified procedure becomes a Metropolis algorithm. Peskun (1973) makes some general comparisons between these two samplers. Besag, Green, Higdon and Mengersen (1995) note that the superiority of Metropolis for binary systems results from its increased mobility around the state space. This rationale applies more generally to the Metropolized Gibbs sampler described here.

6.3.3 Hit-and-run algorithm

Suppose the current state is $\mathbf{x}^{(t)}$. In the hit-and-run (HR) algorithm, one does the following: (a) uniformly select a random direction $e^{(t)}$; (b) sample a scalar $r^{(t)}$ from density $f(r) \propto \pi(\mathbf{x}^{(t)} + re^{(t)})$; and (c) update $\mathbf{x}^{(t+1)} = \mathbf{x}^{(t)} + r^{(t)} e^{(t)}$. This HR algorithm behaves like a random-direction Gibbs sampler and allows for a complete exploration of a randomly chosen direction. It tends to be especially helpful when there are several modes (with comparable sizes) in the target distribution.

A main difficulty in implementing the HR algorithm, however, is that one is rarely able to draw from $f(r)$ in practice. Then, one may end up only using a single step of Metropolis update (Chen and Schmeiser 1993) — which renders the algorithm equivalent to the random-walk Metropolis. The MTM method introduced in Section 5.5 can help achieve the effect of conditional sampling required by the HR algorithm (Liu, Liang and Wong 2000). The following random-ray Monte Carlo scheme is a way of using MTM to achieve the hit-and-run effect. Suppose the current state is $\mathbf{x}^{(t)} = \mathbf{x}^*$; our new algorithm is as follows:

Random-Ray Monte Carlo:

- Randomly generate a direction (a unit vector) e.

- Propose to draw $\mathbf{y}_1,\ldots,\mathbf{y}_k$ from a distribution $T_e(\mathbf{x}^*,\mathbf{y})$ along the direction $\mathbf{e}$. A generic choice is to draw i.i.d. samples $r_1,\ldots,r_k$ from $N(0,\sigma^2)$, where σ can be chosen rather large, and set $\mathbf{y}_j = \mathbf{x} + r_j \mathbf{e}$. Another possibility is to draw $r_j \sim \text{Uniform}(-\sigma,\sigma)$.

- Conduct the MTM step; that is, we choose $\mathbf{y}^*$ from $\mathbf{y}_1,\ldots,\mathbf{y}_k$ with probability proportional to $\pi(\mathbf{y}_j)$; and then draw $\mathbf{x}'_1,\ldots,\mathbf{x}'_{k-1}$ i.i.d. from $T_e(\mathbf{y}^*,\mathbf{x})$. Let $\mathbf{x}^* = \mathbf{x}'_k$. Then, compute the generalized Metropolis ratio

$$r = \min\left\{1, \frac{\sum_{j=1}^{k} \pi(\mathbf{y}_j)T_e(\mathbf{y}_j,\mathbf{x}^*)}{\sum_{j=1}^{k} \pi(\mathbf{x}'_j)T_e(\mathbf{x}'_j,\mathbf{y}^*)}\right\}.$$

In our experience, a much larger σ can be used compared to that in an HR with a single Metropolis update, resulting in a higher acceptance rate for the same computational time. The random-grid Monte Carlo method described in Section 5.5 can also serve as a good alternative to the HR algorithm. In Section 11.1, we will introduce another interesting variation of this algorithm — the adaptive directional sampling (Gilks, Roberts and George 1994) and conjugate gradient Monte Carlo.

6.4 Data Augmentation Algorithm

6.4.1 Bayesian missing data problem

A main reason for statisticians to favor conditional sampling approaches (e.g., the Gibbs sampler) over the more flexible Metropolis algorithm is that in many statistical models, it is not very difficult to derive and to sample from necessary *conditional distributions* and conditional moves are, in general, more "global" than a perturbation-type move. Another reason is that the proposal chain in the Metropolis algorithm seems to be too "random" to be effective in statistical models because the joint space of the parameter $\boldsymbol{\theta}$ and missing data $\mathbf{y}_{\text{mis}}$ in a statistical model (Section 1.9) is often "irregular." For example, $\boldsymbol{\theta}$ may include a discrete component and a continuous component; or different components of $\boldsymbol{\theta}$ (such as a mean vector and a covariance matrix) may have completely different scales. For these problems, defining a reasonable "perturbation" of the current configuration of $(\boldsymbol{\theta}, \mathbf{y}_{\text{mis}})$ is very difficult. However, the Metropolis algorithm is often employed together with a conditional sampling approach (Gelman and Rubin 1992, Liu 1996a, Tierney 1994, Liu, Liang and Wong 2000).

As shown in Sections 1.9 and 3.2, we assume in a Bayesian missing data problem that the "complete-data" model $f(\mathbf{y} \mid \boldsymbol{\theta})$ has a nice analytical form from which we can do all the posterior computations in closed form. Let $\mathbf{y} = (\mathbf{y}_{\text{obs}}, \mathbf{y}_{\text{mis}})$, where $\mathbf{y}_{\text{obs}}$ is observed but $\mathbf{y}_{\text{mis}}$ is missing. The *observed-*

data posterior distribution is

$$p(\boldsymbol{\theta} \mid \mathbf{y}_{\text{obs}}) = \int p(\boldsymbol{\theta} \mid \mathbf{y}_{\text{mis}}, \mathbf{y}_{\text{obs}}) p(\mathbf{y}_{\text{mis}} \mid \mathbf{y}_{\text{obs}}) d\mathbf{y}_{\text{mis}}. \qquad (6.1)$$

If we can draw $\mathbf{y}_{\text{mis}}$ from $p(\mathbf{y}_{\text{mis}} \mid \mathbf{y}_{\text{obs}})$, then a Monte Carlo approximation to (6.1) can be easily obtained. This observation forms the basis of the data augmentation algorithm.

Tanner and Wong (1987) observed that if we start with an approximation $g(\boldsymbol{\theta})$ of the target distribution, $p(\boldsymbol{\theta} \mid \mathbf{y}_{\text{obs}})$, we can draw m independent copies of the missing data, $\mathbf{y}_{\text{mis}}^{(1)}, \ldots, \mathbf{y}_{\text{mis}}^{(m)}$, from

$$\tilde{p}(\mathbf{y}_{\text{mis}}) = \int p(\mathbf{y}_{\text{mis}} \mid \boldsymbol{\theta}, \mathbf{y}_{\text{obs}}) g(\boldsymbol{\theta}) d\boldsymbol{\theta}. \qquad (6.2)$$

This sampling can be achieved by first drawing $\boldsymbol{\theta}^{(j)} \sim g(\boldsymbol{\theta})$ and then drawing $\mathbf{y}_{\text{mis}}^{(j)} \sim p(\mathbf{y}_{\text{mis}} \mid \boldsymbol{\theta}^{(j)}, \mathbf{y}_{\text{obs}})$. The $\mathbf{y}_{\text{mis}}^{(j)}$ so produced are often called *multiple imputations* (Rubin 1987) in statistics literature. With the newly generated copies of $\mathbf{y}_{\text{mis}}$, we can form a hopefully improved approximation of the posterior distribution as

$$g_{\text{new}}(\boldsymbol{\theta}) = \frac{1}{m} \sum_{j=1}^{m} p(\boldsymbol{\theta} \mid \mathbf{y}_{\text{obs}}, \mathbf{y}_{\text{mis}}^{(j)}). \qquad (6.3)$$

Note that if $g(\ )$ were indeed the true posterior distribution, (6.2) would have been the exact predictive distribution of $\mathbf{y}_{\text{mis}}$.

6.4.2 The original DA algorithm

A formal data augmentation scheme starts with a set of imputed missing values $\mathbf{y}_{\text{mis},1}^{(0,1)}, \ldots, \mathbf{y}_{\text{mis}}^{(0,m)}$, providing us with the first approximation of the posterior distribution: $g_0(\boldsymbol{\theta}) = m^{-1} \sum_{j=1}^{m} p(\boldsymbol{\theta} \mid \mathbf{y}_{\text{obs}}, \mathbf{y}_{\text{mis}}^{(0,j)})$. Then, one carries out the following iterations.

Data Augmentation (DA) Algorithm:

- For $t = 1, \ldots N$ (N large):

    For $j = 1, \ldots, m$:

    DA1. Draw l from the set $\{1, \ldots, m\}$ at random (uniformly)
    DA2. Draw a $\boldsymbol{\theta}^*$ from $p(\boldsymbol{\theta} \mid \mathbf{y}_{\text{obs}}, \mathbf{y}_{\text{mis}}^{(t-1,l)})$
    DA3. Draw $\mathbf{y}_{\text{mis}}^{(t,j)}$ from $p(\mathbf{y}_{\text{mis}} \mid \mathbf{y}_{\text{obs}}, \boldsymbol{\theta}^*)$
    End

- End

Steps DA1 and DA2 produce a sample of $\boldsymbol{\theta}$ from the mixture distribution

$$g_{t-1}(\boldsymbol{\theta}) = \frac{1}{m}\sum_{j=1}^{m} p(\boldsymbol{\theta} \mid \mathbf{y}_{\text{obs}}, \mathbf{y}_{\text{mis}}^{(t-1,j)}). \tag{6.4}$$

Thus, the random sample $\mathbf{y}_{\text{mis}}$ produced by Step DA3 follows its updated predictive distribution [i.e., (6.2) with $g(\ )$ substituted by $g_{t-1}(\ )$].

6.4.3 Connection with the Gibbs sampler

After a careful examination, it is observed that imputing multiple copies (m) of $\mathbf{y}_{\text{mis}}$ in each iteration is not really necessary. To illustrate this point, we look only at the first iteration. In order to produce a new imputation $\mathbf{y}_{\text{mis}}^{(1,j)}$, we need to first draw a mixture component (Step DA1), say, the one corresponding to $\mathbf{y}_{\text{mis}}^{(0,l)}$, at random, and then draw

$$\boldsymbol{\theta}^* \sim p(\boldsymbol{\theta} \mid \mathbf{y}_{\text{obs}}, \mathbf{y}_{\text{mis}}^{(0,l)}).$$

Effectively, we can treat $\mathbf{y}_{\text{mis}}^{(1,j)}$ as a "child" of $\mathbf{y}_{\text{mis}}^{(0,l)}$. Because of random sampling, some of the members in the zeroth generation (roughly $m/2.718$ when m is large) will have no children, which means that they will not contribute in anyway to the future approximation of the posterior distribution. In a sense, we can think that 37% of the zeroth-generation imputations are discarded (thus, wasted), completely at random. After a sufficient number of iterations, all the children will come from one ancestor (coalescence), implying that only one member in the zeroth generation contributes to the final approximation. Since the sampling of the mixture component (i.e., the parent) in producing $\boldsymbol{\theta}^*$ is completely at random, the remaining single ancestor bears no selection bias — purely by luck. Consequently, the DA procedure is mathematically equivalent to an algorithm in which one imputes only a single $\mathbf{y}_{\text{mis}}$ (i.e., $k = 1$) in each of its iterations — exactly a Gibbs sampler with two components. From now on, we will just call a two-component Gibbs sampler a *Data Augmentation Scheme*. This procedure can be illustrated more heuristically by the diagram in Figure 6.2:

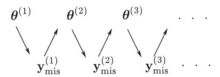

FIGURE 6.2. A graphical illustration of the data augmentation scheme.

An abstract formulation of the data augmentation approach can be summarized as follows. Suppose we are interested in simulating from a distribution $q(\boldsymbol{\theta})$. We can construct an "augmented" system $\pi(\boldsymbol{\theta}, \mathbf{y}_{\text{mis}})$ so that the marginal distribution of $\boldsymbol{\theta}$ under this system is $q(\boldsymbol{\theta})$ [i.e., $\int \pi(\boldsymbol{\theta}, \mathbf{y}_{\text{mis}}) d\mathbf{y}_{\text{mis}} = q(\boldsymbol{\theta})$]. If this augmented system is so nice that it facilitates iterative conditional sampling, then we can simulate from it by the Gibbs sampler and obtain all necessary information regarding $q(\boldsymbol{\theta})$.

6.4.4 An example: Hierarchical Bayes model

We have shown in Section 1.8 that a *hierarchical Bayes* model can be used to improve predictions of students' performances from their LSAT and undergraduate GPA scores (Rubin 1980). Here, we demonstrate by a simple example how the data augmentation scheme can be used to compute with a hierarchical Bayes model.

Efron and Morris (1975) applied an empirical Bayes method to the analysis of a dataset consisting of the first 45 at-bats in the middle of a season for $n = 18$ major league players (shown in column 2 of Table 6.1). They estimated the 18 "true" batting probabilities based on this dataset, and then used them as predictions of each person's batting average for the remainder of the season. Here, we apply a hierarchical Bayes model for the same task. Let Y_i denote the observed batting average (column 2 in the table) in the first 45 at bats of the ith person, and let p_i denote his true batting percentage. A variance-stabilizing transformation of Y_i was first performed in Efron and Morris (1975): We let

$$X_i = \sqrt{45} \arcsin(2Y_i - 1)$$

and let

$$\theta_i = \sqrt{45} \arcsin(2p_i - 1).$$

Then, the X_i can be regarded as Gaussian random variables with mean θ_i and variance 1. Furthermore, a hierarchical structure is assumed on all the θ_i such that $\theta_i \sim N(\mu, \sigma^2)$ independently. Furthermore, we assume that the prior distribution for μ and σ is uniform on $(-\infty, \infty) \times (0, \infty)$, thus improper. As an exercise, the reader may try out other priors, but note that the prior for σ cannot be singular at 0.

With the model and the prior, we can implement a data augmentation scheme as follows:

- Draw θ_i, $i = 1, \ldots, 18$, conditional on μ and σ^2.

- Draw μ and σ^2 conditional on all the values of θ_i.

Figure 6.3 displays the approximated posterior density of μ estimated by Gibbs sampling (almost indistinguishable from its true posterior) as well

Player	Batting average for first 45 at-bats	Batting average for remainder	Stein's estimator	Efron-Morris's estimator
1	.400	.346	.290	.334
2	.378	.298	.286	.313
3	.356	.276	.281	.292
4	.333	.222	.277	.277
5	.311	.273	.273	.273
6	.311	.270	.273	.273
7	.289	.263	.268	.268
8	.267	.210	.264	.264
9	.244	.269	.259	.259
10	.244	.230	.259	.259
11	.222	.264	.254	.254
12	.222	.256	.254	.254
13	.222	.303	.254	.254
14	.222	.264	.254	.254
15	.222	.226	.254	.254
16	.200	.285	.249	.249
17	.178	.316	.244	.233
18	.156	.200	.239	.208

TABLE 6.1. Batting averages and their estimates.

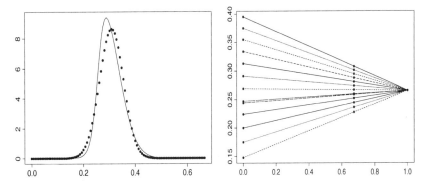

FIGURE 6.3. Left: Posterior density of μ — Gibbs sampling (solid) versus normal (dotted) approximation. Right: a graphical view of how shrinkage estimates are related to their respective MLE's.

as the shrinkage estimates.[1] In fact, this data augmentation procedure can be further modified to improve efficiency. See Liu (1994a), Gelfand, Sahu and Carlin (1995) and Liu and Sabatti (2000) for more discussions.

6.5 Finding Repetitive Motifs in Biological Sequences

In computational biology, it is often of interest to identify common patterns among a diverse class of protein or DNA sequences (Sections 1.5

---

[1] It is seen that the estimates of the p_i are pulled towards their common mean. In this sense, the estimate is "shrunk" toward a common point. The name "shrinkage estimator" is usually used more narrowly for estimators $\hat{p}_i$ that are shrunk towards zero.

and 4.1.3). These common patterns are usually called *"motifs"* in the literature. As illustrated in Figure 4.4, a total of K protein (or DNA) sequences, $\boldsymbol{R} = (R_1, \ldots, R_K)$, with lengths $\boldsymbol{l} = (l_1, \ldots, l_K)$ are given. They are believed to share a common motif as indicated by the blackened region; that is, every sequence in the dataset contains a subsequence of length w that is "similar" to each other. The locations of these subsequences and the motif pattern are unknown. Table 4.2 shows an alignment for a motif in multiple DNA sequences. This alignment matrix (if we know how to align the sequences) can be described by a *product-multinomial* model $\Theta = [\boldsymbol{\theta}_1, \ldots, \boldsymbol{\theta}_w]$ (Section 4.1.3). The base frequency in the background is described by an independent multinomial model $\boldsymbol{\theta}_0$.

By treating the motif locations, called the *alignment variable*, $A = \{a_1, \ldots, a_K\}$, as missing data, we can state our basic *block-motif model* more formally: Every residue/base in the dataset $\boldsymbol{R}$ outside the motif patterns (outside the blackened areas in Figure 4.4) are i.i.d. observations from Multinom($\boldsymbol{\theta}_0$) and the residue/base at position j of the motif pattern of each sequence (the blackened region) follows distribution Multinom($\boldsymbol{\theta}_j$).

6.5.1 A Gibbs sampler for detecting subtle motifs

In Section 4.1.3, we gave a uniform prior on the alignment variable A and independent Dirichlet($N_0\boldsymbol{\alpha}$) priors for all the $\boldsymbol{\theta}_j$, where $\boldsymbol{\alpha}$ is the base frequency in the genome (or other comparable database) and N_0 is the amount of pseudo-counts. Based on these settings, we can easily implement a data augmentation algorithm to compute the posterior distributions of A, $\boldsymbol{\theta}_0$, and Θ.

Data Augmentation Motif Sampler:

(a) Given a realization of the parameter values, $\boldsymbol{\theta}_0$ and Θ, we impute the alignment variable A.

(b) Given the current imputation of the alignment variable A; we sample a new realization of the parameters based on its complete-data posterior.

Step (a) can be easily achieved by "sliding" the pattern matrix Θ along each sequence and computing the relative probability of each position as the start of a motif; that is, we have

$$\pi(a_k = l \mid \Theta, \boldsymbol{\theta}_0, R_k) \propto \boldsymbol{\theta}_0^{h(R_k)} \prod_{j=1}^{w} \left(\frac{\boldsymbol{\theta}_j}{\boldsymbol{\theta}_0}\right)^{h(r_{k,j-1})} \propto \prod_{j=1}^{w} \left(\frac{\boldsymbol{\theta}_j}{\boldsymbol{\theta}_0}\right)^{h(r_{k,l+j-1})}.$$

In words, the probability of $a_k = l$ is proportional to the "signal-to-noise" ratio of the sequence segment $(r_{k,l}, \ldots, r_{k,l+j-1})$. Given A, the posterior distributions of $\boldsymbol{\theta}_0$ and Θ required by Step (b) are trivial to derive provided that their priors are standard Dirichlet or mixture Dirichlet distributions. With Dirichlet priors, these posteriors are, again, Dirichlet distributions.

It is, however, not difficult to see that given A, all the parameters can be analytically integrated out (Liu 1994a, Liu, Neuwald and Lawrence 1995), which results in a joint distribution on A:

$$\pi(A \mid R) \propto \pi(A, R) = \int \int \pi(R, A \mid \boldsymbol{\theta}_0, \boldsymbol{\Theta}) f(\boldsymbol{\theta}_0, \boldsymbol{\Theta}) d\boldsymbol{\theta}_0 d\boldsymbol{\Theta}.$$

This joint distribution can be used to derive a Gibbs sampling algorithm that focuses only on A. Although exact formulas for the conditional distributions required by the Gibbs sampler involve ratios of Gamma functions (Liu, Neuwald and Lawrence 1995), a very simple approximation, the *predictive update* form (Chen and Liu 1996), exists:

$$\pi(a_k = l \mid A_{[-k]}, R) \propto \prod_{j=1}^{w} \left(\frac{\hat{\boldsymbol{\theta}}_{j[-k]}}{\hat{\boldsymbol{\theta}}_{0[-k]}} \right)^{h(r_{k,l+j-1})}, \qquad (6.5)$$

where $\hat{\boldsymbol{\theta}}_{j[-k]}$ is the posterior mean of $\boldsymbol{\theta}_j$ conditioned on the observation R and the current alignment $A_{[-k]}$ (excluding the kth sequence) and $\hat{\boldsymbol{\theta}}_{0[-k]}$ is the posterior mean of $\boldsymbol{\theta}_0$ based on the current non-site positions $R_{\{A_{[-k]}\}^c}$. The formula implies that conditional on the fixed sites of the motif patterns in the remaining sequences, the probability that the motif pattern in sequence k starts at position l is proportional to the likelihood ratio of its being a motif site to its being a nonsite. Equation (6.5) forms the basis of the *site sampler* described in Section 1.5 (used in the second step for the predictive distribution). Both the exact formula and the approximation (6.5) were tested by Liu, Neuwald and Lawrence (1995) and no observable discrepancy between the two results were present.

6.5.2 Alignment and classification

It is often the case in biological applications that the sequences in consideration fall into two (or more) classes and each class has its own motif. To account for this complication, we introduce a *model variable* M: $M = 1$ stands for the one-class model and $M = 2$ for the two-class model, and assume $P(M=1) = P(M=2)$ *a priori*. One of our goals is to compute $P(M \mid R)$ (i.e., whether the data support the one-class model or prefer the two-class model). When $M=2$, we introduce the class indicator vector $C = (c_1, \ldots, c_K)$, with $c_i = 1$ or 2, where K is the total number of sequences. A uniform prior for C is used with the restriction that the minimal class size has to be 3. We let K_1 be the size for class one and $K_2 = K - K_1$ for class two. When $M=1$, we let $c_i=1$ for all i. Assuming that A is independent of M a priori, we have

$$P(R, C, A \mid M = 2)$$
$$= \frac{\prod_{j=1}^{w}[\Gamma(\boldsymbol{h}(r_{i,a_i+j-1}: c_i = 1) + \boldsymbol{\beta})\Gamma(\boldsymbol{h}(r_{i,a_i+j-1}: c_i = 2) + \boldsymbol{\beta})]}{\Gamma(K_1 + \|\boldsymbol{\beta}\|)^w \Gamma(K_2 + \|\boldsymbol{\beta}\|)^w}$$
$$\times \left[\frac{\Gamma(\|\boldsymbol{\beta}\|)}{\Gamma(\boldsymbol{\beta})}\right]^{2w+1} \frac{\Gamma(\boldsymbol{h}(R_{[-A]}) + \boldsymbol{\beta})}{\Gamma(\|\boldsymbol{l}\| - Kw + \|\boldsymbol{\beta}\|)} \frac{1}{[\#A]} \frac{1}{2^{K-1} - \frac{K^2+K}{2} - 1}.$$

Our sampling scheme consists of the following steps:

- **Align:** For a given C, we can use the predictive updating rule (Liu, Neuwald and Lawrence 1995) to update the alignment vector A. Namely, for $\forall\, i$, we update a_i based on $P(a_i \mid C, A_{-i}, R, M)$.

- **Fragment:** Let $A \pm 1 = \{a_1 \pm 1, \ldots, a_K \pm 1\}$; propose a move from A to $A - 1$ or $A + 1$ with equal probability and accept or reject the move based on the Metropolis ratio for $P(A \mid C, R, M)$. This can be seen as a step of *group move*.

- **Classify:** When $M = 2$, we update C by cycling through draws from $P(c_i \mid C_{-i}, A, R, M = 2)$, conditional on A.

- **Jump:** Conditional A, we jump between $M = 1$ and $M = 2$ based on the Metropolis ratio for $P(M, C \mid A, R)$. The proposal distribution from $M = 1$ to $(M = 2, C)$ is uniform on all allowable configuration of C. We use dynamic weighting to help the jump.

In the algorithm, a "cycle" consists of eight rounds of alignment iterations followed by one step of fragmentation and two rounds of classification iterations. The fragmentation step is a group move with the use of a translation group. This step greatly helps the convergence of alignment (Liu 1994a). After every cycle, a model jump step is conducted, with the help of a Q-type dynamic weighting (Section 5.7).

This algorithm was applied to the helix-turn-helix (HTH) dataset of Lawrence et al. (1993), which consists of 30 protein sequences with lengths ranging from 91 to 524. This set represents a large class of sequence-specific DNA binding structures involved in gene regulation. The correct locations of the motif in all the sequences were known from X-ray and nuclear magnetic resonance (NMR) structures or other experiments. The length of the motif was also determined as ~ 20. With $w=15$, our algorithm (with 2500 cycles) correctly identified all the motif locations. It provided a weighted estimate (after truncation of the weights at the 95th percentile; see Section 10.6 for details on the weight truncation method) of the posterior probability of $M=2$ as $\hat{p} < 0.001$.

We also applied the algorithm to another dataset consisting of the first 20 sequences in the HTH dataset and 10 new randomly shuffled sequences.

In each of the 10 random sequences, we inserted a conserved motif of length 15. The motif segment is produced from the pattern "ANHLPEQYTRGI-VAK," with each position having probability 0.3 to be randomly altered. For this new dataset, the weighted estimate (truncation of the weights at the 95th percentile) of the posterior probability of $M=2$ is 0.94, consistent with the simulation. Conditional on $M = 2$, the algorithm (with 5000 cycles) correctly classified the sequences and correctly identified the locations of all the conserved segments. Without using dynamic weighting, the sampler induces a virtually reducible Markov chain. Acceptance probability for the reversible jump between $M = 1$ and $M = 2$ is in the range of 10^{-10}.

6.6 Covariance Structures of the Gibbs Sampler

6.6.1 Data Augmentation

Suppose $\mathbf{x} = (x_1, x_2)$ and the target distribution is $\pi(\mathbf{x})$. As we have redefined in Section 6.4.3, *data augmentation* is equivalent to a two-component Gibbs sampler; that is, with an initial value $(x_1^{(0)}, x_2^{(0)})$, data augmentation iterates as follows:

- Draw $x_1^{(t+1)}$ from the conditional distribution $\pi_{1|2}(\cdot \mid x_2^{(t)})$.

- Draw $x_2^{(t+1)}$ from the conditional distribution $\pi_{2|1}(\cdot \mid x_1^{(t+1)})$.

This sampler has some nice theoretical properties. Under some regularity conditions, it can be shown that the sampler converges geometrically and monotonically (Liu et al. 1994, Liu, Wong and Kong 1995). The convergence rate of the sampler is equal to the *maximal correlation* between the two components, which is closely related to a statistical concept, the *faction of missing information* (Rubin 1987, Liu 1994b) in Bayesian missing data problems (see Chapter 12 for more details).

From the graphical illustration in Figure 6.2, we see that $x_1^{(0)}$ is conditionally independent of $x_1^{(1)}$, given $x_2^{(0)}$. Thus, we have the following theorem.

Theorem 6.6.1 *Suppose the Markov chain resulting from a data augmentation scheme is in stationarity . Then,*

$$\text{cov}\{h(x_1^{(0)}), h(x_1^{(1)})\} = \text{var}_\pi\{E_\pi\{h(x_1) \mid x_2\}\} \tag{6.6}$$

holds for any function h.

Proof: Without loss of generality, we assume that $E_\pi h(x_1) = 0$. Then,

$$\begin{aligned}
\text{cov}\{h(x_1^{(0)}), h(x_1^{(1)})\} &= E\{h(x_1^{(0)})h(x_1^{(1)})\} \\
&= E[E\{h(x_1^{(0)})h(x_1^{(1)}) \mid x_2^{(0)}\}] \\
&= E[E\{h(x_1^{(0)}) \mid x_2^{(0)}\} \cdot E\{h(x_1^{(1)}) \mid x_2^{(0)}\}] \\
&= E_\pi [E_\pi \{h(x_1) \mid x_2\}]^2 \\
&= \text{var}_\pi \{E_\pi \{h(x_1) \mid x_2\}\}.
\end{aligned}$$

The third equation follows from the conditional independence between $x_1^{(0)}$ and $x_1^{(1)}$, given $x_2^{(0)}$; the fourth equation follows from the fact that under the stationarity assumption, both $(x_1^{(0)}, x_2^{(0)})$ and $(x_1^{(1)}, x_2^{(0)})$ follow the target distribution π. $\diamond$

More generally, an explicit expression for lag-n autocovariances can be found:

$$\text{cov}[h(x_1^{(0)}), h(x_1^{(n)})] = \text{var}_\pi[E_\pi[\cdots E_\pi[E_\pi\{h(x_1)|x_2\}|x_1] \mid \cdots]], \quad (6.7)$$

$$\text{cov}[g(x_2^{(0)}), g(x_2^{(n)})] = \text{var}_\pi[E_\pi[\cdots E_\pi[E_\pi\{g(x_2)|x_1\} \mid x_2] \mid \cdots]], \quad (6.8)$$

where the right-hand sides of both (6.7) and (6.8) have n expectation signs conditioned alternately on x_1 and x_2. These identities show that for a two-component Gibbs sampler, the k-lag autocovariances are non-negative and monotone nonincreasing. A similar argument can be applied to show that a random-scan Gibbs sampler also has non-negative and monotone nonincreasing autocovariances.

6.6.2 Autocovariances for the random-scan Gibbs sampler

Suppose $\mathbf{x}$ has d components [i.e., $\mathbf{x} = (x_1, \ldots, x_d)$]. In each iteration step of the random-scan Gibbs sampler, we independently draw an index i according to a preassigned distribution $V = (\alpha_1, \ldots, \alpha_d)$ on the index set $I = \{1, \ldots, d\}$, then replace the value of the random variable x_i corresponding to that index by a new sample drawn from the conditional distribution $\pi\{x_i \mid \mathbf{x}_{[-i]}\}$. The distribution V need not be uniform, but we do require that $\alpha_i > 0$ for all i. One can easily show that π is invariant under the above transition. It is known that the Gibbs sampler with random scanning satisfies the detailed balance relation; thus, it generates a reversible Markov chain. Besides the non-negative even-lag autocovariances guaranteed by the reversibility of the chain, Liu, Wong and Kong (1995) showed that all the autocovariances must be non-negative and monotone decreasing. Furthermore, these autocovariances can be expressed as the variances of some iterative conditional expectations. To establish these properties, we first look at the lag-1 autocovariance.

Lemma 6.6.1 *Let $\mathbf{x}^{(0)}$ and $\mathbf{x}^{(1)}$ be two consecutive realizations of the random-scan Gibbs sampler under stationarity, and let $\mathbf{i}$ be the random variable representing which index is updated at stage one, taking values on $I = \{1, \ldots, d\}$ with distribution V. Then, for any $h(\ ) \in L_0^2(\pi)$,*

$$\mathrm{cov}\{h(\mathbf{x}^{(0)}), h(\mathbf{x}^{(1)})\} = E\left[\sum_{i=1}^{d} \alpha_i E^2\{h(\mathbf{x}) \mid \mathbf{x}_{[-i]}\}\right]$$
$$= E\left[E^2\{h(\mathbf{x}) \mid \mathbf{i}, \mathbf{x}_{[-\mathbf{i}]}\}\right] \geq 0.$$

Proof: From the definition of the scan, it is understood that

$$E\{h(\mathbf{x}^{(0)})h(\mathbf{x}^{(1)})\} = E[E[E\{h(\mathbf{x}^{(0)})h(\mathbf{x}^{(1)}) \mid \mathbf{i}, \mathbf{x}^{(0)}\} \mid \mathbf{x}^{(0)}]]$$
$$= \sum_{i=1}^{d} \alpha_i E[E\{\mathbf{x}^{(0)})h(\mathbf{x}^{(1)}) \mid \mathbf{i} = i, \mathbf{x}_{[-i]}^{(0)}\}]$$
$$= E\left[\sum_{i=1}^{d} \alpha_i E^2\{h(\mathbf{x}^{(1)}) \mid \mathbf{x}_{[-i]}^{(0)}\}\right]$$
$$= E\left[E^2\{h(\mathbf{x}) \mid \mathbf{i}, \mathbf{x}_{[-\mathbf{i}]}\}\right] \geq 0.$$

The second equality follows from our understanding of the random scan; the third equality is true because conditioned on a chosen updating index $\mathbf{i} = i$ and fixed values of the corresponding components, $\mathbf{x}^{(0)}$ and $\mathbf{x}^{(1)}$ are independent and identically distributed under stationarity. ◇

The lemma suggests setting α_i small when $E[E^2\{h(\mathbf{x}) \mid \mathbf{x}_{[-i]}\}]$ is large. Since $h(\mathbf{x})$ has mean zero, we also have

$$E[E^2\{h(\mathbf{x}) \mid \mathbf{x}_{[-i]}\}] = \mathrm{var}\{h(\mathbf{x})\} - E[\mathrm{var}\{h(\mathbf{x}) \mid \mathbf{x}_{[-i]}\}].$$

Hence, α_i should be set small if $E[\mathrm{var}\{h(\mathbf{x}) \mid \mathbf{x}_{[-i]}\}]$ is small, which can be understood as that one should make fewer visits to a component that is less variable.

Theorem 6.6.2 *Let $\mathbf{x}^{(0)}, \mathbf{x}^{(1)}, \ldots$, be consecutive samples generated by the random scan under stationarity, and let $\mathbf{i}$ be the random variable representing the random index in the updating scheme. For $h(\mathbf{x}) \in L_0^2(\pi)$, the lag-$n$ autocovariance between $h(\mathbf{x}^{(0)})$ and $h(\mathbf{x}^{(n)})$ is a non-negative monotone decreasing function of n. It can be written as*

$$\mathrm{cov}\{h(\mathbf{x}^{(0)}), h(\mathbf{x}^{(n)})\} = \mathrm{var}[E[\cdots E[E\{h(\mathbf{x}) \mid \mathbf{i}, \mathbf{x}_{[-\mathbf{i}]}\} \mid \mathbf{x}] \mid \cdots]], \quad (6.9)$$

where there are n conditional expectations taken alternately on $\{\mathbf{i}, \mathbf{x}^{[-\mathbf{i}]}\}$ and $\mathbf{x}$.

Proof: The expression is derived by repeatedly applying Lemma 6.6.1 and the Markov property. The monotonicity is a simple property of conditional expectations. ◇

In Chapter 12, we show how these expressions can be used to compare different sampling schemes and how the the *maximal correlation* among the d variables relates to the convergence rate of the scheme.

6.6.3 More efficient use of Monte Carlo samples

An interesting and immediate consequence of (6.7) and (6.8) is that Rao-Blackwellization *always* increases computational efficiency of Monte Carlo estimates. This problem can be formulated as follows: Suppose we are interested in estimating $I = E_\pi[h(x_1)]$ using the Monte Carlo samples obtained by data augmentation. Then, we have at least two possible estimators:

$$\hat{I} = \frac{1}{m}\left\{h(x_1^{(1)}) + \cdots + h(x_1^{(m)})\right\} \qquad (6.10)$$

$$\tilde{I} = \frac{1}{m}\left\{E[h(x_1) \mid x_2^{(1)}] + \cdots + E[h(x_1) \mid x_2^{(m)}]\right\}. \qquad (6.11)$$

The first estimator $\hat{I}$ is termed the *histogram estimator* and the second is called the *mixture estimator*. The name "mixture" stems from the fact that expression (6.11) is a mixture of complete-data posterior distributions (a natural choice of the kernel densities for a smooth estimate of the density curve). It has been pointed by Gelfand and Smith (1990) that the second estimation $\tilde{I}$ should be preferred — but their argument was based on the assumption that the Monte Carlo samples $\mathbf{x}^{(j)}$ are mutually independent (as explained in Section 3.4.6), which is clearly false in a Gibbs sampler. However, by using (6.7) and (6.8), we see that under stationarity,

$$m^2 \text{var}(\hat{I}) = m\sigma_0^2 + 2(m-1)\sigma_1^2 + \cdots + 2\sigma_{m-1}^2, \qquad (6.12)$$

where $\sigma_k^2 = \text{cov}[h(x_1^{(0)}), h(x_1^{(k)})]$; we have shown from (6.7) that this quantity is non-negative. By using the expression (6.8), we can derive that

$$m^2 \text{var}(\tilde{I}) = m\sigma_1^2 + 2(m-1)\sigma_2^2 + \cdots + 2\sigma_m^2. \qquad (6.13)$$

Comparing the two variances, we see that each term in (6.13) is exactly one lag behind the corresponding term in (6.12). Because of monotonicity of the autocovariances, we conclude that $\text{var}(\tilde{I}) \leq \text{var}(\hat{I})$.

6.7 Collapsing and Grouping in a Gibbs Sampler

Let $\mathbf{x} = (x_1, \ldots, x_d)$ be a random variable that can be partitioned into d components, with density $\pi(\mathbf{x})$. We consider a systematic-scan Gibbs sampler being applied to sample from this target distribution; that is, a

6.7 Collapsing and Grouping in a Gibbs Sampler

Markov chain $\{\mathbf{x}^{(t)} = (x_1^{(t)}, \ldots, x_d^{(t)}), t = 0, 1, \ldots\}$ is constructed with its transition function defined by the d-component Gibbs sampler,

$$K(\mathbf{x}^{(t)}, \mathbf{x}^{(t+1)}) = \prod_{l=1}^{d} \pi\{x_l^{(t+1)} \mid x_1^{(t+1)}, \ldots, x_{l-1}^{(t+1)}, x_{l+1}^{(t)}, \ldots, x_d^{(t)}\}. \quad (6.14)$$

It is easy to check that $\pi(\mathbf{x})$ is invariant under this transition.

Suppose the last two components x_{d-1} and x_d can be drawn together; then, we have a reduced Gibbs sampler on a new partition of the random variable $\mathbf{x}^* = \{x_1, \ldots, x'_{d-1}\}$, where $x'_{d-1} = \{x_{d-1}, x_d\}$, by *grouping*. Furthermore, suppose that the component x_d can be integrated out, then an even more reduced sampler on $\mathbf{x}^- = \{x_1, \ldots, x_{d-1}\}$, with its marginal density $\pi(\mathbf{x}^-) = \int \pi(\mathbf{x}) \, dx_d$, results from *collapsing*. We will compare the three schemes.

In order to argue rigorously, we introduce some concepts concerning a Markov chain and its associated function spaces. Let $L^2(\pi)$ denote the set of all functions $h(\ )$ that are square integrable with respect to π (i.e., has a finite variance). This set is a *Hilbert space* with an inner product defined by $\langle h, g \rangle = E_\pi\{h(\mathbf{x})g(\mathbf{x})\}$. Thus, $\|h\| = \mathrm{var}_\pi(h)$. Let $\mathbf{x}^{(0)}, \mathbf{x}^{(1)}, \ldots$ be a general state-space Markov chain with transition function $K(\mathbf{x}, \mathbf{y}) = P(\mathbf{x}^{(1)} = \mathbf{y} \mid \mathbf{x}^{(0)} = \mathbf{x})$. We define the *forward* operator $\mathbf{F}$ on $L^2(\pi)$ for the Markov chain as

$$\mathbf{F}h(\mathbf{x}) = \int K(\mathbf{x}, \mathbf{y})h(\mathbf{y})d\mathbf{y} = E\{h(\mathbf{x}^{(1)}) \mid \mathbf{x}^{(0)} = \mathbf{x}\}.$$

We observe immediately that the *norm* of the operator is at most 1, where the norm is defined as $\|\mathbf{F}\| = \sup_h \|\mathbf{F}h(\mathbf{x})\|$ with the supremum taken over all functions with $E(h^2) = 1$. On the other hand, since the constant function c is an eigenfunction of the operator corresponding to eigenvalue 1, we know that the norm of $\mathbf{F}$ is exactly 1. When the chain is *reversible* [i.e., the *detailed balance* condition $\pi(\mathbf{x})K(\mathbf{x}, \mathbf{y}) = \pi(\mathbf{y})K(\mathbf{y}, \mathbf{x})$ is satisfied], $\mathbf{F}$ is a self-adjoint operator. When F is compact and self-adjoint, (which is true when the state space is finite and the chain is reversible), the second largest eigenvalue (in absolute value) of F characterizes the mixing rate, or convergence rate, of the Markov chain. Many methods are available for bounding the second largest eigenvalue and finding the actual rate of convergence for this case (Diaconis 1988, Diaconis and Stroock 1991). Methods for dealing with nonreversible chain also exist, although rare (Fill 1991); see Section 12.6 for more details.

Now, we consider $L_0^2(\pi) = \{h(\mathbf{x}) \in L^2(\pi) : E\{h(\mathbf{x})\} = 0\}$, which is a subspace of $L^2(\pi)$ of all mean zero functions. Clearly, this is again a Hilbert space with the same inner product and is invariant under the operator $\mathbf{F}$. We use $\mathbf{F_0}$, called the *forward operator*, to denote the operator on $L_0^2(\pi)$ induced by $\mathbf{F}$. Then, the largest eigenvalue of $\mathbf{F_0}$ is exactly the same as the second largest eigenvalue of $\mathbf{F}$. Typically, the spectral radius of $\mathbf{F_0}$

characterizes the rate of convergence of the Markov chain in both reversible and nonreversible cases. When the chain is reversible, the spectral radius of $\mathbf{F_0}$ is the same as its norm. A general relationship between the norm and the spectral radius of an operator is

$$\lim_{n\to\infty} \|\mathbf{F_0}^n\|^{1/n} = r,$$

where r is the spectral radius. This suggests that one can compare different Markov chains by comparing the norms of the corresponding forward operators. It is interesting to note here that $\|\mathbf{F_0}\|^2$ equals the second largest eigenvalue of the transition operator for the reversiblized chain, which ties in with the method of Fill (1991).

Let $\mathbf{F_s}$ denote the forward operator for the standard Gibbs sampler, corresponding to the transition function (6.14); let $\mathbf{F_g}$ be the forward operator corresponding to the grouping procedure and let $\mathbf{F_c}$ for the collapsed Gibbs sampler with x_d integrated out. The three samplers can be illustrated by the diagrams of their respective visiting schemes:

$$\begin{aligned}
\mathbf{F_s}: &\quad x_1 \to x_2 \to \cdots \to x_d; \\
\mathbf{F_g}: &\quad x_1 \to x_2 \to \cdots \to \{x_{d-1}, x_d\}; \\
\mathbf{F_c}: &\quad x_1 \to x_2 \to \cdots \to x_{d-1}.
\end{aligned} \quad (6.15)$$

Theorem 6.7.1 (Three-schemes theorem) *The norms of the three forward operators are ordered as*

$$\|\mathbf{F_c}\| \leq \|\mathbf{F_g}\| \leq \|\mathbf{F_s}\|.$$

A similar result for the random-scan Gibbs sampler is proved in Section 13.2.2. This theorem can be understood from the the diagrams in Figures 6.4 and 6.5. Consider simulating a three-component random variable $\mathbf{x} = (x_1, x_2, x_3)$ from $\pi(\mathbf{x})$. The deterministic-scan Gibbs sampler is depicted in Figure 6.4 and the samplers resulting from *grouping* and *collapsing* are shown in Figure 6.5.

To illustrate the foregoing theorem, we compared the regular data augmentation and the collapsing approach for the bivariate Gaussian problem with Murray's data (Section 4.4.1). Because the posterior distribution of the unknown covariance matrix Σ is "easy" when given completed data and, given Σ, imputing the missing data is easy, we can implement a standard data augmentation algorithm by iterating between

$$[\Sigma \mid \mathbf{y}_{\text{obs}}, \mathbf{y}_{\text{mis}}]$$

and

$$[\mathbf{y}_{\text{mis}} \mid \mathbf{y}_{\text{obs}}, \Sigma].$$

6.7 Collapsing and Grouping in a Gibbs Sampler   149

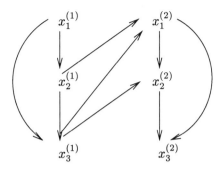

FIGURE 6.4. A graphical illustration of the standard Gibbs sampler with three components. The arrows represents the "causal relationships" in Gibbs sampling: $x_1^{(2)}$ is drawn conditional on $(x_2^{(1)}, x_3^{(1)})$; $x_2^{(2)}$ is drawn conditional on $(x_3^{(1)}, x_1^{(2)})$; and $x_3^{(2)}$ is drawn conditional on $(x_1^{(2)}, x_2^{(2)})$

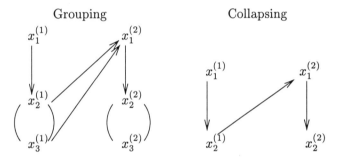

FIGURE 6.5. Graphical representations of grouping and collapsing schemes. The arrows represent the "causal relationships" in Gibbs sampling.

In a collapsing scheme, we can integrate out Σ and iterate only among the missing values; that is, we can iterate the step

$$[y_{\text{mis},i} \mid \mathbf{y}_{\text{obs}}, \mathbf{y}_{\text{mis},[-i]}],$$

which is a noncentral t-distribution whose accurate form can be found in Kong et al. (1994, Section 3.1). More precisely, conditional on the current imputed values for all the missing components $y_{\text{mis},j}$, $j \neq i$, we can easily update $y_{\text{mis},i}$.

To compare the collapsed and the standard schemes, we compute autocovariance curves for each of the eight missing components. Figure 6.6 contains two groups of autocovariance curves; within each group, there are eight curves for eight missing values, respectively. They are estimated from 20,000 iterations for each chain. Since both chains are geometric mixing, we fit the model $\text{auto}(n) = C\rho^n + \epsilon$ to the autocovariances for the two bundles, respectively, where $\text{auto}(n)$ denotes the lag-n autocovariance. It

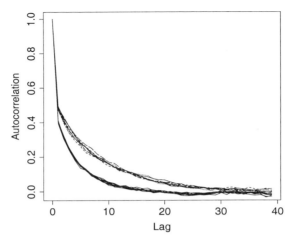

FIGURE 6.6. Autocovariance plot for both the standard and the collapsed Gibbs sampling scheme. Upper group: the standard; lower group: the collapsed.

is seen that the $\hat{\rho}$ estimated from the standard scheme is about 2.5 times larger than that of the collapsed scheme. The ordinary DA has an intuitive appeal for "decoupling" the dependency among the missing data. But our results showed that the collapsing scheme is significantly better in terms of both the convergence rate and the cost per iteration.

As another illustration, we consider a simple Gibbs sampling algorithm for the nonparametric Bayes problem in Section 4.4.2, where we describe a sequential imputation method for the posterior computation. The algorithm described here and its various improvements can be found in Escobar (1994), Liu (1996b), and MacEachern (1994). Recall that we observe $y_i \sim \text{Binom}(l_i, \zeta_i)$, and assume that $\zeta_i \stackrel{\text{i.i.d.}}{\sim} F$ and $F \sim \mathcal{D}(\alpha)$, where $\mathcal{D}(\alpha)$ represents a Dirichlet process. By the Bayes theorem, we obtain the predictive distribution of ζ_i [equivalent to (4.11)]:

$$\pi(\zeta_i \mid y_i, \zeta_{[-i]}) \propto \zeta_i^{y_i}(1-\zeta_i)^{l_i-y_i}\alpha(\zeta_i) + \sum_{j \neq i} \zeta_j^{y_j}(1-\zeta_j)^{l_j-y_j}\delta_{\zeta_j}(\zeta_i).$$

Hence, a collapsed Gibbs sampler (with F collapsed down) can be applied to iteratively sample ζ_i using the foregoing predictive distribution.

Moreover, the Gibbs sampling algorithm for finding repetitive motifs in biological sequences as illustrated by (6.5) is also an application of the collapsing theorem in which the parameters θ_0 and Θ are integrated out from the model (Liu 1994a).

The collapsing theorem suggests that one should avoid introducing unnecessary components into a Gibbs sampler. This is in agreement with the common wisdom for Monte Carlo computation: Do as much analytical work as possible. However, in the next chapter, we show that in the Ising model one can greatly improve computational efficiency by introducing a clever auxiliary variable.

6.8 Problems

1. Implement a Gibbs sampler for simulating from a Ising model defined on a 32×32 grid.

2. Write down the transition matrix of a random-scan Gibbs sampler as defined in Section 6.1. Show that the resulting Markov chain is reversible.

3. Write down the transition matrix of a systematic-scan Gibbs sampler as defined in Section 6.1. Show that the resulting Markov chain is nonreversible.

4. Evaluate the efficiency gain in using the mixture estimate versus the histogram estimate for the bivariate Gaussian example and the hierarchical Bayes example discussed in Section 6.4.4.

5. Why is the random-grid Monte Carlo method described in Section 5.5 a sensible alternative to the hit-and-run algorithm?

6. Explain why the original data augmentation algorithm of Tanner and Wong (1987) is practically equivalent to a Gibbs sampler with two components.

7
Cluster Algorithms for the Ising Model

7.1 Ising and Potts Model Revisit

In Section 1.3, we introduced the Ising model, which is used by physicists to model the magnetization phenomenon and has been studied extensively in statistical physics literature. A closely related model is the *Potts model*. Similar to the Ising model, the Potts model is also defined on the lattice space $\mathcal{L}$ with configurations $\mathbf{x} = (x_l,\ l \in \mathcal{L})$, where

$$\mathcal{L} = \{l = (l_1, l_2) \text{ for } l_1, l_2 = 1, \ldots, N\}.$$

Different from the Ising model, each x_l in the Potts model can take values in an alphabet of size q: $\{1, \ldots, q\}$. The potential energy function for the Potts model can be written as

$$H(\mathbf{x}) = -J \sum_{i \sim j} \delta_{x_i x_j} - \sum_j h_j(x_j),$$

where δ_{ab} is the Kronecker δ-function, equaling 1 when $a = b$ and 0 otherwise. The symbol $i \sim j$ indicates that i and j are neighbors in the lattice space. The function $h_j(\ )$ represents an outside magnet field. The distribution of interest is then the Boltzmann distribution

$$\pi(\mathbf{x}) \propto \exp\{-\beta H(\mathbf{x})\}.$$

Here, $\beta = (kT)^{-1}$ with the Boltzmann constant k and the absolute temperature T. When $q = 2$, the Potts model is equivalent to the Ising model

154   7. Cluster Algorithms for the Ising Model

in that

$$H(\mathbf{x}) = -\frac{1}{2}J\sum_{i\sim j} 2\left(\delta_{x_i x_j} - \frac{1}{2}\right) - \sum_{i\sim j}\frac{1}{2}J - \sum_{j} h_j x_j.$$

For brevity of presentation, in this section we focus on the simulation of the Ising model under the setting $h_j \equiv 0$, which is supposedly the most interesting setting (Newman and Barkema 1999).

An important problem to physicists is the *phase transition phenomena* for such a model; that is, it is observed that when the temperature is high, all the spins behave nearly independently (no long-range correlation), whereas when temperature is below a *critical temperature* c_0, all the spins tend to stay the same (i.e., cooperative performance).

The standard Metropolis algorithm can be easily applied to simulate from this model: First, we randomly pick a spin and turn its value x_l to the opposite $-x_l$; then, we compute the Metropolis ratio to decide whether to accept such a move. However, this single-site update algorithm slows down rapidly once the temperature is approaching or below the critical value c_0, the so-called "critical slowing down." Swendsen and Wang (1987) introduce a powerful clustering algorithm which, together with an implementation modification of Wolff (1989), almost completely eliminates the critical slowing down.

7.2   The Swendsen-Wang Algorithm as Data Augmentation

Conceptually, we can think of the Swendsen-Wang algorithm as a data augmentation scheme (Edwards and Sokal 1988, Higdon 1998, Tanner and Wong 1987). To be precise, we consider augmenting the space of spins by a "bond variable" $\boldsymbol{u} = (u_{l\sim l'})$ with its component variable $u_{l,l'}$ sitting on every edge of the lattice and taking values in $[0, e^{2\beta J}]$:

$$\pi(\mathbf{x}) \propto \exp\left\{\beta J \sum_{l\sim l'} x_l x_{l'}\right\}$$

$$\propto \prod_{l\sim l'} \exp\{\beta J(1 + x_l x_{l'})\}$$

Note that $1 + x_l x_{l'}$ is equal to either 0 or 2. Hence, if we introduce an auxiliary variable $\boldsymbol{u}$ such that

$$\pi(\mathbf{x}, \boldsymbol{u}) \propto \prod_{l\sim l'} I[0 \leq u_{l,l'} \leq \exp\{\beta J(1 + x_l x_{l'})\}],$$

then the marginal distribution of $\mathbf{x}$ is the desirable distribution. Clearly, under this joint distribution, the conditional distribution $[\boldsymbol{u} \mid \mathbf{x}]$ is a product of uniform distributions with ranges depending on two neighboring spins. Conversely, the conditional distribution of **x** given **u** is also easy to figure out: If $u_{l,l'} > 1$, then x_l must be equal to $x_{l'}$; otherwise there is no constraint on x_l's. Thus, **u** affects **x** only through the event $I[u_{l,l'} > 1]$. Based on the configuration of **u**, we "cluster" those lattice sites according to whether they have a "mutual bond" [i.e., whether $u_{l,l'} > 1$]. Then, all the x_l whose site l belongs to a common cluster should take identical value. Conditional on the clusters, every configuration that does not violate the cluster homogeneity is equally likely.

Therefore, we can produce another augmented model that only uses an auxiliary *bonding* variables, $\boldsymbol{b} = (b_{l \sim l'})$, to indicate whether $u(l, l') > 1$ holds. More precisely, we define

$$b_{l,l'} = I[u_{l,l'} > 1].$$

Then, the corresponding augmented model is

$$\pi(\mathbf{x}, \boldsymbol{b}) \propto \prod_{x_l = x_{l'}} \left\{ 1 + b_{l,l'}(e^{2\beta J} - 1) \right\},$$

and $b_{l,l'} = 0$ whenever $x_l \neq x_{l'}$. The clustering of the spins can be achieved by connecting all those neighboring sites whose bond value is 1. Conditional on the realization of $\boldsymbol{b}$, the spin value of one cluster is independent of those of other clusters. The algorithm of Swendsen and Wang (1987) is then a data augmentation scheme that iterates between sampling from $\pi(\boldsymbol{b} \mid \mathbf{x})$ and $\pi(\mathbf{x} \mid \boldsymbol{b})$.

Swendsen-Wang (SW) Algorithm

- For a given configuration of the spins, form the bond variable by giving every edge of the lattice, $\langle l, l' \rangle$, between two "like spins" (i.e., $x_l = x_{l'}$) a bond value of 1 (i.e., $b_{l,l'} = 1$) with probability $e^{-2\beta J}$, and a bond value of 0 otherwise.

- Conditional on the bond variable $\boldsymbol{b}$, update the spin variable **x** by drawing from $p(\mathbf{x} \mid \boldsymbol{b})$, which is uniform on all compatible spin configurations; that is, clusters are produced by connecting neighboring sites with a bond value 1. Every cluster is then flipped with probability 0.5.

7.3 Convergence Analysis and Generalization

Based on Theorem 6.6 and theory in 13, the convergence rate of this algorithm is characterized by the *maximal correlation* between **x** and $\boldsymbol{b}$ (Liu et al. 1994, Liu, Wong and Kong 1995), which is defined as

$$\rho = \sup_{h \in L^2} \frac{\text{var}[E\{h(\boldsymbol{b}) \mid \mathbf{x}\}]}{\text{var}\{h(\boldsymbol{b})\}} \quad (7.1)$$

under the equilibrium distribution. Thus, we can obtain a lower bound of the algorithm's convergence rate by computing $\mathrm{var}[E\{h(b) \mid \mathbf{x}\}]$ for a particular test function. As in Li and Sokal (1989), we can choose $h(b) = \sum b_{l,l'}$. Then, by denoting $p_a = 1 - e^{-2\beta J}$, it is easy to see that

$$E[h(b) \mid \mathbf{x}] = p_a \sum_{l\ l'} \delta_{x_l = x_{l'}},$$

$$\mathrm{var}[h(b) \mid \mathbf{x}] = p_a(1 - p_a) \sum_{l\ l'} \delta_{x_l = x_{l'}}.$$

If we let $U(\mathbf{x}) = \sum_{l\ l'} \delta_{x_l = x_{l'}}$, then $E[U(\mathbf{x})]$ is related to the mean energy and $\mathrm{var}[U(\mathbf{x})]$ is proportional to the specific heat. From definition (7.1) and the theory in Section 13.2.1, we have

$$\lambda_2 = \rho \geq \frac{\mathrm{var}\{E[h(b) \mid \mathbf{x}]\}}{\mathrm{var}[h(b)]} = \frac{p_a^2 \mathrm{var}(U)}{p_a^2 \mathrm{var}(U) + p_a(1 - p_a)E(U)},$$

where λ_2 is the second largest eigenvalue of the transition operator of this data augmentation chain and ρ is as defined in (7.1). This is the key inequality for Li and Sokal (1989) to derive an approximate bound on the critical exponents of the SW algorithm.

A more general formulation of the SW algorithm is given by Edwards and Sokal (1988) and described also in Higdon (1998). In particular, Higdon (1998) successfully applied this approach to tackle a class of image analysis problems. Suppose the target distribution has a form

$$\pi(\mathbf{x}) \propto \pi_0(\mathbf{x}) \prod_k f_k(\mathbf{x}).$$

One can introduce an augmented model with $\mathbf{u} = (u_k)$ so that

$$\pi(\mathbf{x}, \mathbf{u}) \propto \pi_0(\mathbf{x}) I[0 \leq u_k \leq f_k(\mathbf{x})]. \tag{7.2}$$

Then, a data augmentation scheme can be formally implemented by iterating (a) draw $\mathbf{u} \sim [\mathbf{u} \mid \mathbf{x}]$ and (b) draw $\mathbf{x} \sim [\mathbf{x} \mid \mathbf{u}]$. Although step (a) is trivial, step (b) may not be achievable in problems other than the Ising or Potts models. This strategy is also referred to as the *slice sampler* (Section 6.3.1).

A *partial decoupling* method was also given by Higdon (1998) in which he replaces the expression (7.2) by

$$\pi(\mathbf{x}, \mathbf{u}) \propto \pi_0(\mathbf{x}) \prod_k f_k(x)^{1-\delta_k} I[0 \leq u_k \leq f_k(\mathbf{x})^{\delta_k}].$$

One can iterate $[\mathbf{u} \mid \mathbf{x}]$ and $[\mathbf{x} \mid \mathbf{u}]$ as in the previous case. The partial decoupling method is potentially useful when one does not have the nice symmetry as in the Ising or Potts models, which is typically the case in statistical image analysis (a likelihood term will destroy symmetry in the model).

7.4 The Modification by Wolff

Wolff (1989) introduced a modification for the Swendsen-Wang algorithm, which, although both conceptually and operationally simple, significantly outperforms the SW algorithm.

Wolff's Algorithm

- For a given configuration $\mathbf{x}$, one randomly picks a site, say x_l, and grow recursively from it a "bonded set" C as follows:
  - Check all the *unchecked* neighboring sites of a current set $C^{(\text{old})}$; add a bond between a neighboring site and $C^{(\text{old})}$ the same way as in the Swendsen-Wang algorithm.
  - Add those newly bonded neighboring sites to $C^{(\text{old})}$ so as to form a new set $C^{(\text{new})}$.
  - Stop the recursion when there is no unchecked neighbor to add; name the final set C.
- Flip all the spins corresponding to the sites in set C to their opposites (no random sampling here).

The only difference between this algorithm and the SW algorithm is that in each iteration, only *one* cluster is constructed and *all* spins in that cluster are changed to their opposite value. However, the algorithm provides a new insight that is different from the one based on data augmentation. Consider the "move" in the Wolff algorithm from a Metropolis algorithm viewpoint. Suppose the cluster C we have grown has $m + n$ neighboring "links" among which m are linked with $+1$ spins and n with -1 spins. Thus, if the current state of C is all $+1$, by flipping the whole cluster of C to -1, the probability ratio (of the new to old) is $e^{2\beta J(n-m)}$. Now, consider the process of building C (i.e., the proposal). This proposal, in comparison with the reverse proposal, can be viewed as "breaking bonds" along the edge of the cluster C. Since C starts with all $+1$, the probability of breaking m bonds is $e^{-2\beta Jm}$. To propose back, one needs to break n bonds which has a probability $e^{-2\beta Jn}$. Thus, the ratio of the proposal transitions is

$$\frac{T(\mathbf{x}^{(\text{new})} \to \mathbf{x}^{(\text{old})})}{T(\mathbf{x}^{(\text{old})} \to \mathbf{x}^{(\text{new})})} = \frac{\exp\{-2\beta Jn\}}{\exp\{-2\beta Jm\}} = \exp\{2\beta J(m-n)\}.$$

This ratio cancels the probability ratio and, thus, the proposed change is accepted with probability one.

7.5 Further Generalization

There is no essential reason for restricting the growth of the cluster to be among those "like spins." More generally, Niedermayer (1988) suggests that one can allow "bonds" to link neighboring spins of opposite values. Let p_a be the probability of growing a bond between l and l' when $x_l = x_{l'}$, and let p_b be that when $x_l \neq x_{l'}$. After growing the cluster C, we can flip every spin in C to its opposite. First, it is not difficult to see that such a "transition rule" is completely symmetric for the *interior* or the cluster; that is, the energy difference between the two states, $\mathbf{x}_{\text{old}} = (\mathbf{x}_C, \mathbf{x}_{[-C]})$ before the move and $\mathbf{x}_{\text{new}} = (-\mathbf{x}_C, \mathbf{x}_{[-C]})$ after the move, is equal to the energy difference of the two states at the boundary. Now, suppose there are m same-spin links and n different-spin links between C and its complement $\bar{C}$. Then, the transition ratio is

$$\frac{T(\mathbf{x}^{(\text{new})} \to \mathbf{x}^{(\text{old})})}{T(bx^{(\text{old})} \to \mathbf{x}^{(\text{new})})} = \frac{(1-p_a)^m(1-p_b)^n}{(1-p_a)^n(1-p_b)^m} = \left(\frac{1-p_a}{1-p_b}\right)^{m-n}.$$

Therefore, we can choose p_a and p_b so as to cancel the transition ratio with the probability ratio, $e^{2\beta J(n-m)}$, achieving a no-rejection transition in the Metropolis algorithm framework. It is of interest to see if this type of cluster-growth method can be used more generally in statistical computation. One area that might benefit from the method is the statistical image analysis, as shown in Higdon (1998).

7.6 Discussion

The auxiliary variable approach discussed in this chapter seems to be at odds with the theory presented in Section 6.7. This "apparent" conflict seems to suggest that adding "decoupling" variables does not necessarily help in improving convergence rate of the sampler unless the system possesses a special symmetry structure. For example, in the bivariate Gaussian inference problem (Sections 2.2 and 6.7), parameter Σ serves as a "decoupling" variable; that is, given Σ, all the missing components are mutually independent. On the other hand, all the missing components are dependent of each other if we integrate out Σ, as suggested by the collapsing theorem of Section 6.7. Our numerical results clearly showed that this "decoupling" really did not improve convergence rate and is more time-consuming for each iteration. A similar phenomenon was also observed for the Gibbs motif finding algorithm (Section 6.5).

Generally, the Gibbs sampler itself does not specify exactly how the random variable should be augmented or partitioned. This is a decision that users have to make and is where they can apply their ingenuity. There are two conflicting criteria that a good Gibbs sampler algorithm has to meet: (a) Drawing one component conditional on the others is computationally simple; (b) the Markov chain induced by the Gibbs sampler with such partitioning components has to converge reasonably fast to its equilibrium distribution. For example, drawing the variables jointly with no partitioning at all is optimal for convergence, but it is formidable and is the reason why the Gibbs sampler was invented. The above theorem provides a theoretical confirmation of such a conflict. It seems to be a reasonable strategy to "group" or "collapse" when it is computationally feasible. But as a whole, it is left to the reader to make compromises to balance all factors mentioned.

7.7 Problems

1. Implement the SW algorithm for simulating a 64 × 64 Ising model near the critical temperature.

2. Implement Wolff's algorithm for the above task.

3. Implement the Niedermayer's generalization for the above.

4. Experiment with different choices of p_a and p_b in the Niedermayer's algorithm. Can you find a pair of p_a and p_b so that the resulting algorithm outperforms Wolff's algorithm?

5. Is it possible to generalize Wolff's or Niedermayer's algorithms along the line of Li and Sokal (1989) described in Section 7.3?

8
General Conditional Sampling

The fundamental idea underlying all Markov chain Monte Carlo algorithms is the construction of implementable Markov transition rules that leave the target distribution $\pi(\mathbf{x})$ invariant. Although the Metropolis-Hastings algorithm for constructing a desirable Markov chain is very simple and powerful, a potential problem with the Metropolis algorithm, as explained in the previous chapter, is that the proposal function is often chosen out of convenience and is somewhat too "arbitrary." In contrast, the Markov transition rules of the Gibbs sampler are built upon conditional distributions *derived from* the target distribution $\pi(\mathbf{x})$. In this chapter, we describe a more general form of the conditional sampling, *partial resampling*, introduced in Goodman and Sokal (1989) and generalized in Liu and Sabatti (2000).

8.1 Partial Resampling

The basic steps of a Gibbs sampler are (i) decomposing the random vector of interest $\mathbf{x}$ into two components [e.g., $\mathbf{x} = (x_1, \mathbf{x}_{[-1]})$] and (ii) updating one component, x_1, by a new sample x_1' drawn from $\pi(\cdot \mid \mathbf{x}_{[-1]})$. More generally, any transition rule $A(x_1 \to x_1^*)$ that leaves the conditional distribution $\pi(x_1 \mid \mathbf{x}_{[-1]})$ invariant also leaves $\pi(\mathbf{x})$ invariant. Thus, a proper move can be made by drawing a new x_1^* from the transition function $A(x_1 \to \cdot)$ and then updating $\mathbf{x}$ to $\mathbf{x}^* = (x_1^*, \mathbf{x}_{[-1]})$. Additionally, this transition function A can actually *depend* on the current value of $\mathbf{x}_{[-1]}$, in that the invariance

162   8. General Conditional Sampling

is guaranteed by

$$\int \pi(x_1, \mathbf{x}_{[-1]}) A(x_1 \to x_1^* \mid \mathbf{x}_{[-1]}) dx_1$$
$$= \pi(\mathbf{x}_{[-1]}) \int \pi(x_1 \mid \mathbf{x}_{[-1]}) A(x_1 \to x_1^* \mid \mathbf{x}_{[-1]}) dx_1$$
$$= \pi(\mathbf{x}_{[-1]}) \pi(x_1^* \mid \mathbf{x}_{[-1]}) = \pi(\mathbf{x}^*).$$

Although the iterative conditional sampling approach used by the Gibbs sampler is effective in many cases, it can be adversely affected by the parameterization (or, decomposition) of the space. A reparameterization (Gelfand et al. 1995) can sometimes improve the performance of the algorithm significantly. Although a proper decomposition of the sample space (or the random vector) should not be dependent upon the coordinate system employed by the problem, in practice one is confined by the coordinate system imposed on the problem simply because it appears easier to get appropriate conditional distributions.

In general, suppose we can find a partition of the sample space $\mathcal{X}$; that is, we have $\mathcal{X} = \bigcup_{\alpha \in A} \mathcal{X}_\alpha$ and $\mathcal{X}_\alpha \cap \mathcal{X}_\beta = \emptyset$. Then, the corresponding decomposition of π can be constructed:

$$\pi(\mathbf{x}) = \int \nu_\alpha(\mathbf{x}) d\rho(\alpha),$$

where $\nu_\alpha(\mathbf{x})$ is the conditional distribution of $\mathbf{x}$ given that it lies on $\mathcal{X}_\alpha$. Here each $\mathcal{X}_\alpha$ is called a *fiber*. It is seen that any transition of the form $A(\mathbf{x} \to \mathbf{x}')$ *on the fiber* that leaves ν_α invariant also leaves π invariant (Goodman and Sokal 1989, Liu and Sabatti 2000). However, determining the explicit form of ν_α can be a difficult task when an arbitrary fiber construction is given.

To illustrate the idea of partial resampling, we give three "fiber" constructions and the corresponding ν_α for π defined on $\mathcal{X} = \mathbb{R}^2$ with $\mathbf{x} = (x_1, x_2)$.

(i) For $\alpha \in \mathbb{R}^1$, we define the fiber as $\mathcal{X}_\alpha = \{\mathbf{x} : x_1 = \alpha\}$ and, correspondingly,

$$\nu_\alpha(\mathbf{x}) = \begin{cases} \pi(x_2 \mid x_1) & \text{if } x_1 = \alpha \\ 0 & \text{otherwise.} \end{cases}$$

This fiber construction allows one to move along the x_2 direction.

(ii) The fiber is defined as $\mathcal{X}_\alpha = \{\mathbf{x} : x_2 = x_1 + \alpha\}$ and, correspondingly,

$$\nu_\alpha(\mathbf{x}) \propto \begin{cases} \pi(x_1, x_2) & \text{if } x_2 - x_1 = \alpha \\ 0 & \text{otherwise.} \end{cases}$$

This fiber construction allows for a conditional move along the line $x_2 - x_1 = \alpha$.

(iii) For $\alpha \in \mathbb{R}^1 \setminus \{0\}$, we can define $\mathcal{X}_\alpha = \{\mathbf{x} : x_2 = \alpha x_1\}$. By Theorem 8.3.1, the corresponding ν_α is

$$\nu_\alpha(\mathbf{x}) \propto \begin{cases} |x_1| \pi(x_1, \alpha x_1) & \text{if } x_2 = \alpha x_1 \\ 0 & \text{otherwise.} \end{cases}$$

If conditional moves are performed along the fibers, construction (i) results in a standard Gibbs sampling step along the x_2 axis, whereas the latter two constructions correspond to certain reparameterizations of the space. Clearly, different fiber constructions imply different conditional moves for the corresponding Monte Carlo sampler.

The benefits of *partial resampling* are as follows: (a) Similar to a Metropolis move, the conditional move reduces a high-dimensional simulation problem to a series of lower-dimensional ones; (b) it allows the sampler to follow the local dynamics of the target distribution; and (c) it enables "non-axis" moves that cannot be achieved by the Gibbs sampler. Difficulties in applying *partial resampling* in Monte Carlo simulation, however, are also nontrivial. First, there is no clear guidance on how to construct a fiber decomposition of the space. Second, even with a given fiber construction, there is no easy way of obtaining the corresponding conditional measure, $\nu_\alpha(\mathbf{x})$. [Note that $\nu_\alpha(\mathbf{x})$ in case (iii) is not obvious to most people.] In the later part of this chapter, we will describe a way to construct fiber decompositions by using *transformation groups*. We believe that this is perhaps the most general formulation that still permits us to derive some nice and useful mathematical results (Liu and Sabatti 2000).

8.2 Case Studies for Partial Resampling

8.2.1 Gaussian random field model

Consider a Gaussian model defined on an $N \times N$ regular lattice Λ as shown in Figure 8.1. In this model, $\mathbf{x} = (x_s, s \in \Lambda)$ and $x_s \in \mathbb{R}^1$. The target distribution is

$$\pi(\mathbf{x}) \propto \exp\left\{-\frac{1}{2} \sum_{s \sim s'} \beta_{ss'}(x_s - x_{s'})^2 - \frac{1}{2} \sum_{s \in \Lambda} \gamma_s (x_s - \mu_s)^2\right\},$$

where the notation "$s \sim s'$" means that s and s' are neighbors in space Λ, $\beta_{ss'} = \beta_{s's}$, and the $\beta_{ss'}$ and the μ_s are fixed constants. This type of model if often called a *Markov random field* (MRF) model. A single-site Gibbs sampler can be easily applied to simulate from this model: Given the values of all but one site, $\mathbf{x}_{[-s]}$, the conditional distribution of x_s is

$$\pi(x_s \mid \mathbf{x}_{[-s]}) \propto \exp\left\{-\frac{1}{2}\left(\gamma_s + \sum_{s' \sim s} \beta_{ss'}\right)\left(x_s - \frac{\gamma_s \mu_s + \sum_{s' \sim s} \beta_{ss'} x_{s'}}{\gamma_s + \sum_{s' \sim s} \beta_{ss'}}\right)^2\right\}$$

164   8. General Conditional Sampling

(i.e., a Gaussian distribution). However, this single-site updating scheme can be very slow (Goodman and Sokal 1989).

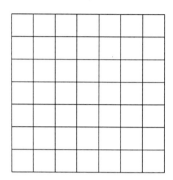
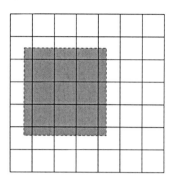

FIGURE 8.1. Left: an 8×8 lattice Λ; Right: a 4×4 sub-lattice S of Λ, as outlined and shaded. There are 16 boundary links that connect pixels in S with pixels out of S.

Alternatively, we can consider updating a "window" of pixels simultaneously. Suppose S is a sub-lattice of size $k \times k$ of Λ (e.g., Figure 8.1). Let $\mathbf{x}_S = \{x_s, \ s \in S\}$. We consider the move

$$\mathbf{x} \longrightarrow (\mathbf{x}_S + \delta, \mathbf{x}_{[-S]}), \tag{8.1}$$

where δ is a real number and we define that $\mathbf{x}_S + \delta = (x_s + \delta, \ s \in S)$. In this "coarsening move," the corresponding fiber construction is $\mathcal{X}_\alpha = \{\mathbf{x} : \mathbf{x}_{[-S]} = \alpha\}$; that is, only those random components indexed by the sites in S are allowed to move.

In order to make the move (8.1) proper, we need to draw δ from some distribution so that the the target distribution π is invariant. By Theorem 8.3.1 in Section 8.3, one can easily derive that δ should be drawn from

$$p(\delta) \propto \pi(\mathbf{x}_S + \delta, \mathbf{x}_{[-S]}),$$

similar to those conditional distributions needed by the regular Gibbs sampler. It is easy to show that distribution $p(\delta)$ is Gaussian with mean

$$\mu_* = -\frac{\sum_{s' \sim s \in \partial S} \beta_{ss'}(x_s - x'_s) + \sum_{s \in S} \gamma_s(x_s - \mu_s)}{\sum_{s' \sim s \in \partial S} \beta_{ss'} + \sum_{s \in S} \gamma_s}$$

and variance

$$\sigma_*^2 = \left(\sum_{s' \sim s \in \partial S} \beta_{ss'} + \sum_{s \in S} \gamma_s \right)^{-1},$$

where the notation $\sum_{s' \sim s \in \partial S}$ means that the summation is over all boundary pixels of S and their respective neighbors outside of S.

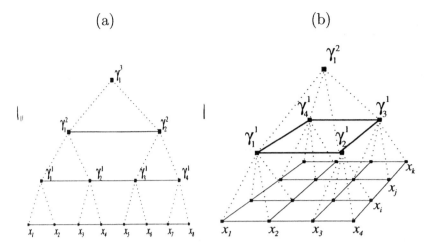

FIGURE 8.2. Graphical representation of the multigrid Monte Carlo schemes. (a) In a one-dimensional Markov random field model, pairs of the variables are moved together at each coarse grid update. (b) In a two-dimensional MRF, quadruples of the variables are moved together.

As shown in Figure 8.2, we can apply the "window move" hierarchically with different "resolution" levels. More precisely, we first update the set of 4-pixel windows; then update the set of 16-pixel windows; and so on. It is helpful to recognize that the *joint distribution* of all non-overlapping 4-pixel window updates form a new lattice model (Goodman and Sokal 1989), which is also a Gaussian random field. Thus, the hierarchical procedure can be understood as a recursive application of the basic "coarsening move." It was shown (Goodman and Sokal 1989) that by properly arranging the coarsening and refining moves (the so-called "W-cycles" and "V-cycles"), one can obtain a Monte Carlo algorithm whose convergence rate (or, equivalently, relaxation time) is nearly independent of the lattice size.

8.2.2 Texture synthesis

An image I as in Figure 8.3 can be represented by an $N \times N$ (e.g., N=128 or 256) lattice Λ with each of its pixels $s \in \Lambda$ having an "intensity value" $I(s)$. It is argued that human perception of texture patterns comes from pre-attentive extraction of certain "summary statistics" such as local contrast, orientation of small patches, etc.

In a sequence of papers (Wu, Zhu and Liu 1999, Zhu, Wu and Mumford 1997, Zhu, Wu and Mumford 1998, Zhu, Liu and Wu 2000), Zhu and collaborators established a mathematical basis for texture modeling. One of their models, *the Julesz ensemble*, can be described as follows. Given a set of K histogram functions (filters), $H = \{h^{(a)} : a = 1, \ldots, K\}$, one can

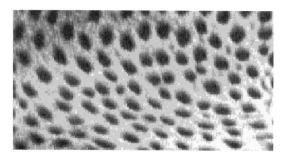

FIGURE 8.3. An observed cheetah's skin texture pattern. (Courtesy of Professors S.C. Zhu and Y.N. Wu.)

compute for any image I the "summary statistics"

$$H(I) = (h^{(1)}(I), \ldots, h^{(K)}(I)).$$

Then, a *Julesz ensemble* with a given summary statistics H_0 is the set of images that match this summary statistics:

$$\Omega_\Lambda(H_0) = \{I : H(I) = H_0\}.$$

Due to limited recording accuracy and intensity quantization, in practice one needs to relax the above definition as

$$\Omega_\Lambda(\mathcal{H}) = \{I : H(I) \in \mathcal{H}\},$$

where $\mathcal{H}$ is an open set around H_0. For example, $\mathcal{H}$ can be defined as $\mathcal{H}(H_0) = \{I : D(H(I), H_0) < \epsilon\}$, where $D(\cdot, \cdot)$ is some distance metric.

For a given observed image whose summary statistics is H_0, it is of interest to "synthesize" new ones whose statistics are as close to H_0 as possible. To achieve this end, Zhu et al. (2000) introduced an energy function defined on the space of all images:

$$G(I) = \begin{cases} 0 & \text{if } D(H(I), H_0) < \epsilon \\ D(H(I), H_0) & \text{otherwise.} \end{cases}$$

Then the distribution

$$q(I; H_0, T) = \frac{1}{Z(T)} \exp\{-G(I)/T\}$$

converges to a uniform distribution on $\Omega_\Lambda(\mathcal{H})$ as $T \to 0$. In their framework, each statistic $h^{(a)}(I)$ is the histogram of the values of the corresponding filter function, $F^{(a)}$, applied to all the windows of size $l \times l$ of Λ. Typical filter functions are shown in Figure 8.4.

It is straightforward to use an iterative single-site update method (i.e., the stanford Gibbs sampler) to simulate from $q(I; H_0, T)$; that is, one randomly picks a site $s \in \Lambda$ and updates the value $I(s)$ by a new value drawn

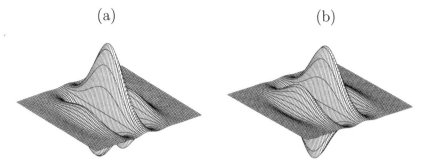

FIGURE 8.4. Two filter functions used in Zhu et al. (2000).

from the conditional distribution $q(\boldsymbol{I}(s) \mid \boldsymbol{I}(-s); \boldsymbol{H}_0, T)$. The efficiency of this algorithm, however, can be exceedingly low because of the special features of the filtering functions, $\boldsymbol{h}^{(a)}$, for $a = 1, \ldots, K$. Recently, Zhu et al. (2000) proposed an efficient method called the *window Gibbs sampler* for sampling from such a model. This new MCMC strategy can be seen as a special way of conducting "coarser moves" in the image space. At each step t of the window Gibbs sampler, one randomly picks a pixel $s \in \Lambda$ and updates

$$\boldsymbol{I}^{(t)}(\cdot) \longrightarrow \boldsymbol{I}^{(t+1)}(\cdot) + \delta^{(a)} W_s^{(a)},$$

where $W_s^{(a)}$ is a window function centered at s and its size is chosen to be the same as the filter $F^{(a)}$ used for extracting texture features. Here, $\delta^{(a)}$ should be chosen from the conditional distribution

$$p(\delta) \propto q(\boldsymbol{I}^{(t)}(\cdot) + \delta W_s^{(a)}; \boldsymbol{H}_0, T).$$

In the computation, δ value is quantized. Otherwise, a random-ray or random-grid Monte Carlo strategy (Section 5.5) can be useful.

For illustration, both the single-site Gibbs and the the window Gibbs were applied to synthesize textures of a cheetah's skin (a true pattern is shown in Figure 8.3). The marginal histograms of eight filters were chosen as the statistics $\boldsymbol{h}$. Both the single-site Gibbs and the window Gibbs algorithms were simulated with two initial conditions: $\boldsymbol{I}_u$, a uniform noise image, and $\boldsymbol{I}_c$, a constant white image. Figure 8.5 displays the results for the first 100 sweeps, where (a) and (c) are the results of the single-site Gibbs starting from $\boldsymbol{I}_u$ and $\boldsymbol{I}_c$, respectively, and (b) and (d) are the results of the window Gibbs starting from $\boldsymbol{I}_u$ and $\boldsymbol{I}_c$, respectively.

The total matching error at each sweep of the Markov chain is defined as

$$E = \sum_{\alpha=1}^{K} ||\boldsymbol{h}^{(\alpha)}(\boldsymbol{I}_{\text{syn}}) - \boldsymbol{h}^{(\alpha)}_{\text{obs}}||_1.$$

The change of E is plotted against the number of sweeps in Figure 8.6. It is seen that the single-site Gibbs starting from both $\boldsymbol{I}_c$ and $\boldsymbol{I}_u$ exhibits a

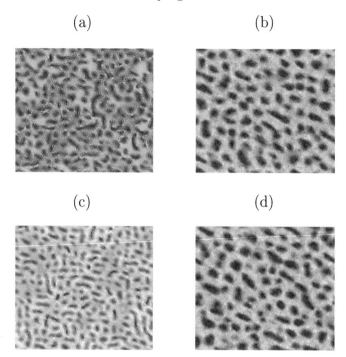

FIGURE 8.5. (a, c): Sampled images from single-site Gibbs sampler; (b, d): sampled images from the window Gibbs sampler. (Courtesy of Professors S.C. Zhu and Y.N. Wu.)

high error floor. In contrast, the matching errors for window Gibbs in both cases dropped under 0.08 (i.e., a less than 1% error on average for each histogram).

In summary, the window Gibbs algorithm outperforms the single-site Gibbs in at least two aspects: (1) The window Gibbs can match statistics faster, particularly when statistics of large features are involved. (2) The window Gibbs moves faster after statistics matching. Thus, it can render texture images of different details.

8.2.3 Inference with multivariate t-distribution

Iterative methods for the inference with t-distribution has been considered (Liu, Rubin and Wu 1998, Liu and Sabatti 2000, Meng and van Dyk 1999). The central component of their algorithms is, in fact, a partial resampling step.

Let $\mathbf{y}_i$, $i = 1, \ldots, n$, be i.i.d. observations from a d-dimensional $t_\nu(\boldsymbol{\mu}, \Sigma)$ distribution, where ν is the known degrees of freedom and $\boldsymbol{\mu}$ and Σ are unknown. Because a t-distributed random variable can be regarded as the

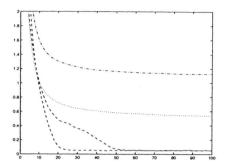

FIGURE 8.6. The matching errors in L_a distance, summed over eight filters, against the number of sweeps in the samplers. Top two curves: single-site Gibbs; bottom two curves: window Gibbs. (Courtesy of Professors S. C. Zhu and Y. N. Wu.)

ratio between a Gaussian random variable and a function of an independent χ^2 random variable, the estimation problem for the t-distribution can be "reconfigured" into a missing data problem. More precisely, we let $q_i \sim \chi^2_\nu/\nu$ be mutually independent and regard them as missing data. Given q_i, we have $[\mathbf{y}_i \mid q_i] \sim \mathcal{N}(\boldsymbol{\mu}, \Sigma/q_i)$. When given an *improper* prior

$$\pi_0(\boldsymbol{\mu}, \Sigma) \propto |\Sigma|^{-(d+1)/2}$$

for $\boldsymbol{\mu}$ and Σ, the joint posterior distribution of $(\boldsymbol{\mu}, \Sigma, q)$ is

$$\pi(\boldsymbol{\mu}, \Sigma, q) \propto \left(\prod_{i=1}^n q_i^{\frac{\nu+d}{2}-1} \right) |\Sigma|^{-\frac{n+d+1}{2}}$$

$$\times \exp\left\{ -\frac{1}{2} \sum_{i=1}^n q_i(\nu + (\mathbf{y}_i - \boldsymbol{\mu})' \Sigma^{-1}(\mathbf{y}_i - \boldsymbol{\mu})) \right\}$$

and a Gibbs sampler/data augmentation scheme can be easily applied:

(a) Conditional on the q_i, the posterior distribution of Σ is an inverse-Wishart distribution (Gelman, Roberts and Gilks 1995) with a scale matrix

$$S = \sum_{i=1}^n q_i (\mathbf{y}_i - \hat{\mathbf{y}})(\mathbf{y}_i - \hat{\mathbf{y}})'$$

where $\hat{\mathbf{y}}$ is a weighted average of the $\mathbf{y}_i$ with weight vector $\mathbf{q} = (q_1, \ldots, q_n)$.

(b) Conditional on the q_i and Σ, $\boldsymbol{\mu}$ is distributed as Gaussian with mean $\hat{\mathbf{y}}$ and covariance matrix $\Sigma/(q_1 + \cdots + q_n)$.

(c) Conditional on Σ and μ, the q_i are mutually independent and distributed as scaled-χ^2 random variables:

$$q_i \sim \chi^2_{\nu+d}/[\nu + (\mathbf{y}_i - \boldsymbol{\mu})'\Sigma^{-1}(\mathbf{y}_i - \boldsymbol{\mu})].$$

We observe that the samples of Σ and $\boldsymbol{q}$ obtained from the above algorithm tend to be tightly "coupled" because of steps (a) and (c). In other words, if the starting values of the q_i are large, the resulting sample of Σ tends to be large and vice versa. Iterative sampling between them is therefore not very efficient. A possible remedy is to consider a potential non-coordinate move:

$$(\Sigma, \boldsymbol{q}) \to \gamma(\Sigma, \boldsymbol{q}) \equiv (\gamma\Sigma, \gamma q_1, \ldots, \gamma q_n),$$

for $\gamma > 0$. Clearly, this move, if achievable, can "damp down" simultaneously the two tightly coupled components, Σ and $\boldsymbol{q}$. Additionally, this move implies a decomposition of the space of $(\Sigma, \boldsymbol{q})$: Each fiber in this decomposition corresponds to one equivalent class, or the *orbit*, under the relationship

$$(\Sigma, \boldsymbol{q}) \equiv (\Sigma', \boldsymbol{q}') \text{ if and only if } \exists \gamma > 0 \text{ such that } (\Sigma, \boldsymbol{q}) = (\gamma\Sigma', \gamma\boldsymbol{q}).$$

To derive the proper sampling distribution for γ, we note that the set of all proper γ's form a *transformation group* (the scale group) on the space of $(\Sigma, \boldsymbol{q})$, a $(n+d(d+1)/2)$-dimensional space. Using Theorem 8.3.1 in the next section, we found that conditional on q and Σ, γ should be drawn from $\chi^2_{n\nu}/(\nu \sum_{i=1}^n q_i)$. Thus, in addition to the regular Gibbs sampling steps, we can add this *partial resampling* step so as to update $(\Sigma, \boldsymbol{q})$ jointly to a new configuration $(\gamma q, \gamma \Sigma)$. This step is central to the success of the so-called "parameter expansion" technique developed recently in statistics literature (Liu, Rubin and Wu 1998, Liu and Wu 1999, Meng and van Dyk 1999).

We have experimented with different dimensional problems ($d = 1, 4, 10$), different numbers of observations, and different degrees of freedom ($\nu = 1, \ldots, 5$). The partial resampling step was helpful for moderate values of ν, but we did not observe a significant increase in efficiency for large ν's, which is consistent with our understanding: The target distribution is almost the same as a Gaussian distribution when ν is large.

8.3 Transformation Group and Generalized Gibbs

A MCMC sampler can be viewed as a scheme, or a plan, for moving a "ghost point" around in its sample space. The output of this sampler is just the trace of this point's movement. For example, at any step of a Metropolis algorithm, a tentative position/configuration that is "close to" the current position of the ghost point is proposed. Then, according to the acceptance-rejection rule, one decides whether to move the ghost point to this new position or to leave it unmoved. In a random-scan Gibbs sampler, one randomly selects a coordinate-direction in the space and moves the ghost point along that direction to a new location, where the new location is drawn from an appropriate conditional distribution.

More generally, suppose that at time t, the "ghost" point is at location $\mathbf{x}^{(t)} = \mathbf{x}$, then at time $t + 1$, this point is "moved" to a new location $\mathbf{x}^{(t+1)} = \mathbf{x}'$. The basic criterion for designing a proper MCMC scheme is to make sure that such a move leaves the target distribution $\pi(\mathbf{x})$ invariant. The partial resampling principle described in Section 8.1 is a very general rule to ensure invariance. However, partial resampling is not sufficiently constructive and is difficult to apply. Here, we seek a more explicit solution under a more specific setting (i.e., when the "moves" can be described by transformation groups).

Suppose the move from $\mathbf{x}$ to $\mathbf{x}'$ in a MCMC sampler can be achieved by selecting a transformation γ (a mover) from a set of transformations (a moving company), Γ, and applying it to $\mathbf{x}$ [so that $\mathbf{x}' = \gamma(\mathbf{x})$]. For example, in a random-scan Gibbs sampler (with the remaining part fixed), the current position $\mathbf{x} = (x_i, \mathbf{x}_{[-i]})$ is moved to a new position $\mathbf{x}' = (x'_i, \mathbf{x}_{[-i]})$. This move can be seen as applying a translation transformation to $\mathbf{x}$:

$$(x_i, \mathbf{x}_{[-i]}) \to (x_i + \gamma, \mathbf{x}_{[-i]}), \quad \gamma \in \mathbb{R}^1,$$

where the set of all eligible γ forms a group under the usual "+" operation. We need to choose γ according to an appropriate distribution so that π is invariant with respect to such a move. Therefore, a general formulation of the problem is as follows: Suppose $\mathbf{x} \sim \pi$ and Γ is a set of transformations, what distribution should one draw $\gamma \in \Gamma$ from so that $\mathbf{x}' = \gamma(\mathbf{x})$ also follows distribution π? An explicit answer to this question is given in Theorem 8.3.1 when Γ forms a locally compact group.

Suppose $\pi(\mathbf{x})$ is a probability distribution of interest defined on sample space $\mathcal{X}$. A set $\Gamma = \{\gamma\}$ of transformations on $\mathcal{X}$ is called a *locally compact group* a topological group or if (i) Γ is a locally compact space, (ii) the elements in Γ form a group with respect to the usual operation for composing two transformations [i.e., $\gamma_1 \gamma_2(\mathbf{x}) = \gamma_1(\gamma_2(\mathbf{x}))$], and (iii) the group operations $(\gamma_1, \gamma_2) \to \gamma_1 \gamma_2$ and $\gamma \to \gamma^{-1}$ are continuous functions (Rao 1987). For any measurable subset $B \subset \Gamma$ and element $\gamma_0 \in \Gamma$, $\gamma_0 B$ defines another subset of Γ resulting from γ_0 "acting" on every element of B. A measure L is called a left- (invariant) Haar measure if for every γ_0 and measurable set $B \in \Gamma$, one has

$$L(B) = \int_B L(d\gamma) = \int_{\gamma_0 B} L(d\gamma) = L(\gamma_0 B).$$

One can similarly define a right-Haar measure. Under mild conditions, these measures exist and are unique up to a positive multiplicative constant (Rao 1987).

Generally speaking, any move from $\mathbf{x}$ to $\mathbf{x}'$ in sample space $\mathcal{X}$ can be achieved by a transformation γ chosen from a suitable group Γ; that is, we can often find a group Γ of transformations so that $\mathbf{x}' = \gamma(\mathbf{x})$ for some $\gamma \in \Gamma$. For example, suppose $\mathbf{x} = (x_1, \ldots, x_d)$ and $\mathbf{x}' = (x_1', x_2, \ldots, x_d)$. Then, the move can be done by a translation group Γ acting on $\mathbf{x}$ in the following way:

$$\Gamma = \{\gamma \in \mathbb{R}^1 : \gamma x = (x_1 + \gamma, x_2, \ldots, x_d)\}.$$

An appropriate sampling distribution for γ gives rise to the Gibbs sampling update of x_1 to x_1'. To obtain a complete Gibbs sampling chain, one needs to use a combination of translation groups (one for each coordinate of $\mathbf{x}$). Provided that this combination is transitive (i.e., one can move from one point to any other in $\mathcal{X}$ via a sequence of transformations selected from the groups), the resulting Markov chain is irreducible.

Given Γ, it is then necessary to determine an appropriate sampling distribution for γ so that π is invariant under the move $\mathbf{x}' = \gamma(\mathbf{x})$. The following theorem, first stated by Liu and Wu (1999) and later extended by Liu and Sabatti (2000), provides an explicit form of the sampling distribution.

Theorem 8.3.1 *Suppose Γ is a locally compact group of transformations on $\mathcal{X}$ and L is its left-Haar measure. Let π be an arbitrary probability measure on space $\mathcal{X}$. Suppose $\mathbf{x} \sim \pi(\mathbf{x})$ and γ is drawn from Γ according to the distribution*

$$p_{\mathbf{x}}(\gamma) \propto \pi(\gamma(\mathbf{x}))|J_\gamma(\mathbf{x})|L(d\gamma),$$

where $J_\gamma(\mathbf{x}) = \det\{\partial\gamma(\mathbf{x})/\partial\mathbf{x}\}$ is the Jacobian of the transformation. Then $\mathbf{x}' = \gamma(\mathbf{x})$ follows π.

The standard Gibbs sampler can be realized by applying this theorem to those Γ's corresponding to the translation group along each coordinate direction. An easy extension of the standard Gibbs sampler is to let Γ be the translation group along an arbitrary direction; that is,

$$\Gamma = \{\gamma \in \mathbb{R}^1 : \gamma(\mathbf{x}) = \mathbf{x} + \gamma e \equiv (x_1 + \gamma e_1, \ldots, x_d + \gamma e_d)\},$$

where $e = (e_1, \ldots, e_d)$ is a fixed vector given in advance. The proper distribution for drawing γ can be derived from Theorem 8.3.1 as $p_{\mathbf{x}}(\gamma) \propto \pi(\mathbf{x} + \gamma e)$. For any locally compact transformation group Γ, we can define a *generalized Gibbs* step as

- draw $\gamma \in \Gamma$ from $p_{\mathbf{x}}(\gamma) \propto \pi(\gamma(\mathbf{x}))|J_\gamma(\mathbf{x})|L(d\gamma);$ \hfill (G1)

- set $\mathbf{x}' = \gamma(\mathbf{x})$. \hfill (G2)

The result described in Theorem 8.3.1 can be understood as a one-step Gibbs sampling updating under a reparameterization guided by a transformation group. More precisely, a point $\mathbf{x} \in \mathcal{X}$ is reparameterized as $(o_x, \boldsymbol{p})$, where o_x indicates the "orbit" point **x** lies on and p is its location on the orbit. In other words, the result of Theorem 8.3.1 can be used to achieve the sampling effect of a reparameterized Gibbs sampler without actually doing the reparameterization.

Groups such as the scale group, the affine transformation group, and the orthonormal transformation group can often be used. If one wishes to update **x** to **x**' as

$$\mathbf{x} = (x_1, \ldots, x_d) \to \mathbf{x}' = (\gamma x_1, \ldots, \gamma x_d), \quad \gamma \in \mathbb{R}^1 \setminus \{0\},$$

for example, one can define a scale-transformation group

$$\Gamma = \{\gamma \in \mathbb{R}^1 \setminus \{0\} : \gamma(x_1, \ldots, x_d) = (\gamma x_1, \ldots, \gamma x_d)\},$$

and sample γ from $p_\mathbf{x}(\gamma) \propto |\gamma|^{d-1} \pi(\gamma \mathbf{x})$.

Compared with the *partial resampling* rule of Goodman and Sokal (1989), the generalized Gibbs step is more constructive and readily implementable. The following rather trivial corollary gives a direct connection with the partial resampling framework.

Corollary 8.3.1 *Suppose $\Gamma = \{\gamma\}$ is a locally compact transformation group on $\mathcal{X}$. Then each orbit of Γ can be treated as a fiber; that is, we can define $\mathcal{X}_\alpha = \{\mathbf{x} \in \mathcal{X} : \mathbf{x} = \gamma(\alpha)\}$, where α is a representative member of the elements in the orbit. The corresponding partial resampling distribution is*

$$\nu_\alpha(\mathbf{x}) \propto \begin{cases} \pi(\gamma(\alpha)) |J_\alpha(\gamma)| L(d\gamma) & \text{if } \mathbf{x} = \gamma(\alpha) \\ 0 & \text{otherwise.} \end{cases}$$

Proof: A direct consequence of Theorem 8.3.1. ◇

Although Theorem 8.3.1 provides us a nice conditional distribution $p_\mathbf{x}(\ )$ for Γ, it is often difficult in practice to sample from this distribution directly. In this case, a Metropolis-type move would be desirable. Suppose $A_\mathbf{x}(\gamma, \gamma')$ is a transition kernel that leaves $p_\mathbf{x}(\gamma)$ invariant. Can we substitute Step G1 with the application of a one-step transition from $A_\mathbf{x}(\gamma, \cdot)$? The complication is as follows. In order to sample a new value γ' with $A_\mathbf{x}(\gamma, \cdot)$, we need a starting value γ. Also, in the event that rejection occurs, we would like to retain $\mathbf{x}' = \mathbf{x}$. Thus, the only way this can be achieved is by applying $A_\mathbf{x}$ to $\gamma = \gamma_{\mathrm{id}}$, the identity transformation. Thus, the question is this: What conditions has $A_\mathbf{x}$ to satisfy so that Step G1 can be replaced by a step $\gamma \sim A_\mathbf{x}(\gamma_{\mathrm{id}}, \gamma)$? The following theorem gives an answer.

Theorem 8.3.2 *Assume that all the conditions in Theorem 8.3.2 hold. Suppose $A_\mathbf{x}(\gamma, \gamma') L(d\gamma)$ is a Markov transition function that leaves*

$$p_\mathbf{x}(\gamma) d\gamma \propto \pi(\gamma(\mathbf{x})) |J_\gamma(\mathbf{x})| L(d\gamma)$$

invariant and satisfies the following transformation-invariant property:

$$A_\mathbf{x}(\gamma, \gamma') = A_{\gamma_0^{-1} \mathbf{x}}(\gamma \gamma_0, \gamma' \gamma_0), \qquad (8.2)$$

174   8. General Conditional Sampling

for all $\gamma, \gamma', \gamma_0 \in \Gamma$. If $\mathbf{x} \sim \pi$ and $\gamma \sim A_{\mathbf{x}}(\gamma_{\mathrm{id}}, \gamma)$, then $w = \gamma(\mathbf{x})$ follows π.

Intuitively, condition (8.2) implies that the local transition function $A_{\mathbf{x}}$ has to be independent of the reference point used (i.e., $\mathbf{x}$) on the "orbit" $\{y : y = \gamma(x), \gamma \in \Gamma\}$ of the group. The condition is satisfied if $A_{\mathbf{x}}(\gamma, \gamma')$ is of the form $g\{p_{\mathbf{x}}(\gamma), p_{\mathbf{x}}(\gamma')\}$. The proof of this theorem is left to the reader as an exercise.

8.4 Application: Parameter Expansion for Data Augmentation

In this section, we consider a special application of Theorem 8.3.1 to the Bayesian missing data framework (Sections 1.9 and A.3.2). Let $\mathbf{y}_{\mathrm{obs}}$ be the observed data, $\mathbf{y}_{\mathrm{mis}}$ be the missing data, and θ be the parameter of interest in model $f(\mathbf{y}_{\mathrm{obs}}, \mathbf{y}_{\mathrm{mis}} \mid \theta)$. With a slight abuse of notations, we let f denote all the probability densities related to the original model. For example, $f(\mathbf{y}_{\mathrm{obs}}, \mathbf{y}_{\mathrm{mis}} \mid \theta)$ represents the complete-data model, $f(\theta)$ denotes the prior, and

$$f(\mathbf{y}_{\mathrm{obs}} \mid \theta) = \int f(\mathbf{y}_{\mathrm{obs}}, \mathbf{y}_{\mathrm{mis}}, \mid \theta) d\mathbf{y}_{\mathrm{mis}}$$

is the observed-data model. As explained in Section 6.4, the data augmentation (DA) algorithm used to simulate from $f(\mathbf{y}_{\mathrm{mis}}, \theta \mid \mathbf{y}_{\mathrm{obs}})$ iterates the following steps :

1. Draw $\mathbf{y}_{\mathrm{mis}} \sim f(\mathbf{y}_{\mathrm{mis}} \mid \theta, \mathbf{y}_{\mathrm{obs}}) \propto f(\mathbf{y}_{\mathrm{obs}}, \mathbf{y}_{\mathrm{mis}} \mid \theta)$.

2. Draw $\theta \sim f(\theta \mid \mathbf{y}_{\mathrm{mis}}, \mathbf{y}_{\mathrm{obs}}) \propto f(\mathbf{y}_{\mathrm{obs}}, \mathbf{y}_{\mathrm{mis}} \mid \theta) f(\theta)$.

In many applications, we can find a "hidden" parameter α in the complete data model $f(\mathbf{y}_{\mathrm{obs}}, \mathbf{y}_{\mathrm{mis}} \mid \theta)$, so that the original model can be embedded into a larger one, $p(\mathbf{y}_{\mathrm{obs}}, \mathbf{w} \mid \theta, \alpha)$, that preserves the observed-data model $f(\mathbf{y}_{\mathrm{obs}} \mid \theta)$. Mathematically, this means that the probability distribution $p(\mathbf{y}_{\mathrm{obs}}, \mathbf{w} \mid \theta, \alpha)$ satisfies

$$\int p(\mathbf{y}_{\mathrm{obs}}, \mathbf{w} \mid \theta, \alpha) d\mathbf{w} = f(\mathbf{y}_{\mathrm{obs}} \mid \theta).$$

We call α an *expansion parameter*.

For notational clarity, we use $\mathbf{w}$ instead of $\mathbf{y}_{\mathrm{mis}}$ to denote the missing data under the expanded model. In order to implement the DA algorithm for the expanded model, we need to give a joint prior distribution $p(\theta, \alpha)$. It is straightforward to show that the posterior distribution of θ is the same for both models if and only if the marginal prior distribution for θ from $p(\theta, \alpha)$ agrees with $f(\theta)$ [i.e., $p(\theta \mid \mathbf{y}_{\mathrm{obs}}) = f(\theta \mid \mathbf{y}_{\mathrm{obs}})$ if and only if

8.4 Application: Parameter Expansion for Data Augmentation

$\int p(\theta, \alpha)d\alpha = f(\theta)]$. Therefore, we only need to specify the conditional prior distribution $p(\alpha \mid \theta)$ while maintaining the marginal prior for θ at $f(\theta)$. It is clear that given θ and $\mathbf{y}_{\text{obs}}$, the posterior distribution of α, $p(\alpha \mid \mathbf{y}_{\text{obs}}, \theta)$, remains as $p(\alpha \mid \theta)$ because α is not identifiable from $\mathbf{y}_{\text{obs}}$.

In many problems, the extra parameter α indexes for a certain set of transformations on the missing data $\mathbf{y}_{\text{mis}}$. In light of the discussions in the previous section and in the latter part of this section, α often corresponds to an element from a transformation group. This step can help achieve more global movements on the missing data space and, therefore, improve efficiency of the algorithm. More precisely, when α indexes a transformation, the expanded likelihood can be derived as

$$p(\mathbf{y}_{\text{obs}}, \mathbf{w} \mid \alpha, \theta) = f(\mathbf{y}_{\text{obs}}, t_\alpha(\mathbf{w}) \mid \theta) |J_\alpha(\mathbf{w})|,$$

where $J_\alpha(\mathbf{w}) = \det\{\partial t_\alpha(\mathbf{w})/\partial \mathbf{w}\}$ is the Jacobian term evaluated at $\mathbf{w}$. When the set of transformations forms a locally compact group, a natural prior for α, as indicated by Theorem 8.3.1, is its left-Haar measure. In this setting, a special form the the *parameter expanded data augmentation* (PX-DA) algorithm can be defined as iterating the following steps:

The PX-DA Algorithm

1. Draw $\mathbf{y}_{\text{mis}} \sim f(\mathbf{y}_{\text{mis}} \mid \theta, \mathbf{y}_{\text{obs}})$.

2. Draw $\alpha \sim p(\alpha \mid \mathbf{y}_{\text{obs}}, \mathbf{y}_{\text{mis}}) \propto f(\mathbf{y}_{\text{obs}}, t_\alpha(\mathbf{y}_{\text{mis}})) \ |J_\alpha(\mathbf{y}_{\text{mis}})| \ H(d\alpha)$.
   Compute $\mathbf{y}'_{\text{mis}} = t_\alpha(\mathbf{y}_{\text{mis}})$.

3. Draw $\theta \sim f(\theta \mid \mathbf{y}_{\text{obs}}, \mathbf{y}'_{\text{mis}})$.

Step 2 in the algorithm can be viewed as an adjustment of the missing data. When the prior for α is proper, Liu and Wu (1999) show that the second step should be of the form

2'. Draw $\alpha_0 \sim p(\alpha)$; compute $\mathbf{w} = t_\alpha^{-1}(\mathbf{y}_{\text{mis}})$.
    Draw $\alpha_1 \sim p(\alpha \mid \mathbf{y}_{\text{obs}}, \mathbf{w}) \propto f(\mathbf{y}_{\text{obs}}, t_\alpha(\mathbf{w})) \ |J_\alpha(\mathbf{w})| \ p_0(\alpha)$.
    Compute $\mathbf{y}'_{\text{mis}} = t_{\alpha_1}(t_{\alpha_0}^{-1}(\mathbf{y}_{\text{mis}}))$.

The interesting result here is that when a Haar measure prior is used for the expansion parameter, one can skip the step of sampling from the prior of α, which is essential when this Haar prior corresponds to an improper probability measure. Note that Steps 2 and 2' are indeed very different.

Based on Theorem 8.3.1, Step 2 leaves the distribution $f(\mathbf{y}_{\text{obs}}, \mathbf{y}_{\text{mis}})$ invariant. As a consequence, the PX-DA leaves the posterior distribution invariant. Liu and Wu (1999) shows that a nice property for using the Haar prior for α is that the resulting adjustment $\mathbf{y}'_{\text{mis}}$ in Step 2 is conditionally *independent* of the previous $\mathbf{y}_{\text{mis}}$, given that they lie on the same *orbit* (fiber). This can leads to a conclusion that using Step 2 is always more preferable to using Step 2', provided that they can be implemented with equal computational cost.

8.5 Some Examples in Bayesian Inference

8.5.1 Probit regression

Let $\mathbf{y} = (y_1, ..., y_n)$ be a set of i.i.d. binary observations from a *probit regression* model:

$$y_i \mid \theta \sim \text{Bernoulli}\{\Phi(X_i'\theta)\},$$

where the X_i ($p \times 1$) are the covariates, θ is the unknown regression coefficient, and Φ is the standard Gaussian cumulative distribution function (cdf). Of interest is the posterior distribution of θ, under, say, a flat prior. A popular way to ease computation is to introduce a "complete-data" model in which a set of latent variables, $z_1, ..., z_n$, is augmented so that

$$[z_i \mid \theta] = N(X_i'\theta, 1) \quad \text{and} \quad y_i = \text{sgn}(z_i),$$

where $\text{sgn}(z) = 1$ if $z > 0$ and $\text{sgn}(z) = 0$ otherwise (Albert and Chib 1993). The standard DA algorithm iterates the following steps:

1. Draw from $[z_i \mid y_i, \theta]$; that is, sample $z_i \sim N(X_i'\theta, 1)$ subject to $z_i \geq 0$ if $y_i = 1$; or draw $z_i \sim N(X_i'\theta, 1)$ subject to $z_i < 0$ for $y_i = 0$.

2. Draw $[\theta \mid z_i] = N(\hat{\theta}, V)$, where $\hat{\theta} = (\sum_i X_i X_i')^{-1} \sum_i X_i z_i$, and $V = (\sum_i X_i X_i')^{-1}$. Both $\hat{\theta}$ and V can be computed using the Sweep operator (e.g., Little and Rubin (1987)).

A problem with this scheme is that the "scales" of the z_i are tightly coupled with the scale of θ. As seen from Step 2 of the foregoing algorithm, conditional on the z_i, the center of θ's distribution is $\hat{\theta}$, a weighted average of the z_i. On the other hand, conditional on θ, the z_i is centered at $X_i'\theta$. Thus, it is helpful to consider a *parameter-expansion* approach.

The complete-data model can be expanded by introducing an expansion parameter α for residual variance, which is originally fixed at 1:

$$[w_i \mid \theta] = N(X_i'\theta\alpha, \alpha^2) \quad \text{and} \quad y_i = \text{sgn}(w_i).$$

It is clear that this model implies a conditional move that can also be derived by using the group of scale transformation

$$\mathbf{z} = t_\alpha(\mathbf{w}) \equiv (w_1/\alpha, \ldots, w_n/\alpha).$$

The corresponding Haar measure is $H(d\alpha) = \alpha^{-1} d\alpha$. The PX-DA algorithm (Section 8.4) has the same first step as in the standard DA, but with slightly different later steps:

2. Draw $\hat{\alpha}^2 \sim \text{RSS}/\chi_n^2$, where $\text{RSS} = \sum_i (z_i - X_i'\hat{\theta})^2$ is a by-product from the SWEEP operator.

3. Draw $\theta \sim N(\hat{\theta}/\hat{\alpha}, V)$.

8.5 Some Examples in Bayesian Inference

This PX-DA scheme also has the following more abstract and general interpretation. The sampling from the target distribution $\pi(\theta, \mathbf{z})$, where $\mathbf{z} = (z_1, \ldots, z_n)$, is achieved by iterating the following steps:

- draw θ from $\pi(\theta \mid \mathbf{z})$;

- update z_i by a sample from $\pi(z_i \mid \mathbf{z}_{[-i]}, \theta)$, $i = 1, \ldots, n$;

- draw γ from $p(\gamma) \propto \gamma^{n-1}\pi(\gamma \mathbf{z})$ and update z_i to γz_i, for $i = 1, \ldots, n$.

Thus, this scheme involves both an integration step and a generalized Gibbs step.

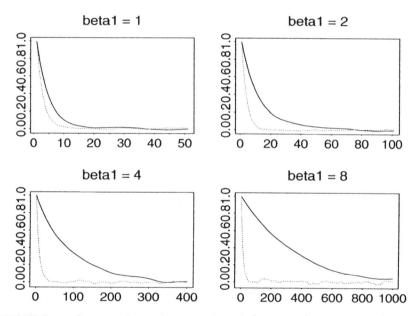

FIGURE 8.7. Autocorrelation functions for DA (solid lines) and PX-DA (dashed lines) with various values of β_1.

We took $n = 100$ and $X_i = (1, x_i)'$, with x_i generated from $N(0, 1)$. The y_i were generated from Bernoulli($\Phi(\beta_0 + \beta_1 x_i)$) with the true values $\beta_0 = 0$ and $\beta_1 = 1, 2, 4, 8$. We implemented both the DA algorithm and the PX-DA algorithm outlined above. Figure 8.7 shows the autocorrelation functions of the draws of β_1 for both algorithms under different true values for β_1. It is clear that when the real value for β_1 gets larger, the improvement of the PX-DA algorithm over the DA algorithm becomes more significant. To put it another way, the PX-DA algorithm is not slowed down very much by the increased value of β_1, whereas the DA algorithm is.

For the DA algorithm, we observe that $\operatorname{var}(\theta \mid \mathbf{y}, \mathbf{z}) = (\sum_i X_i X_i')^{-1}$, whereas for the PX-DA algorithm,

$$\operatorname{var}(\theta \mid \mathbf{y}, \mathbf{z}) = \left(\sum_i X_i X_i'\right)^{-1} + E\left[\hat{\theta}\hat{\theta}'\operatorname{var}(\chi_n)/RSS\right]$$

$$\approx \left(\sum_i X_i X_i'\right)^{-1} + \theta\theta'/2n,$$

which increases with θ. What we observed in Figure 8.7 can be understood from the fact (Chapter 12) that the sample autocorrelation of θ is determined by

$$1 - \frac{E\{\operatorname{var}(\theta|\mathbf{y},\mathbf{z})|\mathbf{y}\}}{\operatorname{var}(\theta|\mathbf{y})}$$

(Liu et al. 1994). The comparison in the rates of convergence should reflect the comparison in real computing time since the implementation of the PX-DA algorithm needs only negligible computing overhead in comparison with the DA algorithm.

8.5.2 Monte Carlo bridging for stochastic differential equation

Consider the stochastic differential equation

$$dY_t = b_t(Y_t, \theta)dt + \sigma_t(Y_t, \theta)dW_t, \tag{8.3}$$

which is frequently used to describe financial quantities such as derivatives and interest rates. In many applications, observations are at discrete time points but an analytical form of the transition function, $p(y_{t+s} \mid y_t)$, is unavailable. To illustrate, we let $y_1, \ldots, y_n$ be observations of the process (8.3) at discrete time $\tau, 2\tau, \ldots, n\tau$, and we assume that b_t and σ_t do not change with t. Of interest is inference about θ. Pederson (1995) proposes a direct Monte Carlo approximation to the transition function via the Chapman-Kolmogorov equation, which tends to be inefficient in complicated problems. Here we describe an alternative approach, Monte Carlo bridging, in combination with the use of multigrid Monte Carlo. The bridging idea has also been independently suggested by C.S. Jones (http://assets.wharton.upenn.edu/~jones13) and Elerian, Chib and Shephard (2001).

To connect two consecutive observations, y_k and y_{k+1}, say, one can impute a large number of "stepping stones," treated as missing data; for example,

$$y_k \to y_{k+1/m} \to \cdots \to y_{k+(m-1)/m} \to y_{k+1},$$

where each intermediate transition can be approximated reasonably well by a Gaussian relationship

$$y_{k+(j+1)/m} = y_{k+j/m} + \delta\, b(y_{k+j/m}, \theta) + \sqrt{\delta}\, \sigma(y_{k+j/m}, \theta)\, W_{k,j},$$

with $\delta = \tau/m$ and $W_{k,j} \sim N(0,1)$. Let us write $\mathbf{y}_{\text{obs}} = \{y_1, \ldots, y_n\}$ and

$$\mathbf{y}_{\text{mis}} = (y_{1,\text{mis}}, \ldots, y_{n-1,\text{mis}})$$
$$\equiv \{y_{k+j/m} : k = 1, \ldots, n-1; \ j = 1, \ldots, m-1\}.$$

The missing data $\mathbf{y}_{\text{mis}}$ is imputed purely for computational convenience and will be called a Monte Carlo bridge of size $M = m - 1$.

Given θ, we can apply a simple Gibbs sampler to cycle through every component of $\mathbf{y}_{\text{mis}}$. Conversely, given $\mathbf{y}_{\text{mis}}$, we can easily update θ from its approximate complete-data posterior. However, as m increases, the efficiency of this standard procedure decreases rapidly because of the high autocorrelation between the neighboring $y_{k+j/m}$. A better approach is to impute $\mathbf{y}_{\text{mis}}$ by the multigrid Monte Carlo. More specifically, consider a segment of the Monte Carlo bridge $y_{k,\text{mis}}$. For simplicity, we denote the $\mathbf{y}_{\text{mis}}$ vector by $(x_1, \ldots, x_M)$, where $M = m - 1$. With $M = 2^L$ and $m_l = 2^{L-l}$, a coarse grid move of level l corresponds to drawing from

$$p(a_1, \ldots, a_{m_l}) \propto \pi(x_1 + a_1, \ldots, x_{2^l} + a_1, \ldots$$
$$\ldots, x_{M-2^l+1} + a_{m_l}, \ldots, x_M + a_{m_l}).$$

For illustration, we consider the Ornstein-Uhlenbeck diffusion $dY_t = (\alpha - \beta Y_t)dt + \sigma dW_t$. Since its transition function $p(y_{t+s} \mid y_t)$ can be obtained analytically, this example serves as a good test case. In our numerical study, we simulated 200 observations from the process with $\alpha = 0$, $\beta = 1$, $\sigma = 1$, and time interval $\tau = 0.5$. With an improper prior $p(\alpha, \beta, \sigma) \propto \sigma^{-1}$, we applied the Monte Carlo bridging idea with bridge sizes $M = 0, 1, 4, 16, 32$, respectively. Figure 8.8 shows the approximated posteriors of σ for these cases and compares them with that resulting from an exact calculation. It is seen that, with 20,000 Monte Carlo samples and a bridge size $M \geq 16$, the approximated posterior is almost indistinguishable from the true one.

We applied level 1, 2 and 3 coarse moves to the case of $M = 16$. A cycle of this multigrid sampler consists of

Standard Gibbs ⇆ Level 2 move ⇆ Level 3 move ⇆ Level 4 move.

In each coarse level move, we did not use the special properties of the Gaussian model; instead, we applied one Metropolis step with its transition kernel satisfying Theorem 8.3.2. The computing time of this multigrid Monte Carlo sampler is 1.8 times that of the standard Gibbs sampler. Figure 8.9 shows that the multigrid Monte Carlo significantly outperforms the simple Gibbs sampler in terms of computational efficiency, with actual cost of computing time taken into consideration.

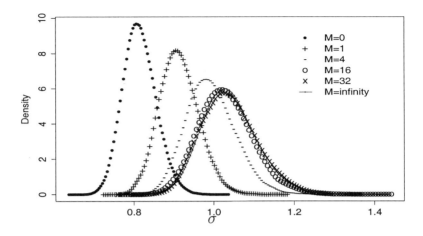

FIGURE 8.8. Comparison between the bridging approximations and the true posterior of σ for the Ornstein-Uhlenbeck process. The discrete observation interval is 0.5.

8.6 Problems

1. Prove Theorem 8.3.2.

2. Find a counterexample to show that the condition in Theorem 8.3.2 is necessary.

3. Given the analytical forms of the two conditional distributions, $\pi_1(x_1 \mid x_2)$ and $\pi_2(x_2 \mid x_1)$, of the joint distribution $\pi(x_1, x_2)$, how to derive the analytical form of $\pi(x_1, x_2)$ up to a normalizing constant?

4. Generalize the above problem to the k-dimensional distribution.

5. Prove Theorem 8.3.1 for the case when L is both the left and the right Haar measure.

6. Prove Theorem 8.3.2.

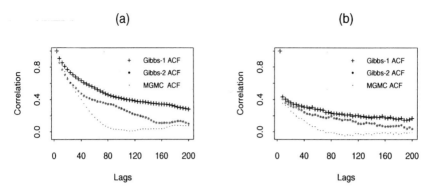

FIGURE 8.9. The Gibbs sampler's lag-1 and lag-2 versus multigrid Monte Carlo's lag-1 autocorrelation functions (a) for σ and (b) for β.

9
Molecular Dynamics and Hybrid Monte Carlo

Molecular dynamics (MD) simulation is a deterministic procedure to integrate the equations of motion based on the classical mechanics principles (Hamiltonian equations). This method was first proposed by Alder and Wainwright (1959) and has become one of the most widely used research tools for complex physical systems. In a typical MD simulation study, one first sets up the quantitative system (model) of interest under a given condition (e.g., fixed number of particles and constant total energy). Then, successive configurations of the system, as a function of time, are generated by following Newton's laws of motion. After a period of time for "equilibration," one can start to collect "data" from this computer experiment — the data consist of a sequence of snapshots that record the positions and velocities of the particles in the system during a period of time. Based on these records, one can estimate "typical characteristics," which can often be expressed as the time average of a function of the realized configurations, of the simulated physical system.

Ideally, one would like to generate continuous trajectories (video shots instead of snapshots), but in computer realizations, one is forced to use discrete time steps (a discretization of the Hamiltonian equations). Since the time step used in MD simulations is constrained by the need to conserve the total energy (at least approximately), how to choose a good step size has always been an art in the field. On one hand, the system will evolve very slowly if the time step is too small, whereas on the other hand, the simulation result will be very inaccurate if the time step used in simulation is too large.

Molecular dynamics simulation is a deterministic procedure and has to move on a hyper-surface where a certain Hamiltonian is conserved. In contrast, the goal of a Monte Carlo simulation is to sample "typical configurations" from a Boltzmann distribution which is determined by the system's potential energy and temperature. Because the "time average" will converge to the "configuration average" in a large system (the ergodicity theorem[1]), estimations from MC simulations often correspond very well with those from MD simulations. Although many problems can be treated by both means, it is also the case that in some problems one method is easier to implement than the other.

It is noted in Duane, Kennedy, Pendleton and Roweth (1987) that, besides the ergodicity theorem, there is also a very close technical connection between the MD and the MC simulations. The basic idea behind the new technique they proposed, hybrid Monte Carlo (HMC), is that one can use MD to generate trial moves in a MCMC sampler. Thus, in a certain sense, a bad MD move can be a good MCMC proposal. The advantage of the MD proposal is that the resulting MCMC moves follow the dynamics of the target distribution more closely. As a consequence, much of the randomness in an "unbiased" random-walk proposal is suppressed. Even for a non-physical system, using an artificial Hamiltonian equation to generate MD proposals is very helpful in making good MCMC transitions (Neal 1996). This chapter will illustrate the basic idea behind the HMC and discuss some useful generalizations.

9.1 Basics of Newtonian Mechanics

Let $\mathbf{x}(t)$ denote the d-dimensional position vector of a body of particles at time t (e.g., $d = 3N$ for an N-particle system in a three-dimensional space). We also assume that there is a d-dimensional mass vector $\mathbf{m} = (m_1, \ldots, m_d)$. Let $\mathbf{v}(t) \equiv \dot{\mathbf{x}}(t)$ denote the speed vector of the particles and let $\dot{\mathbf{v}}(t)$ be its acceleration vector. Suppose $\vec{F}$ is the force exerted on the particle; then, the Newton's law of motion states that

$$\vec{F} = \mathbf{m}\dot{\mathbf{v}}(t),$$

---

[1] The ergodicity theorem states that the average over a period of time of a function of the system configuration, as the time period goes to infinity, is equal to the average of that function over all configurations weighted by the Boltzmann factor $\exp\{-U(\mathbf{x})/\beta T\}$, where $U(\mathbf{x})$ is the potential energy of the system and T is the temperature; that is,

$$\lim_{t \to \infty} \frac{1}{t} \int_0^t h(\mathbf{x}_s) ds = Z^{-1} \int h(\mathbf{x}) \exp\{-U(\mathbf{x})/\beta T\} d\mathbf{x}.$$

where the product between two vectors is assumed to be component-wise:

$$\mathbf{m}\dot{\mathbf{v}} = (m_1\dot{v}_1, \ldots, m_d\dot{v}_d).$$

Instead of using the velocity vector **v**, a more convenient but equivalent variable, the *momentum* vector, defined as

$$\mathbf{p} = \mathbf{mv} \equiv (m_1 v_1, \ldots, m_d v_d),$$

is most frequently seen in classical mechanics. The *kinetic energy* of the system is usually defined as

$$k(\mathbf{p}) = \frac{1}{2}\sum_{i=1}^{d} m_i v_i^2 = \frac{1}{2}\sum_{i=1}^{d} \frac{p_i^2}{m_i} \equiv \frac{1}{2}\left\|\frac{\mathbf{p}}{\sqrt{\mathbf{m}}}\right\|^2.$$

Here, the ratio between two vectors (e.g., $\mathbf{p}/\sqrt{\mathbf{m}}$) is a component-wise operation which results in a new vector $(p_1/m_1, \ldots, p_d/m_d)$. The *phase space* of a system is defined as the product space of **x** and **p**, and vector $(\mathbf{x}, \mathbf{p})$ is often referred to as a point in the phase space.

Let $U(\mathbf{x})$ be the potential energy field of the system. Then the total energy of the particle system at a given time is

$$H(\mathbf{x}, \mathbf{p}) = U(\mathbf{x}) + k(\mathbf{p}). \tag{9.1}$$

The law of the conservation of energy says that the total energy remains constant in a closed system. Differentiating both sides of (9.1) with respect to t, we derive the Newton's law of motion from the law of energy conservation. It is mathematically more convenient to write Newton's equation in the form of *Hamiltonian equations*:

$$\dot{\mathbf{x}}(t) = \frac{\partial H(\mathbf{x}, \mathbf{p})}{\partial \mathbf{p}} \tag{9.2}$$

$$\dot{\mathbf{p}}(t) = -\frac{\partial H(\mathbf{x}, \mathbf{p})}{\partial \mathbf{x}} \tag{9.3}$$

Clearly, the first equation describes the definition of **p** and the second equation is essentially Newton's law. This classical formulation through Hamiltonian equations can be generalized to quantum mechanics, in which the energy function is replaced by a Hamiltonian operator.

9.2 Molecular Dynamics Simulation

The main task of MD simulation is the integration of the equations of motion over a given period of time and to study the physical and chemical properties of the system during a particular period (such as the effect of water in the process of protein folding). Because one can only operate discretely on a computer, the continuous-time equations of motion have to be discretized and a difference method has to be used. By standard Taylor expansion, the Hamiltonian equations can be approximated as

$$\mathbf{x}(t+dt) = \mathbf{x}(t) + \frac{\mathbf{p}(t)}{\mathbf{m}}dt + \frac{\dot{\mathbf{p}}(t)}{2\mathbf{m}}dt^2 + \cdots, \tag{9.4}$$

$$\mathbf{p}(t+dt) = \mathbf{p}(t) + \dot{\mathbf{p}}(t)dt + \frac{\ddot{\mathbf{p}}(t)}{2}dt^2 + \cdots. \tag{9.5}$$

This type of approximation forms the basis for all finite-difference methods. One of the most widely used algorithms for integrating the equations of motion is the so-called Verlet algorithm (Verlet 1967), which is the simplest and perhaps the best. The *Verlet algorithm* is based on the observation that

$$\mathbf{x}(t+dt) + \mathbf{x}(t-dt) = 2\mathbf{x}(t) + \frac{\dot{\mathbf{p}}(t)}{\mathbf{m}}dt^2 + O(dt^4), \tag{9.6}$$

$$\mathbf{x}(t+dt) - \mathbf{x}(t-dt) = 2\frac{\mathbf{p}(t)}{\mathbf{m}}dt + O(dt^3). \tag{9.7}$$

For a chosen small time increment Δt, Equation (9.6) gives rise to the position update

$$\mathbf{x}(t+\Delta t) = 2\mathbf{x}(t) - \mathbf{x}(t-\Delta t) - \frac{1}{\mathbf{m}}\frac{\partial H}{\partial \mathbf{x}}\bigg|_t (\Delta t)^2,$$

and Equation (9.7) gives rise to the momentum or, equivalently, the speed update

$$\mathbf{p}(t+\Delta t) = \mathbf{m}\frac{\mathbf{x}(t+\Delta t) - \mathbf{x}(t-\Delta t)}{2\Delta t}.$$

All the foregoing vector operations involving **m** are component-wise.

Another commonly used method for the MD simulation is the *leap-frog* method (Hockney 1970), which is equivalent to the Verlet algorithm (i.e., giving identical trajectories). The distinctive feature of the leap-frog algorithm is that it updates the momentum variable $\mathbf{p}(t)$ at half-time intervals:

$$\mathbf{x}(t+\Delta t) = \mathbf{x}(t) + \Delta t \frac{\mathbf{p}\left(t+\frac{1}{2}\Delta t\right)}{\mathbf{m}}, \tag{9.8}$$

$$\mathbf{p}\left(t+\frac{1}{2}\Delta t\right) = \mathbf{p}\left(t-\frac{1}{2}\Delta t\right) + \frac{\partial H}{\partial \mathbf{x}}\bigg|_t \Delta t. \tag{9.9}$$

Note that $\partial H/\partial \mathbf{x}$ is a function of $\mathbf{x}$ alone and does not depend on the value of $\mathbf{p}$. The momentum at time t can be computed as $[\mathbf{p}(t+\frac{1}{2}\Delta t) + \mathbf{p}(t-\frac{1}{2}\Delta t)]/2$ afterward.

To illustrate, we consider a small "ball" with mass 1 in a one-dimensional space, with Hamiltonian $H(x,p) = U(x) + k(p)$, where

$$U(x) = x^2 + a^2 - \log[\cosh(ax)], \quad k(p) = p^2/2.$$

9.2 Molecular Dynamics Simulation   187

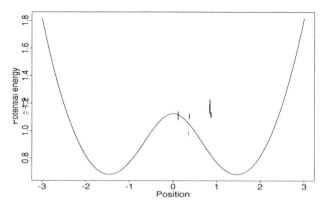

FIGURE 9.1. The shape of the potential energy $U(x)$, showing that two energetically favorable positions are at $x = a$ and $x = -a$.

For $a = 1.5$, the potential function $U(x)$ is shown in Figure 9.1.

We let the initial speed $p(0)$ be 2.0, 1.5, 1.1, 0.7, and 0.1, respectively, and let $x(0) = 1$ for all the five cases. The leap-frog algorithm was performed for 200 time steps with step size $\Delta t = 0.1$. Figure 9.2 shows that the total energy in all the cases is not exactly preserved (which is due to the discrete approximation), whereas the trajectories in the phase space are well behaved. As one can see from Figure 9.2(b), the particle travels across the two energy wells when the initial speed $p(0)$ is large (greater than 1.1), and the particle is "trapped" in the mode it is started from when the initial speed is small.

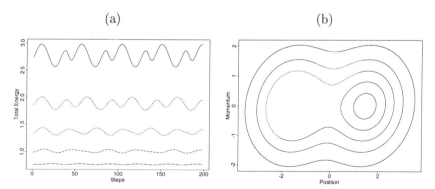

FIGURE 9.2. Molecular dynamics simulation for a toy example: (a) the serial plot for the total energy; (b) the trajectories in the phase space.

An important property of the Verlet or leap-frog algorithm is, as in the exact Hamiltonian equations, that the volume is preserved from one step to another; that is, suppose $V(0)$ is a subset in the phase space $\mathcal{X} \times \mathcal{P}$. One

can define
$$V(t) = \{(\mathbf{x}(t), \mathbf{p}(t)) : (\mathbf{x}(0), \mathbf{p}(0)) \in V(0)\}.$$
Then, the "volume preservation" property says that
$$|V(t)| \stackrel{\text{def}}{=} \int\int_{V(t)} d\mathbf{x}d\mathbf{p} = \int\int_{V(0)} d\mathbf{x}d\mathbf{p} = |V(0)|.$$

For the Hamiltonian equation, this property is called *Liouville's theorem* (Arnold 1989). To see why the volume preservation property holds in a Hamiltonian system, we define the *phase flow* mapping
$$g^t : (\mathbf{x}(0), \mathbf{p}(0)) \longmapsto (\mathbf{x}(t), \mathbf{p}(t)).$$
Then, $\{g^t : t \in (-\infty, \infty)\}$ is a group of transformations. A simple application of the Taylor expansion in combination with the Hamiltonian equations shows that, for a small t,
$$\begin{aligned} g^t(\mathbf{x}, \mathbf{p}) &= (\mathbf{x}, \mathbf{p}) + (\dot{\mathbf{x}}, \dot{\mathbf{p}})t + O(t^2) \\ &= (\mathbf{x}, \mathbf{p}) + \left(\frac{\partial H}{\partial \mathbf{p}}, -\frac{\partial H}{\partial \mathbf{x}}\right)t + O(t^2). \end{aligned}$$

Let $\mathbf{y} = (\mathbf{x}, \mathbf{p})$ and $\mathbf{f}(\mathbf{y}) = \left(\frac{\partial H}{\partial \mathbf{p}}, -\frac{\partial H}{\partial \mathbf{x}}\right)t + O(t^2)$. Then, the Hamiltonian mapping becomes
$$g^t(\mathbf{y}) = \mathbf{y} + \mathbf{f}(\mathbf{y})t + O(t^2).$$

A key argument in proving Liouville's theorem is the observation that for small t,
$$\begin{aligned} |V(t)| &= \int_{V(0)} \det\left(\frac{\partial g^t(\mathbf{y})}{\partial \mathbf{y}}\right) d\mathbf{x}d\mathbf{p} \\ &= \int_{V(0)} \det\left\{I + \frac{\partial \mathbf{f}(\mathbf{y})}{\partial \mathbf{y}}t + O(t^2)\right\}. \end{aligned}$$

By Lemma 2 of Arnold (1989), we have
$$\det(I + At) = 1 + t\,\text{tr}(A) + O(t^2).$$
Thus,
$$\begin{aligned} |V(t)| &= \int_{V(0)} \left(1 + \text{div}\left(\frac{\partial H}{\partial \mathbf{p}}, -\frac{\partial H}{\partial \mathbf{x}}\right)t + O(t^2)\right) d\mathbf{x}d\mathbf{p} \\ &= |V(0)| + O(t^2). \end{aligned}$$

The last equation follows from the fact that the divergence
$$\text{div}\left(\frac{\partial H}{\partial \mathbf{p}}, -\frac{\partial H}{\partial \mathbf{x}}\right) = \frac{\partial}{\partial \mathbf{x}}\left(\frac{\partial H}{\partial \mathbf{p}}\right) + \frac{\partial}{\partial \mathbf{p}}\left(-\frac{\partial H}{\partial \mathbf{x}}\right) = 0.$$

Thus, we have shown that $\lim_{t\to 0}(|V(t)| - |V(0)|)/t \to 0$, which implies that the volume as a function of t is constant.

It is much easier to show that the leap-frog algorithm is volume-preserving as well. From (9.9), the half-step update for the momentum gives rise to a map in the phase space:

$$(\mathbf{x}, \boldsymbol{p}) \longmapsto (\mathbf{x}, \boldsymbol{p} + f(\mathbf{x})\Delta t),$$

which results in an identity Jacobian. Therefore, this mapping is volume-preserving. The other half-step update for the position variable $\mathbf{x}$, following (9.8), results in a mapping

$$(\mathbf{x}, \boldsymbol{p}) \longmapsto (\mathbf{x} + g(\boldsymbol{p})\Delta t, \boldsymbol{p}),$$

which is also volume-preserving.

9.3 Hybrid Monte Carlo

A major advantage of molecular dynamics simulation in physical systems is its reliance on basic physics principles (e.g. Newton's equation), which has been shown by nature to work well. Typically, the trajectory of a MD simulation in the position space follows the dynamics of the potential function. As we have seen in Figure 9.2, however, the total energy cannot be exactly preserved in a MD simulation. The fluctuation can be rather substantial if the step size of MD moves is large in comparison with the initial momentum given to the ball. Therefore, a main problem with MD simulation is the stringent requirement of a small time-step size. For example, the protein folding process takes about 10^{-3} seconds in nature. A proper MD simulation of such a process needs a step size of order 10^{-12} and will take about 10^6 days using a current computer. Another potential problem with the MD simulation is its unquantified error due to discretization of the Hamiltonians.

In contrast, in a standard Metropolis-type Monte Carlo simulation, the proposal distribution cannot be easily adapted to "local dynamics" of the target distribution. For example, if the system of interest consists of closely packed particles, a random proposal for moving a particle is most likely rejected because the proposed new position has been partially occupied by others. Or, if the target distribution π is "banana shaped" (or "snake shaped"), a random proposal in the configuration space tends to waste a lot of effort poking around in wrong directions. To overcome some of these difficulties, Duane et al. (1987) introduced the method of *hybrid Monte Carlo* (HMC) which combines the basic idea of MD (i.e., proposing new positions based on Hamiltonian equations) and the Metropolis acceptance-rejection rule to produce Monte Carlo samples from a given target distribution.

Suppose of interest is to draw samples from $\pi(\mathbf{x}) \propto \exp[-U(\mathbf{x})]$. Two basic observations are important for the HMC method: (a) If we can simulate $(\mathbf{x}, \boldsymbol{p})$ from the distribution $\pi(\mathbf{x}, \boldsymbol{p}) \propto \exp[-H(\mathbf{x}, \boldsymbol{p})]$, then, marginally, $\mathbf{x} \sim \pi$ and $\boldsymbol{p}$ follows the Gaussian distribution $\phi(\boldsymbol{p}) \propto \exp(-\|\boldsymbol{p}/\sqrt{\mathbf{m}}\|^2/2)$; (b) the Hamiltonian path is "time reversible." Property (b) needs some more explanation: if we run the leap-frog algorithm, starting from $(\mathbf{x}, \boldsymbol{p})$ for t steps to reach $(\mathbf{x}', \boldsymbol{p}')$, then we can start from $(\mathbf{x}', -\boldsymbol{p}')$ and run t steps of the same algorithm to get back to the starting position but with opposite momentum [i.e., to $(\mathbf{x}, -\boldsymbol{p})$]. This property is easy to check for the leap-frog algorithm because each half-step can be reversed by negating the momentum. The importance of property (b) will made clearer later in this section.

Duane et al. (1987) introduced a "fictitious" momentum variable conjugate to the configuration variable $\mathbf{x}$ and a *guide* Hamiltonian

$$H'(\mathbf{x}, \boldsymbol{p}) = U'(\mathbf{x}) + k(\boldsymbol{p}),$$

where $\boldsymbol{p}$ is an auxiliary variable with the same dimensionality as $\mathbf{x}$, $k(\boldsymbol{p}) = \sum_{i=1}^{d} p_i^2/m_i$, and the m_i are positive quantities representing the "masses" of the components. Vector $\boldsymbol{p}$ plays the role of "momentum variable." The function $U'(\mathbf{x})$ is allowed to be different from the target one, $U(\mathbf{x})$. In the next section, we generalize this formulation to accommodate a larger class of "kinetic energy" functions, $k(\boldsymbol{p})$. The guide Hamiltonian is used to generate a proposal state. Another Hamiltonian, *the acceptance Hamiltonian*, is

$$H(\mathbf{x}, \boldsymbol{p}) = U(\mathbf{x}) + k(\boldsymbol{p}),$$

which is used to decide acceptance and rejection. The reasons that one might use two different Hamiltonians, H' and H, are two: (i) since one can only run a discrete time process on a computer, the resulting discretization of the "exact" Hamiltonian can be treated as a "guide Hamiltonian" and (ii) one sometimes wants to be flexible in proposing the new position in the phase space. For example, if the original potential energy $U(\mathbf{x})$ has steep energy barriers, having a smoother or "tempered" potential energy function [e.g., $U'(\mathbf{x}) = U(\mathbf{x})/T$ with $T > 1$] may help the sampler get out of local energy wells.

The HMC algorithm is an iterative procedure and can be implemented as follows. Suppose at time t we are at position $\mathbf{x}$ (i.e., $\mathbf{x}^{(t)} = \mathbf{x}$) of the configuration space $\mathcal{X}$ (one needs not record the momentum information). Then, at time $t + 1$, we do the following:

- generate a new momentum vector $\boldsymbol{p}$ from the Gaussian distribution [i.e., from $\phi(\boldsymbol{p}) \propto \exp\{-k(\boldsymbol{p})\}$];

- run the leap-frog algorithm (or any deterministic time-reversible and volume-preserving algorithm), starting from $(\mathbf{x}, \boldsymbol{p})$, for L steps, to obtain a new state in the phase space, $(\mathbf{x}', \boldsymbol{p}')$;

- accept the proposed state $(\mathbf{x}', \mathbf{p}')$ (i.e., let $\mathbf{x}^{(t+1)} = \mathbf{x}'$) with probability

$$\min\{1, \exp\{-H(\mathbf{x}', \mathbf{p}') + H(\mathbf{x}, \mathbf{p})\}\},$$

and let $\mathbf{x}^{(t+1)} = \mathbf{x}$ with the remaining probability.

To see why this algorithm works, we define the leap-frog move at the second step by a mapping g^L on the phase space:

$$g^L : (\mathbf{x}(0), \mathbf{p}(0)) \mapsto (\mathbf{x}(L), \mathbf{p}(L)).$$

A heuristic argument (Duane et al. 1987, Neal 1993) goes as follows: (a) If we let $(\mathbf{x}', \mathbf{p}') = g^L(\mathbf{x}, \mathbf{p})$, we have $(\mathbf{x}, -\mathbf{p}) = g^L(\mathbf{x}', -\mathbf{p}')$; (b) $\pi(\mathbf{x}, \mathbf{p}) = \pi(\mathbf{x}, -\mathbf{p})$ for any pair of $(\mathbf{x}, \mathbf{p})$; and (c) because of volume preservation, we have $d\mathbf{x}d\mathbf{p} = d\mathbf{x}'d\mathbf{p}'$. These three properties suggest that the proposal is "symmetric," as required by the Metropolis algorithm, implying that the algorithm is valid.

Here, we provide a more rigorous mathematical proof of invariance. Suppose $\mathbf{x}$ follows the target distribution $\pi(\mathbf{x})$ at time 0; after generating a new $\mathbf{p}$, making the MD transition, and deciding on rejection-acceptance, at time 1, we have a new point $\mathbf{x}^*$ with density $f(\mathbf{x}^*)$. To show that $f(\ )$ is in fact the same as $\pi(\ )$, we will show that for any square integrable function $h(\ )$, the equality $E_\pi h(\mathbf{x}) = E_f h(\mathbf{x})$ holds. To proceed, we let g^t be the Hamiltonian mapping on the phase space [i.e., $(\mathbf{x}', \mathbf{p}') = g^L(\mathbf{x}, \mathbf{p})$] and let g^{-L} denote the inverse mapping [i.e., $(\mathbf{x}, \mathbf{p}) = g^{-L}(\mathbf{x}', \mathbf{p}')$]. Then

$$E_f h(\mathbf{x}^*) = \int\int h(\mathbf{x}^*)\left[\pi(g^{-L}(\mathbf{x}^*, \mathbf{p}))\min\left\{1, \frac{\pi(\mathbf{x}^*, \mathbf{p})}{\pi(g^{-L}(\mathbf{x}^*, \mathbf{p}))}\right\}|J_{g^{-L}}|\right.$$
$$\left. + \pi(\mathbf{x}^*, \mathbf{p})\left(1 - \min\left\{1, \frac{\pi(g^k(\mathbf{x}^*, \mathbf{p}))}{\pi(\mathbf{x}^*, \mathbf{p})}\right\}\right)\right]d\mathbf{x}^*d\mathbf{p}$$

where the Jacobian term $|J_{g^{-L}}|$ is equal to 1 because of the volume preservation property of the Hamiltonian map g^L and g^{-L}. The first part of the right-hand side comes from acceptance and the second part comes from rejection.

Continuing with the equation,

$$E_f h(\mathbf{x}^*) = \int h(\mathbf{x}^*)\pi(\mathbf{x}^*, \mathbf{p})d\mathbf{x}^*d\mathbf{p}$$
$$+ \int h(\mathbf{x}^*)[\min\{\pi(g^{-L}(\mathbf{x}^*, \mathbf{p})), \pi(\mathbf{x}^*, \mathbf{p})\}$$
$$- \min\{\pi(\mathbf{x}^*, \mathbf{p}), \pi(g^L(\mathbf{x}^*, \mathbf{p}))\}]d\mathbf{x}^*d\mathbf{p} \quad (9.10)$$
$$= \int h(\mathbf{x}^*)\pi(\mathbf{x}^*, \mathbf{p})d\mathbf{x}^*d\mathbf{p} = \int h(\mathbf{x}^*)\pi(\mathbf{x}^*)d\mathbf{x}^*. \quad (9.11)$$

The first equality follows from the volume-preserving property of the mapping g^L (so that the Jacobian term disappears). Finally, because of symmetry, we have

$$\pi(\mathbf{x}, -\mathbf{p}) = \pi(\mathbf{x}, \mathbf{p}). \quad (9.12)$$

192   9. Molecular Dynamics and Hybrid Monte Carlo

If we let $(\mathbf{x}', \mathbf{p}') = g^L(\mathbf{x}, \mathbf{p})$, then time-reversibility means that
$$g^L(\mathbf{x}', -\mathbf{p}') = (\mathbf{x}, \mathbf{p}). \tag{9.13}$$
Combining (9.12) and (9.13), we have
$$\pi(\mathbf{x}, -\mathbf{p}) = \pi(\mathbf{x}, \mathbf{p}) = \pi(g^{-L}(\mathbf{x}', \mathbf{p}')) = \pi(g^L(\mathbf{x}', -\mathbf{p}')).$$
Thus, the equality between (9.10) and (9.11) follows from
$$\int h(\mathbf{x}') \min\{\pi(g^{-L}(\mathbf{x}', \mathbf{p}')), \pi(\mathbf{x}', \mathbf{p}')\} d\mathbf{x}' d\mathbf{p}'$$
$$= \int h(\mathbf{x}') \min\{\pi(g^{-L}(\mathbf{x}', -\mathbf{p}')), \pi(\mathbf{x}', -\mathbf{p}')\} d\mathbf{x}' d\mathbf{p}'$$
$$= \int h(\mathbf{x}') \min\{\pi(g^{-L}(\mathbf{x}', -\mathbf{p}')), \pi(\mathbf{x}', \mathbf{p}')\} d\mathbf{x}' d\mathbf{p}'$$
$$= \int h(\mathbf{x}) \min\{\pi(\pi(\mathbf{x}, \mathbf{p}), g^L(\mathbf{x}, \mathbf{p}))\} d\mathbf{x} d\mathbf{p}.$$

At this point, we have proved that $f(\mathbf{x}') = \pi(\mathbf{x}')$, meaning that the HMC transition leaves the target distribution $\pi(\mathbf{x})$ invariant. Besides sharing the same volume-preserving and time-reversible properties as the guide Hamiltonian, the actual Hamiltonian also preserves the value of $H(\mathbf{x}, \mathbf{p})$, thus incurring no rejections. It is also clear from the discussion in Section 9.2 that the leap-frog moves are valid for proposing moves in HMC (i.e., volume-preserving and time-reversible).

9.4 Algorithms Related to HMC

9.4.1 Langevin-Euler moves

As in the previous subsection, we let $\pi(\mathbf{x}) \propto \exp\{-U(\mathbf{x})\}$ be the target distribution of interest. The Langevin diffusion refers to the following result: The solution of the stochastic differential equation
$$d\mathbf{x}_t = -\frac{1}{2}\frac{\partial U(\mathbf{x}_t)}{\partial \mathbf{x}} dt + dW_t,$$
where W_t the standard Brownian motion, follows the target distribution π. Thus, a discretization of the equation is of the form
$$\mathbf{x}_{t+1} = \mathbf{x}_t - \frac{1}{2}\frac{\partial U(\mathbf{x}_t)}{\partial \mathbf{x}} h + \sqrt{h} Z_t,$$
where Z_t follows a standard Gaussian distribution. It we let $\sqrt{h} = dt$ in Equation (9.4) and let $\mathbf{p}_t$ be a refreshed momentum drawn from the Gaussian distribution $\phi(\mathbf{p})$, the Langevin update is exactly the same as the second order Taylor expansion of the Newton's law. Thus, a Langevin update is equivalent to a single-step hybrid Monte Carlo move Neal (1993).

9.4.2 Generalized hybrid Monte Carlo

Suppose we can augment the sample space of **x** to include a "momentum" variable, say p. So now the pseudo-phase space consists of $\phi = (\mathbf{x}, p)$ and the target distribution $\pi(\mathbf{x})$ is augmented to $\pi(\phi)$. Furthermore, suppose we can find a irreducible transition rule on the phase space, $T(\phi, \phi')$, so that the detail balance $\pi^*(\phi)T(\phi, \phi') = \pi^*(\phi)T(\phi', \phi)$ is maintained for some non-negative function $\pi^*(\phi)$ (unique up to a normalizing constant). Let the current position in the original space be $\mathbf{x}^{(0)}$. Then we can supplement it with a momentum realization $p^{(0)}$ in a suitable way and then evolve the phase space point $\phi^{(0)} = (\mathbf{x}^{(0)}, p^{(0)})$ according to the surrogate transition $T(\cdot, \cdot)$. After k steps, we accept the new point with probability

$$\min\left\{1, \frac{\pi(\phi^{(1)})/\pi^*(\phi^{(k)})}{\pi(\phi^{(0)})/\pi^*(\phi^{(0)})}\right\}.$$

In hybrid Monte Carlo, one augment the original target distribution $\pi(\mathbf{x})$ to a distribution defined on the phase space; that is, $\pi(\phi) = \pi(\mathbf{x}) \times f(p)$, where $\log f(p) = k(p) = -\|p/\sqrt{\mathbf{m}}\|^2/2$. The leap-frog algorithm leads to a uniform π^*. From the proof in the previous section and the construction of the leap-frog moves, we see that the kinetic energy function $k(p)$ can be rather arbitrary. The basic requirements for $k(p)$ are that (a) it is symmetric in p [i.e., $k(p) = k(-p)$] and (b) it is bounded from below. In this case, the leap-frog algorithm becomes

$$\mathbf{x}(t + \Delta t) = \mathbf{x}(t) + \Delta t \left.\frac{\partial H}{\partial p}\right|_{t+\frac{\Delta t}{2}},$$

$$p\left(t + \frac{\Delta t}{2}\right) = p\left(t - \frac{\Delta t}{2}\right) + \Delta t \left.\frac{\partial H}{\partial \mathbf{x}}\right|_t.$$

By the same argument as in Section 9.2, we can clearly see that this generalized leap-frog move is volume-preserving. To see why it is time reversible when the kinetic energy function $k(p)$ is symmetric, we let $\mathbf{x}'(t) = \mathbf{x}(t+\Delta t)$, $p'(t - \Delta t/2) = -p(t + \Delta t/2)$, and apply one leap-frog step. Then, we will end at position $\mathbf{x}(t)$ with the negated momentum $-p(t - \Delta t/2)$. It is not clear, though, how to choose an effective kinetic energy function.

An interesting special case of the generalized hybrid Monte Carlo (GHMC) for the one-dimensional sampler was proposed by Gustafson (1998). First, he augmented the original state space x to (x, P) where P is either -1 or 1 and has a marginal uniform distribution. Suppose that currently we are at $(x^{(t)}, p^{(t)})$; then, the proposed new configuration is $y = x^{(t)} + p^{(t)}|Z|$, where $Z \sim N(0, \sigma^2)$, and let

$$(x^{(t+1)}, p^{(t+1)}) = \begin{cases} (y, p^{(t)}) & \text{with probability } r \\ (x^{(t)}, -p^{(t)}) & \text{with probability } 1 - r, \end{cases}$$

where r is the usual Metropolis ratio. It is very easy to generalize the algorithm to higher-dimensional cases: The auxiliary variable P is a uniformly distributed random direction in a d-dimensional space and $Z = \sqrt{X^2}$, where X^2 has a χ^2-distribution with d degrees of freedom. Although effective for one-dimensional problems, this method may not be useful in multidimensional cases. The reason is that negating direction P_n in a high-dimensional space typically will not lead to a better search direction.

9.4.3 Surrogate transition method

In some Monte Carlo simulation problems (e.g., simulation of polarized liquid water), evaluation of the energy function $h(\mathbf{x}) = -\log \pi(\mathbf{x})$ involves expensive computation (such as inverting a 2000×2000 matrix to compute the polarization vector). It is often very inexpensive, however, to obtain a reasonably good approximation $h^*(\mathbf{x})$ of $h(\mathbf{x})$. For example, instead of solving a large linear equation completely, one can opt to perform a few rounds of iterative updates. Then, the Metropolis algorithm needs to be adjusted to accommodate this variation.

Mathematically, we assume that one can conduct a *reversible* Markov transition $S(\mathbf{x}, \mathbf{y})$ (surrogate) which leaves $\pi^*(\mathbf{x}) \propto \exp\{-h^*(\mathbf{x})\}$ invariant; that is, the detailed balance

$$\pi^*(\mathbf{x})S(\mathbf{x}, \mathbf{y}) = \pi^*(\mathbf{y})S(\mathbf{y}, \mathbf{x})$$

is satisfied. A valid surrogate transition can be devised by making use of the Metropolis principle on $\pi^*(\mathbf{x})$.

Suppose our current sample is $\mathbf{x}^{(t)}$. We let $\mathbf{y}_0 = \mathbf{x}^{(t)}$ and recursively sample $\mathbf{y}_i \sim S(\mathbf{y}_{i-1}, \cdot)$ for $i = 1, \ldots, k$. Then, we update $\mathbf{x}^{(t+1)} = \mathbf{y}_k$ with probability

$$\min\left\{1, \frac{\pi(\mathbf{y}_k)/\pi^*(\mathbf{y}_k)}{\pi(\mathbf{x}^{(t)})/\pi^*(\mathbf{x}^{(t)})}\right\} \tag{9.14}$$

and let $\mathbf{x}^{(t+1)} = \mathbf{x}^{(t)})$ with the remaining probability.

To show that the foregoing procedure is valid, we see that the proposal transition function from $\mathbf{y}_0$ to $\mathbf{y}_k$ can be formally written as

$$S^{(k)}(\mathbf{y}_0, \mathbf{y}_k) = \int \cdots \int S(\mathbf{y}_0, \mathbf{y}_1) \cdots S(\mathbf{y}_{k-1}, \mathbf{y}_k) d\mathbf{y}_1 \cdots d\mathbf{y}_{k-1}.$$

In words, $S^{(k)}(\cdot, \cdot)$ is the k-step transition function for the surrogate Markov chain defined by S. It is easily seen that $\pi^*(\mathbf{x})S^{(k)}(\mathbf{x}, \mathbf{y}) = \pi^*(\mathbf{y})S^{(k)}(\mathbf{y}, \mathbf{x})$. Thus, the actual transition function from $\mathbf{x}^{(t)} = \mathbf{x}$ to $\mathbf{x}^{(t+1)} = \mathbf{y} \neq \mathbf{x}$ has the form

$$A(\mathbf{x}, \mathbf{y}) = S^{(k)}(\mathbf{x}, \mathbf{y}) \min\left\{1, \frac{\pi(\mathbf{y})/\pi^*(\mathbf{y})}{\pi(\mathbf{x})/\pi^*(\mathbf{x})}\right\}.$$

Hence,

$$\pi(\mathbf{x})A(\mathbf{x},\mathbf{y}) = \pi^*(\mathbf{x})S^{(k)}(\mathbf{x},\mathbf{y})\min\left\{\frac{\pi(\mathbf{x})}{\pi^*(\mathbf{x})}, \frac{\pi(\mathbf{y})}{\pi^*(\mathbf{y})}\right\}$$
$$= \pi^*(\mathbf{y})S^{(k)}(\mathbf{y},\mathbf{x})\min\left\{\frac{\pi(\mathbf{x})}{\pi^*(\mathbf{x})}, \frac{\pi(\mathbf{y})}{\pi^*(\mathbf{y})}\right\}$$
$$= \pi(\mathbf{y})A(\mathbf{y},\mathbf{x}),$$

which is the detailed balance condition. A surrogate transition procedure with $k=1$ is given by Liu and Chen (1998) under the sequential importance sampling framework. We find that the surrogate procedure can be more generally applicable in Monte Carlo simulation of complicated systems. At a conceptual level, the HMC can be seen as a surrogate transition method in which a discretized Hamiltonian is used as a surrogate to guide for the dynamical moves in the phase space (joint space of positions and momentums). Because of the time reversibility and volume preservation properties, the corresponding π^* in HMC is the uniform distribution.

9.5 Multipoint Strategies for Hybrid Monte Carlo

9.5.1 Neal's window method

In a standard HMC algorithm, each Monte Carlo update involves L (often between 40 and 70) steps of deterministic Hamiltonian leap-frog moves. The acceptance-rejection decision is made, however, based only on the "energy" comparison between the starting and the ending configurations of the leap-frog trail. Neal (1994) suggested that some middle steps in the trail can also be used in order to increase the acceptance rate in HMC. Suppose the current position vector is $\mathbf{x}^{(t)}$ and we generate a renewed momentum vector $\boldsymbol{p}^{(t)}$ from its Gaussian distribution. Then, the "window algorithm" can be stated as follows:

- Choose a window size $W < L$ (either deterministically or from a fixed distribution).

- Draw K from $\{0, 1, \ldots, W-1\}$ uniformly.

- Starting from the current point in the phase space $\phi(0) = (\mathbf{x}^{(t)}, \boldsymbol{p}^{(t)})$, run the leap-frog steps backward for K steps and forward for $L-K$ steps to result in a trajectory
$$\phi(-K), \ldots, \phi(-1), \phi(0), \phi(1), \ldots, \phi(L-K),$$

- Place the "acceptance window" of size W at the end of the trajectory:
$$\mathcal{A} = \{\phi(L-K-W+1), \ldots, \phi(L-K)\};$$

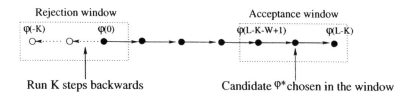

FIGURE 9.3. Illustration of Neal's "window" hybrid Monte Carlo method.

and the "rejection window" at the beginning

$$\mathcal{R} = \{\phi(-K), \ldots, \phi(-K+W-1)\}.$$

- Define the "free energy" of a window $\mathcal{W}$ as

$$F(\mathcal{W}) = -\log\left(\sum_{\phi(j) \in \mathcal{W}} \exp\{-H(\phi(j))\}\right).$$

Go to the acceptance window $\mathcal{A}$ with probability

$$\min\{1, \exp\{F(\mathcal{A}) - F(\mathcal{R})\}\}.$$

and stay in the rejection window $\mathcal{R}$ with the remaining probability.

- Having decided on the window $\mathcal{W}$, a particular state $\phi = \phi(j)$ within that window is selected according to probability

$$P(\phi(j)) = \exp\{-H(\phi(j)) + F(\mathcal{W})\}.$$

The new state is updated as $(\mathbf{x}^{(t+1)}, \mathbf{p}^{(t+1)}) = \phi$.

A graphical illustration is given in Figure 9.3. (Neal 1994) showed by examples that this approach, not surprisingly, can improve the acceptance rate of the HMC algorithm. However, it is less clear if the window method actually improves the computational efficiency of the algorithm (i.e., resulting in a more rapidly mixing Markov chain).

9.5.2 Multipoint method

The multipoint Metropolis method described in Section 5.5 allows one to choose a good candidate among multiple random proposals. Its basic principle can be further extended to accommodate the multiple leap-frog steps in the HMC algorithm.

Similar to the description of the window method in the previous section, we define $\phi(0) = (\mathbf{x}^{(t)}, \mathbf{p}^{(t)})$, where $\mathbf{p}^{(t)}$ is a "renewed" momentum vector, and conduct the following procedures:

Multipoint HMC

9.5 Multipoint Strategies for Hybrid Monte Carlo

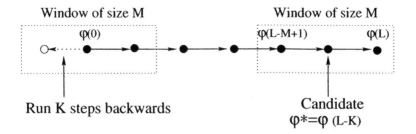

FIGURE 9.4. Illustration of the multiple-trial HMC method.

- From the starting states, $\phi(0)$, run L leap-frog iterations to obtain $\phi(1), \ldots, \phi(L)$.

- Select one candidate ϕ' from the last M configurations [i.e., ϕ' is chosen from $\phi(L - M + 1), \ldots, \phi(L)$] according to their respective Boltzmann probabilities; that is,

$$\Pr(\phi' = \phi(L - M + k)) \propto w_k \exp\{-H(\phi(L - M + k))\},$$

for $k = 1, \ldots, M$. Suppose we have selected $\phi' = \phi(L - K)$.

- Run K reversed leap-frog steps from $\phi(0)$ (by using the negated momentum) to get $\phi(-1), \ldots, \phi(-K)$.

- Let $(\mathbf{x}, \mathbf{p}) = \phi'$ with probability

$$p = \min\left\{1, \frac{\sum_{j=1}^{M} w_j \exp\{-H(\phi(L - M + j))\}}{\sum_{j=1}^{M} w_j \exp\{-H(\phi(M - K - j))\}}\right\},$$

and let $(\mathbf{x}, \mathbf{p}) = \phi(0)$ with probability $1 - p$.

- Let $\mathbf{x}^{(t+1)} = \mathbf{x}$ and sample (renew) $\mathbf{p}^{(t+1)}$ from $\mathcal{N}(\mathbf{0}, \Sigma)$, where $\Sigma = \text{diag}(m_1^{-1}, \ldots, m_d^{-1})$.

The w_j in the algorithm are non-negative numbers are completely controlled by the user. The role of this weighting vector to give a prior preference for certain points on the leap-frog trajectory. For example, one may wish to emphasize those points that are farther to the end of the leap-frog trajectory more. In this case, we may choose $w_j \propto \sqrt{j}$. A graphical view of the multipoint method is given in Figure 9.4. One can see that the multipoint approach is very similar to the window method, but is more flexible and more efficient in general (Qin and Liu 2001). The correctness of this algorithm can be shown by combining a similar argument in validating the general multipoint approach with the volume preservation and time reversibility properties of the leap-frog moves.

9.6 Application of HMC in Statistics

Using the HMC to solve statistical inference problems was first introduced by Neal (1996). This effort was only 10 years behind that in physics and theoretical chemistry. In contrast, statisticians were 40 years late in using the Metropolis algorithm. The connection between statistical problems, especially Bayesian inference problems, and physics problems is, in fact, quite simple: Since all unknowns in a probabilistic model, be they tuning parameters, missing observations, latent structures, or observed data, can be treated as joint random variables in a Bayesian framework, all the inference tasks can be reduced to the evaluation of certain expectation with respect to the *posterior distribution* of unknown variables. This target posterior distribution can always be written out explicitly, up to a normalizing constant, as

$$\pi(\boldsymbol{\theta}) \propto f(\mathbf{y} \mid \boldsymbol{\theta})\pi_0(\boldsymbol{\theta}) \equiv c\exp\{-U(\boldsymbol{\theta})\},$$

where $f(\cdot)$ is the probabilistic model that connects data with unknown parameters, $\pi_0(\ )$ is the prior distribution of $\boldsymbol{\theta}$, and

$$U(\boldsymbol{\theta}) = -\log f(\mathbf{y} \mid \boldsymbol{\theta}) - \log \pi_0(\boldsymbol{\theta}).$$

In order to use the HMC to sample from this posterior distribution, we need to introduce an auxiliary "momentum" variable $\boldsymbol{p}$ and construct the "guide Hamiltonian" $H(\boldsymbol{\theta}, \boldsymbol{p}) = U(\boldsymbol{\theta}) + k(\boldsymbol{p})$.

Although it is conventional to let

$$k(\boldsymbol{p}) = \sum_{i=1}^{d} p_i^2/2m_i,$$

this pseudo-energy function can be chosen more flexibly (Section 9.4.2). Even with the conventional choice of $k(\boldsymbol{p})$, a tricky question is how to choose appropriate m_i's so as to make the algorithm more efficient. Intuitively, m_i is the "mass" of the ith component. Thus, the larger an m_i is, the "slower" that component moves. The efficiency of HMC can be improved by setting m_i differently for each p_i according to the properties of component x_i. A related important issue is the step-size choice for the leap-frog moves in HMC. In the following subsections, we describe a few statistical inference problems in which the HMC has an obvious edge over other MCMC strategies. We hope that these expository descriptions will motivate other researchers to study theoretical properties of HMC and to investigate strategies for tuning the algorithm.

9.6.1 Indirect observation model

Consider the following statistical inference problem. Let θ be a parameter vector in some parameter space Ω and let $\mathbf{x}(\theta)$ be a random vector whose distribution is known completely if θ is given. Suppose, however, that we cannot directly observe $\mathbf{x}$, but observe, instead, a vector $\mathbf{y}$ whose relationship with $\mathbf{x}$ and θ can be described as

$$\mathbf{y} = g(\mathbf{x}(\theta), \theta), \qquad (9.15)$$

where the functional form of $g(\cdot)$ is known. Of interest is the Bayesian inference on the parameter vector θ. We further assume that inverting the function g is difficult; thus, obtaining the likelihood function of θ is infeasible. The standard approach in literature is the *simulated method of moment* (McFadden 1989), but it may be inefficient when the moment functionals are not chosen properly. We take a likelihood-based approach here.

To resolve the difficulty that the likelihood function cannot be evaluated easily, we modify model (9.15) by introducing a Gaussian noise and pretend that $\mathbf{y}$ is drawn from the modified model

$$\mathbf{y} = g(\mathbf{x}(\theta), \theta) + \epsilon \qquad (9.16)$$

where $\epsilon \sim N(0, \sigma^2 \mathbf{I})$, $\mathbf{I}$ is the identity matrix, and σ is a tuning parameter controlled by the user. Therefore, under model (9.16), the joint posterior distribution of $\mathbf{x}$ and θ can be derived as follows:

$$\pi_\sigma(\mathbf{x}, \theta \mid \mathbf{y}) \propto f_\sigma(\mathbf{y} \mid \mathbf{x}, \theta; \sigma^2) f(\mathbf{x} \mid \theta) \pi_0(\theta), \qquad (9.17)$$

where π_0 is the prior for θ. When the data vector $(\mathbf{x}, \mathbf{y})$ can be decomposed as $(x_1, y_1), \ldots, (x_n, y_n)$ and they are i.i.d. given θ, then we have

$$\pi_\sigma(\mathbf{x}, \theta \mid \mathbf{y}) \propto \prod_{i=1}^{n} \left[f_\sigma(y_i \mid x_i, \theta; \sigma^2) \right) f(x_i \mid \theta) \right] \pi_0(\theta),$$

where f_σ stands for the density function of the modified model (9.16). The posterior distribution $\pi_\sigma(\theta \mid \mathbf{y})$ is then a marginal of $\pi_\sigma(\mathbf{x}, \theta \mid \mathbf{y})$. It can be shown that under mild conditions, $\pi_\sigma(\theta \mid \mathbf{y})$ converges to the posterior density $\pi(\theta \mid y)$ almost surely as $\sigma \to 0$.

Our strategy (Chen, Qin and Liu 2001) for the estimation in indirect observation model is to choose a sequence of σ's:

$$\sigma_1 > \sigma_2 > \cdots > \sigma_l,$$

and generate Monte Carlo samples from all the π_{σ_j}, $j = 1, \ldots, l$. If we are interested in the posterior mean $\hat{\theta}$ of θ, for example, we compute the posterior means $\hat{\theta}_j$ under the jth modified distribution and then fit a quadratic function

$$\theta_j = \beta_0 + \beta_1 \sigma_j + \beta_2 \sigma_j^2 + e_j.$$

The estimate $\hat{\beta}_0$ of β_0 then corresponds to the unmodified model and serves as our final estimate of $\hat{\theta}$.

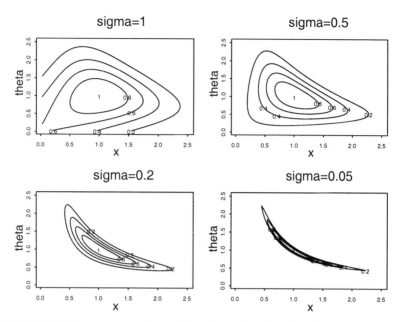

FIGURE 9.5. The contour plots of density $\pi_\sigma(x,\theta)$ in (9.18) with a flat prior $\pi_0(\theta) \equiv c$.

A technical difficulty remains in our approach: Monte Carlo sampling from (9.17) is generally difficult to do. In particular, as $\sigma \to 0$, the posterior distribution of π_σ lives on an almost degenerate subspace, making the random-walk-type Metropolis algorithm very inefficient. For example, a trivial example is

$$y = \theta x, \quad x \sim N(\theta, 1).$$

Under the modified model, we have

$$\pi_\sigma(x,\theta) \propto \exp\left\{-\frac{(y-x\theta)^2}{2\sigma^2} - \frac{(x-\theta)^2}{2}\right\} \pi_0(\theta). \tag{9.18}$$

When σ is small, the sample is forced to lie around the curve $x\theta = y$. See the contour plots in Figure 9.5. Thus, a random directional proposal will almost surely fail. To overcome this computational difficulty, we can apply the HMC method to draw samples from each π_{σ_i}.

Having to simulate from multiple related systems is also a common problem in physics and chemistry where it is often of interest to estimate a certain system property at different temperatures. Several "tempering" strategies (Chapter 10) have been developed in the physics and statistics literature for improving MCMC efficiencies in these problems. We found that the parallel tempering method (Geyer 1991) is especially useful for our problem. The basic idea of parallel tempering is to run the l sampling processes (each corresponding to a π_{σ_i}) in parallel and to allow the sampler to switch between different configurations corresponding to different distributions.

More precisely, suppose φ_i and φ_j are the current states for two HMC chains corresponding to π_{σ_i} and π_{σ_j} ($i \neq j$). An exchange of the two configurations is proposed with a small probability δ. If the exchange move is proposed, it is accepted with probability

$$\min\left\{1, \frac{\exp\{-H(\varphi_i;\sigma_j^2) - H(\varphi_j;\sigma_i^2)\}}{\exp\{-H(\varphi_i;\sigma_i^2) - H(\varphi_j;\sigma_j^2)\}}\right\}. \qquad (9.19)$$

Since the detailed balanced condition is never violated, each HMC chain converges to the stationary distribution with its particular value of σ.

9.6.2 Estimation in the stochastic volatility model

The stochastic volatility (SV) model is a nonlinear state-space model (Sections 1.6, 3.3, and 4.5) and can be considered as a generalization of the celebrated Black-Scholes formula (Hull and White 1987). A simple stochastic model has the form

$$y_t = \epsilon_t \beta \exp(x_t/2), \quad x_{t+1} = \phi x_t + \eta_t, \quad t = 1, \ldots, T, \qquad (9.20)$$

where $\epsilon_t \sim N(0,1)$ and $\eta_t \sim N(0,\sigma^2)$. One can see that the $\log\{\text{var}(y_t)\}$ in (9.20) follows an AR(1) process. Because of its nonlinear nature, the model parameters are difficult to estimate. Shephard and Pitt (1997) suggested a way to use the Gibbs sampler to obtain Bayes estimates. They noted that the usual Gibbs sampler converges extremely slowly and developed an improved MCMC algorithm based on conditional sampling of a block of variables (Liu and Sabatti 2000). An even more efficient algorithm was suggested lately in Chib, Nardari and Shephard (2002).

We recently reported some promising results for using a HMC-based algorithm to compute the Bayes estimates in a SV model (Chen et al. 2001). Our dataset consists of daily exchange rates of pound/dollar from 10/1/1981 to 6/28/1985 (a total of $T=946$ observations). Let r_t be the daily exchange rate and let $dr_t = \log r_{t+1} - \log r_t$. Since the mean of y_t is zero in the SV model, we define y_t as

$$y_t = 100\left(dr_t - \sum dr_t/T\right). \qquad (9.21)$$

for $t = 1, \ldots, T$ and fit a SV model (9.20) with these y_t.

Let $\mathbf{x} = (x_1, \ldots, x_T)$ and $\mathbf{y} = (y_1, \ldots, y_t)$, and let the prior for β be $p(\beta^2) \propto \beta^{-2}$ (improper); for σ^2, Inv-$\chi^2(10, 0.05)$; and for $(\phi+1)/2$, a beta prior with shape parameters 20 and 1.5. Then the following conditional distributions can be easily sampled from:

$$\beta^2 \mid \mathbf{y}, \mathbf{x} \sim \text{Inv-}\chi^2\left(T, \frac{1}{T}\sum_{t=1}^{T}\frac{y_t^2}{\exp(x_t)}\right),$$

$$\sigma^2 \mid \phi, \mathbf{x} \sim \text{Inv-}\chi^2(T+10, V),$$

$$\pi(\phi \mid \sigma^2, \mathbf{x}) \propto \exp\left\{-\frac{x_1^2(1-\phi^2) + \sum_{t=2}^{T}(x_t - \phi x_{t-1})^2}{2\sigma^2}\right\}(1+\phi)^{19.5}(1-\phi),$$

where

$$V = \frac{1}{T+10}\left[0.5 + x_1^2(1-\phi^2) + \sum_{t=2}^{T}(x_t - \phi x_{t-1})^2\right].$$

Once the parameter values are given, the negative log-density is

$$U(\mathbf{x}) = \sum_{t=1}^{T}\left\{\frac{x_t}{2} + \frac{y_t^2}{2\beta^2 \exp(x_t)}\right) + \frac{x_1^2(1-\phi^2)}{2\sigma^2}\right\} + \sum_{t=1}^{T-1}\frac{(x_{t+1} - \phi x_t)^2}{2\sigma^2}.$$

The posterior density of $\mathbf{x}$, given the parameter values, is proportional to $\exp\{-U(x)\}$.

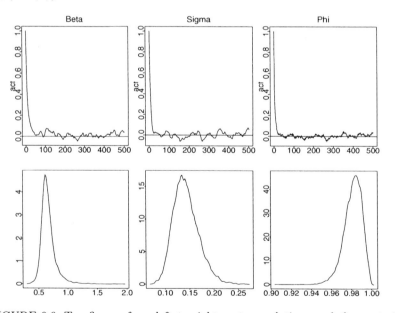

FIGURE 9.6. Top figures from left to right: autocorrelations and the posterior density estimates or β, σ and ϕ; bottom figures from left to right: their respective posterior densities.

We implemented the following iterative sampling algorithm: Given $\mathbf{x}$, we drew the parameters β, ϕ, and σ^2 from the above conditional distributions;

whereas given the realized values of the parameters, we drew the state variable $\mathbf{x}_t$ by the HMC. The step sizes for the HMC moves of $x_2, \ldots, x_T$ were chosen as 0.03 and that for x_1 was 0.06.

This Gibbs-HMC method were run for 28,000 iterations and the results from the last 20,000 iterations are reported in Figure 9.6 and Table 9.1. It can be seen from Figure 9.6 that the efficiency of the Gibbs-HMC algorithm is comparable to that of the "grouping" method in Shephard and Pitt (1997). A relative advantage of the HMC approach is that no special Gaussian approximations of certain likelihood function is needed, which makes our method more automatic.

Parameter	Mean	Standard Error	Covariance		
β	0.6433	0.0958	9.17e−03	−2.16e−04	1.48e−04
σ	0.1481	0.0273	−2.16e−04	7.44e−04	−1.64e−04
ϕ	0.9801	0.0093	1.48e−04	−1.64e−04	8.64e−05

TABLE 9.1. Bayes estimates of the parameters in the stochastic volatility model.

10
Multilevel Sampling and Optimization Methods

In this chapter, we describe a few innovative ideas in using auxiliary distributions and multiple Markov chains (in parallel) to improve the efficiency of Monte Carlo simulations. Roughly speaking, in order to improve the mixing property of an underlying Monte Carlo Markov chain, one can build a few "companion chains" whose sole purpose is to help bridging parts of the sample space that are separated by very high energy (or low probability) barriers in the original distribution. This bridging idea is achieved most simply by incorporating an auxiliary "temperature" parameter, which is based on the following simple but important observation: A "tempered" distribution

$$\pi_T(\mathbf{x}) \propto \exp\{-h(\mathbf{x})/T\}$$

becomes "flatter" (more uniform) as the temperature T increases and becomes "spikier" as T decreases toward 0. The augmented system with companion chains often consists of a collection of tempered distributions with different temperature parameters. When $T \to 0$, all of the probability mass will be concentrated around the global minimum of $h(\mathbf{x})$.

If a Monte Carlo sampler can move freely among the augmented system according to the Metropolis rule, then satisfactory results will be obtained for the distribution with the lowest temperature. In practice, however, this is not easy to achieve. A major difficulty concerns the number of the auxiliary distributions to be used. Since the performance of the method depends on how freely the sampler can move in the augmented system, one must make sure that the waiting time for a "traversal" of all the distributions in the system is not too large. Another interesting topic has to do with the choice of auxiliary distributions. Although tempered distributions are most frequently used, other forms with reduced structures can often be more effective (Liu and Sabatti 1998, Wong and Liang 1997).

10.1 Umbrella Sampling

The essential idea of using multiple "scaled" (or "tempered") densities to improve the Monte Carlo estimation seems to appear first in Torrie and Valleau (1977), where they proposed a method, the *umbrella sampling*, for estimating the ratio of two partition functions (or normalizing constants). The problem can be stated generally as follows: Suppose we can evaluate analytically two non-negative and integrable functions, $q_0(\mathbf{x})$ and $q_1(\mathbf{x})$. We are interested in knowing the ratio

$$A = \frac{\int q_1(\mathbf{x})d\mathbf{x}}{\int q_0(\mathbf{x})d\mathbf{x}} \equiv \frac{Z_1}{Z_0}.$$

Clearly, $\pi_i(\mathbf{x}) = q_i(\mathbf{x})/Z_i$ defines a probability density. A rather simple idea is to simulate $\mathbf{x}^{(1)}, \ldots, \mathbf{x}^{(m)}$ from π_0 and to estimate A by averaging the likelihood ratio:

$$\hat{A} = \frac{1}{m}\left\{\frac{q_1(\mathbf{x}^{(1)})}{q_0(\mathbf{x}^{(1)})} + \cdots + \frac{q_1(\mathbf{x}^{(m)})}{q_0(\mathbf{x}^{(m)})}\right\}.$$

This estimate is legitimate because, by ergodicity, $\hat{A}$ converges to

$$E_0\left\{\frac{q_1(\mathbf{x})}{q_0(\mathbf{x})}\right\} = \int \frac{q_1(\mathbf{x})}{q_0(\mathbf{x})}\frac{q_0(\mathbf{x})}{c_0}d\mathbf{x} = A.$$

However, this method can be very inefficient when the two distributions π_1 and π_0 differ too much, in which case the variance of the likelihood ratio is huge. To overcome this difficulty, Torrie and Valleau (1977) noticed that we can, in fact, simulate the Monte Carlo samples $\mathbf{x}^{(1)}, \ldots, \mathbf{x}^{(m)}$ from another "umbrella distribution" $\pi_u(\mathbf{x})$ (known up to a normalizing constant). Note that for any $\pi_u(\mathbf{x})$, with respect to which π_0 and π_1 are absolutely continuous, we have the identity

$$\frac{E_u\{q_1(\mathbf{x})/q_u(\mathbf{x})\}}{E_u\{q_0(\mathbf{x})/q_u(\mathbf{x})\}} = A, \tag{10.1}$$

where $q_u(\mathbf{x}) = c_u\pi_u(\mathbf{x})$ and $E_u(\ )$ stands for the expectation with respect to π_u. Thus, we can use the ratios of two average likelihood ratios to estimate A. If one chooses a proper π_u which covers simultaneously the "importance regions" of the both target distributions, then a very accurate estimate of A can be obtained. This idea is generally termed as *umbrella sampling*. According to Valleau (1999), an umbrella sampler

... is designed to span a substantial range of different physical situations in a single MC run, sampling on a distribution quite unlike an ordinary ensemble distribution.

A challenge in designing a useful umbrella sampler is to actually find a good "umbrella distribution." The first suggestion of Torrie and Valleau (1977) is to consider the class of distributions

$$\pi_u(\mathbf{x}) \propto w[\Delta h(\mathbf{x})]\pi_0(\mathbf{x}),$$

where $\pi_0(\mathbf{x})$ and $\pi_1(\mathbf{x})$ are of the Boltzmann form

$$\pi_i(\mathbf{x}) \propto \exp\{-h_i(\mathbf{x})/kT_i\}, \quad i = 0, 1$$

and $\Delta h(\mathbf{x}) = h_1(\mathbf{x})/kT_1 - h_0(\mathbf{x})/kT_0$. The weight function $w(\ )$ can be chosen by trial and error in a pilot study. In particular, one may want to choose $w(\ )$ to favor those configurations with relatively large contributions to the estimate of A (i.e., smaller Δh). In the case when $h_0(\mathbf{x}) = h_1(\mathbf{x})$, the two distributions correspond to two systems with the same "internal energy function" but at different temperatures. In this case, the weighting function is a function of the *energy variable* $U_0 = h_0(\mathbf{x})$. Thus, the reweighting idea of Torrie and Valleau (1977) can be seen as a precursor to the idea of multicanonical sampling (Berg and Neuhaus 1991) and $1/k$-ensemble sampling (Hesselbo and Stinchcombe 1995), both of which attempt to choose the weight function by a recursive procedure so as to achieve a certain marginal distribution for the energy variable U_0 (see Section 10.5 for more details).

When $h_0(\mathbf{x}) = h_1(\mathbf{x})$ but $T_0 \neq T_1$ (without loss of any generality, we let $T_0 > T_1$), Torrie and Valleau suggested the second idea which can be seen as a precursor to various tempering approaches. In particular, they consider intermediate systems of the "temperature-scaling" form

$$\pi_{\alpha_i}(\mathbf{x}) \propto \exp\{-h_0(\mathbf{x})/T_{\alpha_i}\},$$

for $T_0 > T_{\alpha_1} > \cdots > T_{\alpha_{k-1}} > T_1$, where $0 < \alpha_1 < \cdots < \alpha_{k-1} = 1$. A particular choice of the temperature sequences is $T_{\alpha_i} = T_0 + \alpha_i(T_1 - T_0)$. With these intermediate systems, one can estimate the ratios of the normalizing constants for all pairs of neighboring systems, $c_{\alpha_{i+1}}/c_{\alpha_i}$, by MC sampling from the relatively hotter system, π_{α_i}. Then, the ratio A can be estimated as

$$\hat{A} = \frac{\hat{c}_{\alpha_1}}{\hat{c}_0} \times \cdots \times \frac{\hat{c}_1}{\hat{c}_{\alpha_{k-1}}}.$$

The idea of using temperature-scaled distributions to aid Markov chain Monte Carlo sampling has become one of the major techniques for modern-day Monte Carlo computation.

The problem of estimating the ratio of two normalizing constants is given a new twist in some recent articles (Meng and Wong 1996, Gelman and Meng 1998). In particular, Meng and Wong (1996) note that the identity,

$$A = \frac{c_1}{c_0} = \frac{E_0\{q_1(\mathbf{x})\alpha(\mathbf{x})\}}{E_1\{q_0(\mathbf{x})\alpha(\mathbf{x})\}}$$

holds for an arbitrary function $\alpha(\mathbf{x})$ defined on the common support of q_1 and q_2. Interestingly, this identity is "opposite" to (10.1) used in umbrella sampling. Suppose one can generate m_i Monte Carlo samples from π_i, for $i = 0, 1$. Then, one can estimate A by

$$\hat{A}_{\text{BS}} = \frac{\sum_{j=1}^{m_0} q_1(\mathbf{x}_0^{(j)}) \alpha(\mathbf{x}_0^{(j)})/n_0}{\sum_{j=1}^{m_1} q_0(\mathbf{x}_1^{(j)}) \alpha(\mathbf{x}_1^{(j)})/n_1}. \quad (10.2)$$

Meng and Wong found that an optimal choice of α that minimizes the variance of $\hat{A}_{\text{BS}}$ is

$$\alpha(\mathbf{x}) \propto \{m_0 \pi_0(\mathbf{x}) + m_1 \pi_1(\mathbf{x})\}^{-1}.$$

This formula is not directly usable because its form depends on A. An iterative procedure was suggested to overcome the difficulty. A more serious problem with the bridge sampling method is that it becomes very inefficient when π_0 and π_1 have very little overlap (Chen, Shao and Ibrahim 2000).

The bridge sampling idea was further extended to *path sampling* by Gelman and Meng (1998), in which they defined an augmented system $q(\mathbf{x} \mid \lambda)$, with normalizing constants $c(\lambda)$, such that q_0 is equal to $q(\mathbf{x} \mid \lambda = 0)$ and q_1 is equal to $q(\mathbf{x} \mid \lambda = 1)$. Then, one can draw random samples from an augmented distribution,

$$\pi(\mathbf{x}, \lambda) \propto q(\mathbf{x} \mid \lambda) f(\lambda),$$

from which the ratio A can be estimated. Gelman and Meng pointed out that their approach is closely related to the methods of simulated tempering and parallel tempering (Sections 10.3 and 10.4).

Chen and Shao (1997) showed that if one estimates r using umbrella sampling with one intermediate $\pi_u(\mathbf{x})$, the optimal choice of π_u that minimizes a relative mean square error in estimating A is

$$\pi_u(\mathbf{x}) \propto |\pi_0(\mathbf{x}) - \pi_1(\mathbf{x})|. \quad (10.3)$$

They further proved that this optimal umbrella sampling is *always* better (in terms of the same relative mean square error) than both the optimal bridge sampling and the optimal path sampling estimates. Clearly, the optimal umbrella distribution in (10.3) is not really usable because its form depends on the unknown ratio A. It is conceivable, however, that one can employ an iterative procedure to obtain a sequence of π_u that converges to the optimal one.

10.2 Simulated Annealing

The idea of using a "temperature" parameter to control the simulation of the target density or the optimization of a target function started to become popular with the introduction of the *simulated annealing* (SA) method by Kirkpatrick et al. (1983). Although SA is an *optimization* instead of simulation method, we will spend some time to explain the basic SA operation in order to let the reader get familiar with the use of the (artificial) temperature parameter. The SA algorithm has had tremendous impact in the computer industry (e.g. circuit and silicon chip designs) and many other scientific research areas.

In condensed matter physics, *annealing* is known as a thermal process for obtaining low-energy states of a solid in a *heat bath*. The process has two steps:

- Raise the temperature of the heat bath high enough for the solid (metal) to melt.

- Decrease the temperature of the heat bath slowly to near zero so that the particles in the system can arrange themselves in the ground state of the solid (i.e., crystallize).

At the high-temperature phase, the solid metal becomes a liquid and all its particles can "flow" around relatively freely. In this phase, the particles may be able to find better settling positions. When the temperature is decreased, the particles' movements are more and more confined due to the high energy cost. Eventually, they are forced to "line up" so as to attain the (locally) lowest-energy state.

Realizing that the Metropolis algorithm can be used to simulate particle movements at various temperature to reach *thermal equilibrium*, Kirkpatrick et al. (1983) proposed a computer imitation of the thermal annealing process, the SA, and applied it to solve combinatorial optimization problems.

Suppose our task is to find the *minimum* of a target function $h(\mathbf{x})$. This is equivalent to finding the maximum of $\exp\{-h(\mathbf{x})/T\}$ at any given temperature T. Let $T_1 > T_2 > \cdots > T_k > \cdots$ be a sequence of monotone decreasing temperatures in which T_1 is reasonably large and $\lim_{k\to\infty} T_k = 0$. At each temperature T_k, we run N_k steps of the Metropolis-Hastings (M-H) or Gibbs sampling scheme with $\pi_k(\mathbf{x}) \propto \exp\{-h(\mathbf{x})/T_k\}$ as the equilibrium distribution. An important mathematical observation is that for any system in which $\int \exp\{-h(\mathbf{x})/T\}d\mathbf{x} < \infty$ for all $T > 0$, distribution π_k, as k increases, puts more and more of its probability mass (converging to 1) into a vicinity of the global maximum of h. Hence, a sample drawn from π_k would almost surely be in a vicinity of the global minimum of $h(\mathbf{x})$ when T_k is is close to zero. Theoretically, at least, we should be able to obtain good samples from π_k if we let the number of M-H iterations N_k be sufficiently large. The foregoing procedure can be summarized as the following:

SA Algorithm

- Initialize at an arbitrary configuration $\mathbf{x}^{(0)}$ and temperature level T_1.

- For each k, we run N_k iterations of an MCMC scheme with $\pi_k(\mathbf{x})$ as its target distribution. Pass the final configuration of $\mathbf{x}$ to the next iteration.

- Increase k to $k+1$.

It can be shown that the global minimum of $h(\mathbf{x})$ can be reached by SA with probability 1 if the temperature variable T_k decreases sufficiently slowly [i.e., at the order of $O(\log(L_k)^{-1})$, where $L_k = N_1 + \cdots + N_k$ (Geman and Geman 1984)]. In practice, however, no one can afford to have such a slow annealing schedule. Most frequently, people use a linear or even exponential temperature decreasing schedule, which can no longer guarantee that the global optimum will be reached. It was shown by Holley, Kusuoka and Stroock (1989) that no cooling schedule that is faster than a logarithmic rate can be guaranteed to find the global optimum. However, many researchers' experiences during the past 15 years have testified that SA is a very attractive general-purpose optimization tool. See Aarts and Korst (1989) for further analysis.

10.3 Simulated Tempering

In order to let a MCMC scheme move more freely in the state space, Marinari and Parisi (1992) and Geyer and Thompson (1995) proposed a technique, *simulated tempering* (ST), in the same spirit as SA. To implement ST, one first constructs a family of distributions $\Pi = \{\pi_i(\mathbf{x}) \ i \in I\}$ by varying a single parameter, the *temperature*, in the target distribution π; that is, $\pi_i(\mathbf{x}) \propto \exp\{-h(\mathbf{x})/T_i\}$ for an appropriate temperature T_l. The original target distribution π corresponds to the member of this family with the lowest temperature.

A new target distribution, $\pi_{\text{st}}(\mathbf{x}, \mathbf{i}) \propto c_\mathbf{i} \exp\{-h(\mathbf{x})/T_\mathbf{i}\}$, is defined on the augmented space $(\mathbf{x}, \mathbf{i}) \in \mathcal{X} \times I$. Here, the c_i are constants that can be controlled by the user and they should be tuned so that each tempered distribution in the system should have a roughly equal chance to be visited. Ideally, the c_i should be proportional to the reciprocal of the ith partition function, $Z_i = \int \exp\{-h(\mathbf{x})/T_i\}$. But in practice, one needs to tune these parameters via some pilot studies. An interesting tuning procedure called the *reverse logistic regression* is described in Geyer and Thompson (1995). After setting up the augmented tempering system, a MCMC sampler can be used to draw samples from π_{st}. The intuition behind ST is that by heating up the distribution repeatedly, the new sampler can escape from local modes and increase its chance of reaching the "main body" of the distribution. Initiated with $\mathbf{i}^{(0)} = 0$ and any $\mathbf{x}^{(0)}$ in the configuration space, the ST algorithm consists of the following steps:

ST Algorithm

- With the current state $(\mathbf{x}^{(t)}, \mathbf{i}^{(t)}) = (\mathbf{x}, \mathbf{i})$, we draw $u \sim \text{Uniform}[0,1]$.

- If $u \leq \alpha_0$, we let $\mathbf{i}^{(t+1)} = i$ and let $\mathbf{x}^{(t+1)}$ be drawn from a MCMC transition $T_i(\mathbf{x}, \mathbf{x}^{(t+1)})$ that leaves π_i invariant.

- If $u > \alpha_0$, we let $\mathbf{x}^{(t+1)} = \mathbf{x}$ and propose a level transition, $i \to i'$, from a transition function $\alpha(i, i')$ (usually a nearest-neighbor simple random walk with reflecting boundary), and let $\mathbf{i}^{(t+1)} = i'$ with probability

$$\min\left\{1, \frac{c_{i'} \pi_{i'}(\mathbf{x}) \alpha(i', i)}{c_i \pi_i(\mathbf{x}) \alpha(i, i')}\right\};$$

otherwise let $\mathbf{i}^{(t+1)} = i$.

Even in the ideal situation when the sampler behaves like a symmetric random walk along the temperature space, the expected waiting time for a traversal of all the distributions in the augmented system is of the order of L^2. This puts a severe limitation on how many temperature levels we can afford to employ. On the other hand, in order for ST to work well, the two adjacent distributions π_i and π_{i+1} need to have sufficient overlap, requiring that the temperature difference be small enough and the α_i be tuned properly. Otherwise, there will be large probability barriers for the moves between the two adjacent temperature levels, making it practically impossible to accept temperature transitions based on the Metropolis rule. Thus, although the idea of sampling from a set of tempered distributions is attractive, its usefulness is still severely limited by the waiting time dilemma. The same criticism also applies to the more efficient *parallel tempering* algorithm in the next section.

A possible remedy to this dilemma is to use the dynamic weighting rule Wong and Liang (1997) described in Section 5.7 to overcome steep energy barriers encountered in temperature transitions (Section 10.6). For optimization problems, we have designed a *relaxed* version of ST in which the detailed balance condition is not strictly followed (Cong, Kong, Xu, Liang, Liu and Wong 1999). Although our experiences showed that this strategy worked well for a number of very-large-scale integration (VLSI) design problems, the relaxed ST can no longer be used to sample from a target distribution.

For a given sequence of temperatures, $T_1 < \cdots < T_L$, ST simulates from the L tempered distribution, $\pi_l(\mathbf{x}) \propto \exp\{-h(\mathbf{x})/T_l\}$, simultaneously. By doing so, one hopes to "borrow" information across different chains to achieve a better sampling result. There is no reason why the distributions at different levels have to be generated by varying only the temperature. We may think of varying the dimensionality and the structure of the original state space as well (Liu and Sabatti 1998). For example, we may approximate the original distribution by a sequence of "coarser" distributions with reduced dimensions. These distributions of different "resolution levels" constitute an augmented system and this system, as in ST, can be represented by the state vector $(\mathbf{i}, \mathbf{x_i})$, where $\mathbf{i}$ is the indicator for the complexity level and the dimensionality of $\mathbf{x}_i$ increases as i increases. The reversible jumping rule (Green 1995) described in Section 5.6 has to be used to guide for the transitions between spaces of different dimensions.

10.4 Parallel Tempering

The *parallel tempering* is an (PT) interesting and powerful twist of the *simulated tempering* (ST) first proposed by Charles Geyer in a conference proceedings (Geyer 1991). The same technique was reinvented later by Hukushima and Nemoto (1996) under the name *exchange Monte Carlo*. Instead of augmenting $\mathcal{X}$ to $\mathcal{X} \times I$ as in the ST, Geyer (1991) suggested directly dealing with the product space $\mathcal{X}_1 \times \cdots \times \mathcal{X}_I$, where the $\mathcal{X}_i$ are identical copies of $\mathcal{X}$. Suppose $(\mathbf{x}_1, \ldots, \mathbf{x}_I) \in \mathcal{X}_1 \times \cdots \times \mathcal{X}_I$. For the family of distributions $\Pi = \{\pi_i, i = 1, \ldots I\}$, we define a joint probability distribution on the product space as

$$\pi_{\text{pt}}(\mathbf{x}_1, \ldots, \mathbf{x}_I) = \prod_{i \in I} \pi_i(\mathbf{x}_i)$$

and run parallel MCMC chains on all of the $\mathcal{X}_i$. An "index swapping" operation is conducted in place of the temperature transition in ST. The PT algorithm can be more rigorously defined as follows:

- Let the current state be $(\mathbf{x}_1^{(t)}, \ldots \mathbf{x}_I^{(t)})$; draw $u \sim \text{Uniform}[0,1]$.

- If $u \leq \alpha_0$, we conduct the *parallel step*; that is, we update every $\mathbf{x}_i^{(t)}$ to $\mathbf{x}_i^{(t+1)}$ via their respective MCMC scheme.

- If $u > \alpha_0$, we conduct the *swapping step*; that is, we randomly choose a neighboring pair, say i and $i+1$, and propose "swapping" $\mathbf{x}_i^{(t)}$ and $\mathbf{x}_{i+1}^{(t)}$. Accept the swap with probability

$$\min\left\{1, \frac{\pi_i(\mathbf{x}_{i+1}^{(t)})\pi_{i+1}(\mathbf{x}_i^{(t)})}{\pi_i(\mathbf{x}_i^{(t)})\pi_{i+1}(\mathbf{x}_{i+1}^{(t)})}\right\}.$$

This scheme is very powerful in simulating complicated systems such as bead polymers and other molecular structures. It has also been very popular in dealing with statistical physics models (Hukushima and Nemoto 1996). Compared with the ST, PT does not need fine-tuning (to adjust normalizing constants α_i) and can utilize information in multiple MCMC chains.

Typically, the auxiliary distributions π_i are chosen as the tempered distributions; that is,

$$\pi_i(\mathbf{x}) \propto \exp\{-U(\mathbf{x})/T_i\},$$

for $1 = T_1 < T_2 < \ldots < T_I$. As we have discussed in Section 10.3, it is very important to choose a proper number of temperature levels. Since the acceptance probability for an exchange operation is controlled by both the "typical" energy difference and the temperature difference, a rough guideline is to choose the T_i so that

$$\left(\frac{1}{T_i} - \frac{1}{T_{i+1}}\right)|\Delta U| \approx -\log p_a,$$

where $|\Delta U|$ is the "typical" energy difference (e.g., the mean energy change under the target distribution) and p_a is the lower bound for the acceptance rate. Another useful guiding formula is given in (11.11) on page 233.

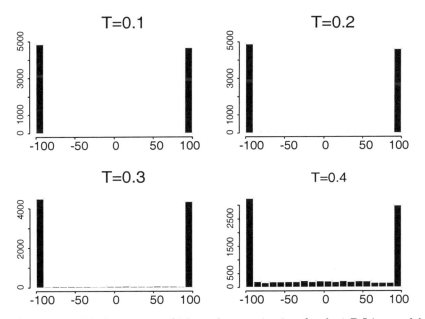

FIGURE 10.1. The histograms of M, total magnetization, for the 1-D Ising model with four different temperatures at $T = 0.1, 0.2, 0.3$, and 0.4, respectively.

214   10. Multilevel Sampling and Optimization Methods

A simple illustration. We consider the simulation of a one-dimensional Ising model of the form

$$\pi_T(\mathbf{x}) \propto \exp\left\{\frac{1}{T}\sum_{i=1}^{d-1} x_t x_{i+1}\right\},$$

where $\mathbf{x} = (x_1, \ldots, x_d)$ and $x_j = \pm 1$. The goal here is to simulate from this distribution with the temperature at $T_0 = 0.1$. As shown in Section 2.4, it is easy to do the exact simulation from this distribution, which can be used to double check our results (it is easy to make coding errors for these simulation problems). Without using parallel tempering, the single-site update algorithm was trapped in one of the modes in 200,000 iterations.

As an initial test, we introduced three tempered distributions corresponding to $T_1 = 0.2$, $T_2 = 0.3$, and $T_3 = 0.4$, respectively. We then implemented parallel tempering for this system. The acceptance rates for the exchange proposals between neighboring pairs were 0.95, 0.88, and 0.65, respectively. The histograms of the total magnetization $M = \sum_{i=1}^{d} x_i$ for the four distributions are shown in Figure 10.1 and the autocorrelation plots are shown in Figure 10.2.

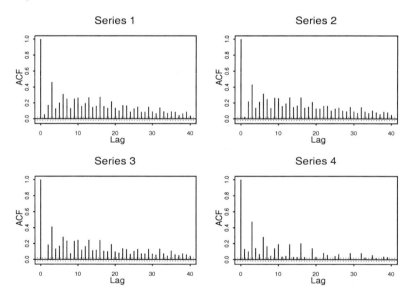

FIGURE 10.2. The autocorrelation plots of M for the 1-D Ising model with temperatures at $T= 0.1$, 0.2, 0.3, and 0.4, respectively.

We felt that the acceptance rates for the exchange between the first 3 distributions were too high. After some pilot studies, we set a new tempering system with $T_1=0.33$, $T_2=0.40$, and $T_3=0.46$. With this setting, the acceptance rates for the exchange moves were 0.74, 0.75, and 0.74 respectively.

The new autocorrelation plots are shown in Figure 10.3. The total sum of the autocorrelations [i.e., the integrated autocorrelation time (Section 5.8)] for π_1 is 7.88 under the first setting and is 5.15 under the new setting.

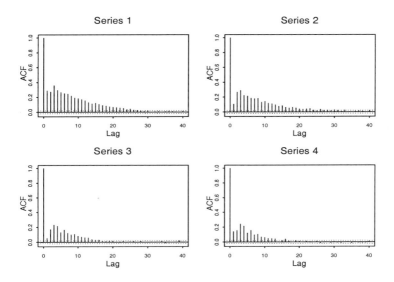

FIGURE 10.3. The autocorrelation plots of M for the 1-D Ising model with temperatures at $T=$ 0.1, 0.33, 0.40, and 0.46, respectively.

10.5 Generalized Ensemble Simulation

In statistical physics, simulating from the *Boltzmann* distribution $\pi(\mathbf{x}) = Z^{-1}(\beta)e^{-\beta H(\mathbf{x})}$, where $\beta = 1/kT$, is often called the *canonical ensemble* simulation. One can think of this "canonical" distribution as giving each configuration $\mathbf{x}$ a weighting factor

$$w_B(\beta, \mathbf{x}) = e^{-\beta H(\mathbf{x})},$$

which is derived under the *canonical ensemble* assumption (also called the constant-NVT ensemble; N represents the number of particles, V the total volume, and T the temperature of the system), where the system has a fixed number of particles, a fixed volume, and a fixed temperature.

The two methods to be introduced in this section, the *multicanonical sampling* and *1/k-ensemble sampling* suggest using weighting functions different from w_B, which, in physics, implies that one has to simulate systems under different "ensemble" conditions simultaneously. The basic idea of these methods is to sample from a modified distribution $\pi'(\mathbf{x})$ which is much "smoother" than π but also possesses key characteristics of π. A way to achieve this is to choose a particular π' so that the "energy variable," $U = H(\mathbf{x})$, or some other variable such as the entropy variable V (to be defined later), has an approximately uniform distribution under π'. The advantage of sampling in the generalized ensemble is that the resulting distribution is "easier" to explore in comparison to the canonical ensemble distribution.

10.5.1 Multicanonical sampling

Multicanonical sampling (Berg and Neuhaus 1991) seeks to sample from a modified distribution $\pi'(\mathbf{x}) \propto w(\mathbf{x})\pi(\mathbf{x})$ under which the energy variable $U = H(\mathbf{x})$ is approximately uniformly distributed. With a slight abuse of notations, we see that the marginal distribution of U, when $\mathbf{x} \sim \pi$, is

$$\pi_U(u) = Z^{-1}(\beta)\Omega(u)e^{-\beta u},$$

where $\Omega(u)$ is called the *density of the states* (or spectral density). Now, if we can draw $\mathbf{x}$ from

$$\pi'(\mathbf{x}) \propto e^{-S\{h(\mathbf{x})\}}, \quad \text{where } S(u) = \log \Omega(u),$$

then the resulting distribution for U is $\pi'_U(u) \propto c$. Therefore, the central idea of multicanonical sampling is to iteratively update the approximation, $\hat{\Omega}(u)$, to the spectral density $\Omega(u)$ and then produce Monte Carlo samples from an approximated version of π'.

The procedure starts with a Monte Carlo sampling with the tempered distribution $\pi_\beta(\mathbf{x}) \propto e^{-\beta H(\mathbf{x})}$, and initializing the histogram function $N_\beta(\ )$ of U which tabulates the realized U values of the starting Monte Carlo run. In other words, the function $N_\beta(\cdot)$ is defined on the set of integer $\{0,\ldots,B\}$. Suppose the range of U is discretized as $-\infty < u_0 < u_1 < \cdots < u_B < \infty$, with $u_b = u_{b-1} + \Delta u$. Suppose we have Monte Carlo samples $\mathbf{x}^{(1)},\ldots,\mathbf{x}^{(N)}$ drawn from π_β. A *histogram function* for U can be produced as

$$N_\beta(i) = \#\{\mathbf{x}^{(j)} : u_{i-1} < H(\mathbf{x}^{(j)}) \leq u_i\}.$$

Clearly,

$$N_\beta(i)/N \stackrel{n\to\infty}{\approx} Z^{-1}(\beta)\Omega(u_i)e^{-\beta u_i}\Delta u.$$

Thus, we can estimate the spectral density as

$$\hat{\Omega}_\beta(u) = \frac{N_\beta(i)e^{\beta u_i}}{\sum_{b=1}^{B} N_\beta(b)e^{\beta u_b}} \quad \text{for } u_{i-1} < u \leq u_i. \qquad (10.4)$$

A first-order approximation (ignoring the variation in the denominator) to the variance of this estimate, with fixed u, is

$$\mathrm{var}\{\hat{\Omega}_\beta(u)\} \approx \frac{1}{N|\mathcal{S}|^2} \Omega(u) \left\{ Z(\beta) e^{\beta u} - \Omega(u) \right\}, \qquad (10.5)$$

where $|\mathcal{S}|$ is the total number of states. If u corresponds to the minimum energy, then the larger the β, the larger this variance.

It is important to note that the function $\Omega(u)$ is independent of β, although the estimate $\hat{\Omega}_\beta$ depends on β explicitly. In other words, we can simulate from a set of distributions $\pi_{\beta_k}(\mathbf{x}) \propto e^{-\beta_k \mathbf{x}}$ corresponding to different temperature levels, as in the simulated tempering or parallel tempering, and then use *all* the samples to estimate $\Omega(u)$, up to a normalizing constant. It is of interest to investigate how to optimally combine the estimates $\hat{\Omega}(u)$ produced at different temperature levels. More precisely, suppose we have Monte Carlo samples $\{\mathbf{x}_k^{(j)}, j=1,\ldots,N\}$ drawn from π_k for $k=1,\ldots,K$; then, an estimate of $\Omega(u)$ can be

$$\tilde{\Omega}(u) = \sum_{k=1}^{K} \alpha_k \hat{\Omega}_{\beta_k}(u) \quad \text{for} \quad u_{i-1} < u \leq u_i,$$

where $\sum_k \alpha_k = 1$ and they can be chosen to minimize the total variance. In particular, these coefficients can be chosen differently for different u's. Expression (10.5) suggests that one should give low-temperature estimates a larger weight when u is close to the minimum energy, and vice versa for u far above the minimum energy.

With an initial estimate of the spectral density $\tilde{\Omega}(u)$, the multicanonical sampling algorithm can be implemented as follows:

- Sample states $\mathbf{x}$ sufficiently long according to the current estimate $p_j(\mathbf{x}) \propto e^{-S_j(H(\mathbf{x}))}$.

- Obtain the new estimate of the spectral density $\tilde{\Omega}_j(u)$.

- Update $S_{j+1}(u) = S_j(u) - \log\{\tilde{\Omega}_j(i) + c_i\}$ for $u \in (u_{i-1}, u_i]$ and let $S_{j+1}(u)$ equal infinity (the maximum) when u is outside of the range (u_0, u_K). The c_i serves as a "prior counts" to smooth out the estimation in Ω.

Note that the estimate of $S_{j+1}(u)$ is invariant when both $\tilde{\Omega}_j$ and c_j are multiplied by a constant. It is conceivable that a more robust updating method for the last step may improve the performance of the algorithm.

10.5.2 The 1/k-ensemble method

Similar in spirit to multicanonical sampling, the 1/k-ensemble method (Hesselbo and Stinchcombe 1995) seeks to produce a modified distribution

$\pi^*(\mathbf{x})$ so that the *entropy variable*

$$S \equiv S(H(\mathbf{x})) = \log\{\Omega(H(\mathbf{x}))\}$$

is approximately uniform. If S is indeed uniform, then the distribution of the energy variable U is

$$\pi_s^*(u) \propto \frac{d\log \Omega(u)}{du}. \tag{10.6}$$

To achieve uniformity in S, Hesselbo and Stinchcombe suggested sampling from

$$\pi^*(\mathbf{x}) \propto 1/k(H(\mathbf{x})), \tag{10.7}$$

where $k(H)$ is the number of configurations "with smaller or equal energy." This is exactly how the name of the method was coined. In our previous notation,

$$k \equiv k(H) = \int_{-\infty}^{H} \Omega(H')dH'.$$

Thus, under π^*, the distribution of U is

$$P_{1/k}(u) \propto \frac{\Omega(u)}{k(u)} = \frac{d\log k(u)}{du}.$$

Since in many physics systems $k(u)$ is a rapidly increasing function of energy, we have $\log k(u) \approx \log \Omega(u)$ for a wide range of u. Hence, sampling from (10.7) can make the entropy variable S almost uniformly distributed.

In practice, we will not be able to know the function $k(u)$ beforehand. The same iterative strategy as the one described in the previous subsection is also applicable here. First, the energy range is discretized and the integrals replaced by summations. Then, the current sampling distribution is updated by reweighting using the histogram of S. Obviously, if one has a good estimation to the spectral density function $\Omega(u)$, one would be able to get a good approximation to $k(u)$, and vice versa.

10.5.3 Comparison of algorithms

Recently, Hansmann and Okamoto (1997) conducted a study to compare the performances of multicanonical sampling, $1/k$-sampling, and simulated tempering for a protein folding problem. The protein segment under their consideration is *Met-enkephalin* which has five amino acid residues: Tyr-Gly-Gly-Phe-Met. The energy function they used is similar to the one in Section 1.4 and is given by the sum of four terms: the electrostatic term, the 12-6 Lennard-Jones potential term, the hydrogen-bond term, and the torsion angle term.

Repeats× Iterations	MU	$1/k$	ST	SA
10×100,000	10	9	9	5
20×50,000	18	15	16	8
50×20,000	21	22	17	10
100×10,000	28	29	20	13

TABLE 10.1. Comparison of the four annealing methods in finding minimum energy states. From left to right: multicanonical sampling; $1/k$-sampling, simulated tempering, and simulated annealing.

In their test, all three algorithms were given sufficient tuning time to find their best settings (e.g., the weight function for both multicanonical sampling and $1/k$-sampling and the temperature ladder for simulated tempering). Then, a production run of one million MC sweeps for each of the three methods were made after a burn-in period of 10,000 sweeps. The starts of all the runs were completely random. Hansmann and Okamoto (1997) concluded after a careful study of their simulation results that all the three methods do not differ much from each other, but they are all much more efficient than traditional methods.

Hansmann and Okamoto (1997) also modified the three sampling methods to suit the global optimization task [i.e., finding the ground (minimum energy) states]. In an annealing version of multicanonical sampling, for example, they introduced an upper bound, H_{wall}, in energy, above which the proposed new configuration is not accepted. The weight is updated so that the energy distribution is made flat in the interval $(H_{\text{wall}} - \Delta H, H_{\text{wall}})$. The upper bound H_{wall} is lowered once after each sweep so that $H_{\text{wall}} = H_0 + \Delta H$, where H_0 is the lowest energy found in the preceding iteration. A simulated tempering annealing method is similar to the above procedure except that the energy bounds are replaced by temperature bounds. Their simulation experiments showed that all three methods significantly outperformed the traditional simulated annealing. Table 10.1 summarizes the number of times each method finds the ground states under different settings.

10.6 Tempering with Dynamic Weighting

If a ST sampler can traverse the temperature ladder of the augmented system freely according to the Metropolis rule, then satisfactory results will be obtained for the "coldest" distribution. In difficult simulation problems, however, it is not easy to achieve this goal because, as discussed as the end of Section 10.3, one has to either employ many tempered distributions in the augmented system, which apparently affects computational efficiency, or suffer from repeated rejections for temperature transition proposals due to the high-energy barriers between adjacent tempered distributions. Here, we describe how to use dynamic weighting rule introduced in Section 5.7 to alleviate this difficult situation.

The new *simulated tempering with dynamic weighting* (STDW) algorithm is essentially the same as the ST algorithm (Section 10.3) except that a weighting variable w is now associated with the configuration $(\mathbf{x}, \mathbf{i})$ and the dynamic weighting rule is used to guide the transitions between adjacent temperature levels. Let $0 < \alpha_0 \leq 1$ be given in advance and suppose the current state is $(\mathbf{x}^{(t)}, \mathbf{i}^{(t)}, w^{(t)}) = (\mathbf{x}, i, w)$. The *STDW algorithm* iterates as follows.

- Draw $u \sim$ Uniform[0,1].

- If $u \leq \alpha_0$, we let $\mathbf{i}^{(t+1)} = i$, $w^{(t+1)} = w$, and let $\mathbf{x}^{(t+1)}$ be a MCMC update of $\mathbf{x}$ with respect to π_i (this update can be a result of more than one MCMC step).

- If $u > \alpha_0$, we let $\mathbf{x}^{(t+1)} = \mathbf{x}$ and propose a level transition, $i \to i'$, from a transition function $\alpha(i, i')$ (usually a nearest-neighbor simple random walk). Conduct an R-type (or Q-type) move (with $\theta = 1$) to update $(\mathbf{i}, w)$; that is,

    - compute the Metropolis ratio
    $$r(i, i') = \frac{c_{i'} \pi_{i'}(\mathbf{x}) \alpha(i', i)}{c_i \pi_i(\mathbf{x}) \alpha(i, i')};$$

    - update
    $$(\mathbf{i}^{(t+1)}, w^{(t+1)}) = \begin{cases} (i', wr(i, i') + 1) & \text{if } U \leq \dfrac{wr(i, i')}{wr(i, i') + 1} \\ (\mathbf{x}^{(t)}, w[wr(i, i') + 1]) & \text{otherwise.} \end{cases}$$

After we have obtained a set of weighted samples, $(\mathbf{x}^{(t)}, \mathbf{i}^{(t)}, w^{(t)})$, from this method, we first retain those $(\mathbf{x}^{(t)}, w^{(t)})$ corresponding to $\mathbf{i}^{(t)} = 1$, the target distribution. Then, we estimate the quantity of interest by a weighted average after a *stratified truncation*. Suppose the estimation of $\mu = E_\pi \rho(\mathbf{x})$ of interest. First, the samples are stratified into subsets of comparable sizes according to the value of $\rho(\mathbf{x})$. The highest $k\%$ (usually $k = 1$ or 2) of the weights within each stratum are then trimmed down to be equal to their $(100 - k)$th percentile of that stratum. The resulting weights after modification are then used to produce a weighted average of the $\rho(\mathbf{x}^{(t)})$.

For example, one may be interested in estimating the expected value of the spontaneous magnetization $\mu = E[\rho(\mathbf{x})] = E|\sum_i x_i|/d^2$, where $\mathbf{x} = (\sigma)$ in a Ising model simulation (Section 1.3). Then, after STDW sampling, we divide the range of $\rho(\mathbf{x})$ into small intervals, $b_0 < b_1 < \cdots < b_k$, and stratify the weighted samples $(\mathbf{x}^{(t)}, w^{(t)})$, $t = 1, \ldots, m$, according to the values of $\rho(\mathbf{x})$; that is, we construct $S_j = \{(\mathbf{x}^{(t)}, w^{(t)}) : \rho(\mathbf{x}^{(t)}) \in (b_{j-1}, b_j)$. The $w^{(t)}$ in each strata S_j is truncated to $\tilde{w}^{(t)} = w^{(t)} \wedge w_j^*$, where w_j^* is the $(100-k)$th percentile of the weights in S_j. Finally, an estimation of μ is given by (13.19), with the $w^{(t)}$ replaced by the $\tilde{w}^{(t)}$.

10.6.1 Ising model simulation at sub-critical temperature

Simulating a 2-D Ising model on a large lattice space and investigating the phase transition phenomena have always been a favorite test for a new Monte Carlo strategy. As explained in Section 1.3, a 2-D Ising model on a $L \times L$ lattice is defined as a Boltzmann distribution for the spin configuration $\mathbf{x} = \{x_s,$ with $s = (a, b)$ and $1 \leq a, b \leq L\}$:

$$\pi(\mathbf{x}) = \frac{1}{Z(K)} \exp \left\{ K \sum_{i \sim j} x_i x_j \right\}.$$

Here, each spin x_i only takes value in $\{-1, 1\}$, the notation $i \sim j$ denotes the nearest neighbors on the lattice, the symbol K is the coupling constant (inverse temperature), and $Z(K)$ is the partition function. Liang and Wong (1999) obtained some interesting results on Ising model simulations by STDW with R-type moves. The simulations were done on lattices of sizes 32^2, 64^2, and 128^2, respectively. Similar to ST, they treated the inverse temperature K as a dynamic variable. This variable takes values in a set of levels uniformly spaced in the interval $[0.4, 0.5]$ (the critical point is known to be 0.44). The R-type moves were applied to cross various temperature levels, whereas the Metropolis-type moves (heat-bath algorithm) were used within each temperature level. In each of the three lattice sizes, they started a single run with the configuration that all spins are $+1$. The run continued until 10,000 configurations were obtained at the lowest temperature level. Figure 10.4 plots the estimate of the expected absolute value of the spontaneous magnetization (defined as $E|\sum x_i|/d^2$, where d is the lattice size) at various inverse temperatures K for the different sizes of lattices. Estimation was done by weighted averaging with the weights stratified according to spontaneous magnetization and then truncated at 99%.

The smooth curve in Figure 10.4 is the celebrated infinite lattice result (i.e., the "truth" when the lattice size is infinite) discovered by Onsager (1949) and proved by Yang (1952). It is seen that the critical point (0.44) can be estimated quite well from the STDW simulations by the crossing of the curves for the 64^2 and 128^2 models. A main strength of the STDW algorithm is that a single run of the process can yield accurate estimates over the entire temperature range, extending well below the critical point. In contrast, simulated tempering encountered a serious difficulty in sampling even the smallest, 64×64, lattice model (Liang and Wong 1999).

222   10. Multilevel Sampling and Optimization Methods

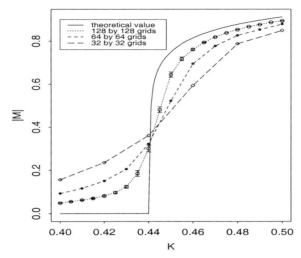

FIGURE 10.4. Ising model simulations via dynamic weighting. The expected absolute value of the spontaneous magnetization (defined as $E|\sum_i x_i|/d^2$, where d is the lattice size) is plotted against inverse temperature K, with lattices of size 32^2, 64^2, 128^2, and infinite, respectively. The smooth curve corresponds to the theoretical infinite lattice result.

10.6.2 Neural network training

The artificial neural network is a simple mathematical model motivated by neuron functions and has been a widely used tool in learning and classification problems (Hopfield 1982, Rumelhart and McClelland 1986). The most popular among these networks is the *multi-layer perceptrons* (MLP), in which all the units (nodes) are grouped into ordered layers (typically consisting of input, hidden, and output layers). The units in the lower layer (input) only connects with the units in the one above it (Ripley 1996). Each node in a higher layer independently processes the values fed to it by nodes in the lower layer in the form

$$y_k = f_k\left(\alpha_k + \sum_{j \sim k} w_{jk} x_j\right),$$

where the x_j are inputs, and then present the output y_k as an input for the next layer. Here, we take f_k as the same sigmoidal function [i.e., $f(s) = 1/(1 + \exp(-s))$] throughout the network.

It can be shown that a neural network with a sufficient number of hidden units and adjustable connection weights can approximate any bounded continuous function arbitrarily well (Cybenko 1989, Hornik, Stinchcombe and White 1989, White 1992). Hence, it has been hoped that such a "universal approximator" can solve important prediction problems. Neural network "learning" is accomplished by choosing the connection strengths w_{jk}

10.6 Tempering with Dynamic Weighting

so that the network's outputs match the desired outputs in the training data as closely as possible. In the last several decades, many algorithms have been proposed to train neural networks, including conjugate gradient, back-propagation (BP) and their variants (Rumelhart, Hinton and Williams 1986), and Bayesian methods (Buntine and Weigend 1991, Neal 1996). However, these methods can fail badly in some cases, one of which is the two-spiral problem (Lang and Witbrock 1988). By using the STDW algorithm, Wong and Liang (1997) treated the two-spiral problem with considerable success (both the 2-25-1 and 2-14-4-1 networks have been fitted and the results were close to be perfect, whereas the error rate for BP is generally greater than 40%).

In training programs such as back-propagation, the total mean square error

$$H_p = \sum_p \|O_p - T_p\|^2$$

is used as the cost function, where T_p is the pth training case's ideal output and O_p is the output of the network. We can define a target distribution jointly for the connection strengths w_{jk} and the temperature parameter T so that

$$\pi(w_{jk}, \text{ all } j, k; T) \propto \alpha(T) \exp(-H_p/T).$$

Here, T represents a finite number of temperature levels: $t_1 > t_2 > \cdots > t_L$ (Wong and Liang used $L = 4$ for the two-spiral problem). Then, a standard Metropolis algorithm can be used for modifying local connection strengths at a fixed temperature, and a dynamic weighting rule is used for temperature transition. After a reasonable configuration is obtained from the STDW algorithm, a *post-optimization* procedure is applied to zoom in for the local optimum. The often-used post-optimization methods include steepest-gradient-decent and conjugate gradient.

Both the encoder problem (Ackley, Hinton and Sejnowski 1985) and the parity problem (Rumelhart et al. 1986), two classic examples in the neural network community, were considered by Wong and Liang. The input in the *encoder problem* is a length-d binary sequence and the output is desired to be *identical to the input*. A requirement for the network designed for the task is that the hidden layers cannot have more than $\log_2(d)$ nodes. Obviously, a network with a hidden layer of d nodes is trivially perfect. A three-layer network with five hidden units were trained for $d = 32$ (so that this constitutes a 32-5-32 network). Note that one is dealing with a $5 \times 32 \times 2 = 320$-dimensional optimization problem in this example. The STDW algorithm with post-optimization achieved perfect learning in about 5 minutes on a Sparc-20 Sun Workstation. With a longer running time, perfect learning was also achieved on the more difficult 32-4-32 (there are 4 hidden units, 256 scalar parameters involved) encoder problem.

10. Multilevel Sampling and Optimization Methods

The input of a *d-parity problem* is also a binary sequence of length d. The output is required to be 1 if the input sequence contains an odd number of 1's, and is 0 otherwise. So this exercise is meant to show how a "black-box" network can mimic a very nonlinear and discontinuous function. Rumelhart et al. (1986) show that *at least d* hidden units are required for a three-layer MLP to solve this problem. The STDW method had no difficulty solving this problem with a d-d-1 ($2 \leq d \leq 8$) network. A perfect solution for $d = 8$ (a 72-dimensional optimization problem) was obtained in Liang (1997).

11
Population-Based Monte Carlo Methods

In parallel tempering (Section 10.4), the target distribution is embedded into a larger system which hosts a number of similar distributions differing with each other only in a temperature parameter. Then, parallel Monte Carlo Markov chains are conducted to sample from these distributions simultaneously. An important step which makes PT effective and which connects the multiple distributions in the augmented system is to propose configuration exchanges between two adjacent sampling chains. The attractiveness of this configuration-swap step can be loosely attributed to a population-based "learning" strategy; that is, in high-temperature states, radically different new configurations are allowed to arise, whereas in lower-temperature states, a configuration is given opportunities to refine itself. By making exchanges, we can retain and improve those good configurations generated in the population by putting them into low-temperature chains. However, one may feel that this "exchange" step is a rather minimal interaction among the multiple chains in the "population." More active interactions such as those employed in a genetic algorithm might be more helpful. In this chapter, we will follow this thought to venture into population-based Monte Carlo strategies.

Genetic algorithm has been very successfully applied in various optimization problems. Its basic principle of mutations and crossovers are so attractive that researchers have made attempts to employ them in guiding Markov chain Monte Carlo (MCMC) moves. However, such attempts have met many obstacles because of the difficulties in designing Markov chain moves that are both evolutionarily oriented (good ones live and bad ones die) and reversible (satisfying the detailed balance). Some recent work along the line of adaptive directional sampling (Gilks et al. 1994), multipoint Metropolis principle (Section 5.5), and parallel tempering heuristics seems to have opened the door for designing proper and effective population-based MCMC algorithms (Liang and Wong 2000).

11.1 Adaptive Direction Sampling: Snooker Algorithm

Adaptive direction sampling (ADS) is a very interesting idea first proposed by Gilks et al. (1994). In this method, one runs multiple MCMC chains in parallel and adapts the future movement of one chain along a direction generated by other chains. ADS is clearly a population-based method. But its learning strategy, which is directional instead of configurational, is quite different from that of PT. A special form of the ADS is the *snooker algorithm*, which will be explained below.

At iteration t of the snooker algorithm, one keeps a population of samples, say $\mathcal{S}^{(t)} = \{\mathbf{x}^{(t,1)}, \ldots, \mathbf{x}^{(t,m)}\}$, of size m. Each of the $\mathbf{x}_t^{(t,j)}$ is also called a "stream." Then, the next-generation population $\mathcal{S}^{(t+1)}$ is generated as follows:

(a) A stream $\mathbf{x}^{(t,c)}$ is chosen at random from $\mathcal{S}^{(t)}$.

(b) An anchor point $\mathbf{x}^{(t,a)}$ is chosen at random from $\mathcal{S}^{(t)} \setminus \{\mathbf{x}^{(t,c)}\}$ and a direction e is generated as $e = (\mathbf{x}^{(t,c)} - \mathbf{x}^{(t,a)})/\|\mathbf{x}^{(t,c)} - \mathbf{x}^{(t,a)}\|$.

(c) A scalar r is drawn from an appropriate distribution $f(r)$.

(d) Update $\mathbf{x}^{(t+1,c)} = \mathbf{x}^{(t,a)} + re$, and leave others unchanged [i.e., we let $\mathbf{x}^{(t+1,j)} = \mathbf{x}^{(t,j)}$ for $j \neq c$].

Gilks et al. (1994) and Roberts and Gilks (1994) show that $f(r)$ should be of the form

$$f(r) \propto |r|^{d-1} \pi(\mathbf{x}^{(t,a)} + re), \tag{11.1}$$

where d is the dimensionality of the state space. Note that if direction e is generated from a distribution that is independent of $\mathbf{x}^{(t,c)}$, as in the hit-and-run algorithm, then the proper distribution is for sampling r is $f(r) \propto \pi(\mathbf{x}^{(t,c)} + re)$.

To see why (11.1) is true, we need to show that at the equilibrium the new point $\mathbf{x}^{(t+1,c)}$ is independent of the $\mathbf{x}^{(t,j)}$ for $j \neq c$ and is distributed as π, provided that $\mathbf{x}^{(t,c)}$ is independent of the $\mathbf{x}^{(t,j)}$ for $j \neq c$ and follows distribution π. This fact follows from the following lemma (Liu, Liang and Wong 2000), which is the generalization of a result of Gilks et al. (1994).

Lemma 11.1.1 *Suppose* $\mathbf{x} \sim \pi$ *and* $\mathbf{y}$ *is any fixed point in a d-dimensional space. Let* $\mathbf{e} = (\mathbf{x} - \mathbf{y})/\|\mathbf{x} - \mathbf{y}\|$ *be a unit vector. If r is drawn from distribution in (11.1), then* $\mathbf{x}' = \mathbf{y} + r\mathbf{e}$ *follows distribution* π. *If* $\mathbf{y}$ *is generated from a distribution* $D(\mathbf{y})$ *independent of* $\mathbf{x}$, *then* $\mathbf{x}'$ *is independent of* $\mathbf{y}$ *and has density* $\pi(\mathbf{x}')$.

Proof: Without loss of generality, we need only to show the case when $\mathbf{y} = 0$. Then, the Markovian move is of the form

$$\mathbf{x}' = r^* \mathbf{x}, \quad r^* \neq 0,$$

and is a *scale* group transformation of $\mathbf{x}$. Now, the question becomes this: What distribution should we impose on this transformation group so that an element r^* drawn from it leaves π invariant? This general question is answered by Theorem 8.3.1 in Section 8.3, which gives rise to expression (11.1) under our current setting. ◇

Although ADS is a powerful framework enabling interactions among multiple chains, the method alone has not been very effective in improving sampling efficiency. A main reason is that the direction generated in ADS is still rather arbitrary. It is not clear why one wants to let the current chain move in the direction of other chains. An important question is, thus, (a) how one can select a meaningful direction $\mathbf{e}_t$. A related question is (b) how to sample from $f(r)$ in (11.1)? In the next section, we describe a novel algorithm of Liu, Liang and Wong (2000) that combines the ADS framework with deterministic mode finding procedures and the multipoint method (Section 5.5).

11.2 Conjugate Gradient Monte Carlo

The main features of this algorithm are (i) the local optimality information revealed by a deterministic local-search scheme is explicitly used for adaptation and (ii) the companion MTM algorithm allows for a very large step-size in searching along a "promising" direction, which partially resolves issue (b) raised at the end of the previous section. Some numerical examples show that the new sampler offers significant improvement over the traditional Metropolis sampler, especially in difficult problems (Liu, Liang and Wong 2000).

We follow the ADS approach of evolving a population of samples, say, $\mathcal{S}^{(t)} = \{\mathbf{x}^{(t,1)}, \ldots, \mathbf{x}^{(t,m)}\}$, at each iteration. To update one of the samples, say, $\mathbf{x}^{(t,c)}$, we use the other samples to construct a good "anchor" point $\mathbf{y}$ and then update $\mathbf{x}_t^{(t,c)}$ by a multiple-try Metropolis (MTM) transition along the direction defined by $\mathbf{x}^{(t,c)}$ and $\mathbf{y}$. With Lemma 11.1 proved in the previous section, one can see that essentially *any* way of choosing the anchor point **y** is appropriate provided that **y** is independent of $\mathbf{x}^{(t,c)}$ and that the distribution along the line $f(r)$ is properly adjusted. For example, we can use a conjugate gradient search to construct the anchor point. In summary, at iteration $t+1$, we do the following steps.

Conjugate Gradient Monte Carlo (CGMC)

1. Choose $\mathbf{x}^{(t,a)} \in \mathcal{S}^{(t)}$ at random. Obtain either the gradient or the conjugate gradient direction of $\log \pi(\ )$ at **x**; and conduct a *deterministic* search to find a local mode of π or any point with a higher density value, **y**. This point **y** is called an *anchor point*.

2. Choose another member $\mathbf{x}^{(t,c)}$ from $\mathcal{S}^{(t)} \setminus \{\mathbf{x}^{(t,a)}\}$ at random.

3. Let $e = (\mathbf{y} - \mathbf{x}^{(t,c)})/\|\mathbf{y} - \mathbf{x}^{(t,c)}\|$ and sample along the line $\mathbf{y} + r\mathbf{e}$ by the MTM method; that is, we draw $r \in (-\infty, \infty)$ from

$$f(r) \propto |r|^{d-1} \pi(\mathbf{y} + r\mathbf{e}), \qquad (11.2)$$

where d is **x**'s dimension, and let $\mathbf{x}^{(t+1,c)}$ be $\mathbf{y} + r\mathbf{e}$.

4. We can update other members in $\mathcal{S}^{(t)}$ in the same way as we update $\mathbf{x}^{(t,c)}$ in the previous step. However, we can also leave them unchanged. The new population $\mathcal{S}^{(t+1)}$ consists of both updated and unchanged samples in the previous population.

The gradient/conjugate gradient procedure in Step 1 can be iterated for any number of times and can also be replaced by *any* effective local optimization method, such as the iterative conditional maximization or a few EM steps. In all of our examples, we have used the conjugate gradient directional method coupled with a one-dimensional minimization algorithm taken from the *Numerical Recipe* (Press and Vetterling 1995). It should be noted, however, that the generation of **y** cannot be dependent upon any members in generation $t-1$. For example, both of the conjugate directions in a CG search have to be generated based solely on $\mathbf{x}^{(t,a)}$.

The population size m needs not be too large. In fact, we found that having $m = 2$ to 5 was good enough for many of our examples. However, it should be a worthwhile topic to study the effect of m on the convergence of the algorithm. By using Lemma 11.1, it is not difficult to prove that the new algorithm is proper (Liu, Liang and Wong 2000).

11.3 Evolutionary Monte Carlo

The snooker algorithm and CGMC are two useful attempts to incorporate a population-based learning capability in Monte Carlo simulations. However, in comparison with a typical genetic algorithm, the interactions among the members in the population are still rather minimal and the effects of these interactions are only modest. It is thus desirable to incorporate more powerful interactions, such as the crossover operations used in genetic algorithms. A partial support for this desire is that the genetic algorithm is known to make very efficient use of distributed information across states of a population, which may serve as a good basis for a MCMC sampler with "learning." But designing a useful crossover move in a MCMC sampler has been difficult mainly because of the stringent requirement of the detailed balance condition. For example, if we propose to conduct a random crossover between two members in the current population, we need to retain two offsprings (otherwise one would not be able to maintain the detailed balance). Usually, however, at least one of the offsprings does not have a good "fitness" value. Thus, when applying the Metropolis acceptance-rejection rule to the proposal, we will almost surely reject it. In this section, we will explain how the crossover operation can be effectively used in a tempering framework (Liang and Wong 2000).

Suppose the target distribution of interest is, again,

$$\pi(\mathbf{x}) \propto \exp\{-H(\mathbf{x})\},$$

where $\mathbf{x}$ takes a value in the sample space $\mathcal{X}$. Let $\mathbf{X} = \{\mathbf{x}_1, \mathbf{x}_2, \ldots, \mathbf{x}_m\}$ denote a population, where $\mathbf{x}_i$ takes values in the original sample space $\mathcal{X}$ and m is the population size. A set of N different temperatures, $T = (t_1, t_2, \ldots, t_m)$, are given and ordered as $t_1 > t_2 > \cdots > t_m$. Each individual (or chromosome) $\mathbf{x}_i$ in the population is attached to a temperature t_i for $i = 1, \ldots, m$. The fitness function of $\mathbf{x}_i$ is defined as the "Hamiltonian" function $H(\mathbf{x}_i)$ given by the original target distribution. The corresponding Gibbs distribution for the ith member in the population is

$$\pi_i(x_i) = \frac{1}{Z_i(t_i)} \exp\{-H(\mathbf{x}_i)/t_i\}, \qquad (11.3)$$

where $Z_i(t_i)$ is the normalizing constant for the distribution; that is,

$$Z_i(t_i) = \sum_{\text{all } \mathbf{x}_i} \exp\{-H(\mathbf{x}_i)/t_i\}. \qquad (11.4)$$

If one lets the lowest temperature $t_m = 1$, then π_m corresponds to the target distribution $\pi(\mathbf{x})$. Similar to *parallel tempering*, one considers a Markov chain sampler on the augmented state space defined by $\mathbf{X} = (\mathbf{x}_1, \mathbf{x}_2, \ldots, \mathbf{x}_m)$. The target distribution of this augmented system is defined as the *augmented* Boltzmann distribution as in *parallel tempering*:

$$\pi(\mathbf{X}) = \frac{1}{Z(T)} \exp\left\{-\sum_{i=1}^{m} H(\mathbf{x}_i)/t_i\right\}, \qquad (11.5)$$

$$Z(T) = \prod_{i=1}^{m} Z_i(t_i). \tag{11.6}$$

We now discuss how various aspects of the genetic algorithm can be incorporated into this framework.

11.3.1 Evolutionary movements in binary-coded space

In this subsection, we will focus on the binary-coded state space. In other words, our $\mathbf{x}_i$ is defined as a vector, $\mathbf{x}_i = (b_{i,1}, \ldots, b_{i,d})$, where $b_{i,j}$ is either 0 or 1. Needless to say, many discrete problems can be coded in this way. Even some problems with continuous components can be reduced to a space of binary strings. Examples include the variable selection problem (Section 11.5.3), the change-point problem, and the sequence alignment problem in biology (Sections 1.5, 4.1.3, and 6.5).

Mutation. This operation can be achieved by a standard Metropolis step: A "chromosome," say $\mathbf{x}_k$, is first selected at random from the current population $\boldsymbol{X}$. Then $\mathbf{x}_k$ is mutated to $\mathbf{y}_k$ by flipping the values at some random positions of the binary string of $\mathbf{x}_k$. The new population is proposed to be $\boldsymbol{Y} = \{\mathbf{x}_1, \ldots, \mathbf{y}_k, \ldots, \mathbf{x}_N\}$. According to the Metropolis rule, the proposal is accepted with probability $\min(1, r_m)$, where

$$r_m = \exp\{-(H(\mathbf{y}_k) - H(\mathbf{x}_k))/t_k\}. \tag{11.7}$$

If the proposal is accepted, the current population $\boldsymbol{X}$ is replaced by $\boldsymbol{Y}$, otherwise, the population $\boldsymbol{X}$ is kept unchanged.

The possible choices for mutation operator include 1-point mutation and 2-point mutation. One can also consider *uniform mutation,* in which each digit of $\mathbf{x}_k$ has a probability p_m of flipping its value. All of these mutation operators are symmetric (i.e., the transition probability from $\boldsymbol{X}$ to $\boldsymbol{Y}$ is equal to that from $\boldsymbol{Y}$ to $\boldsymbol{X}$).

Crossover. One chromosome pair, say $\mathbf{x}_i$ and $\mathbf{x}_j$ ($i \neq j$), are selected from the current population $\boldsymbol{X} = \{\mathbf{x}_1, \ldots, \mathbf{x}_i, \ldots, \mathbf{x}_j, \ldots, \mathbf{x}_N\}$ according to a selection procedure (e.g., roulette wheel or random selection). The new "offsprings" $\mathbf{y}_i$ and $\mathbf{y}_j$ are generated by one crossover operator, which is discussed below. A new population is proposed as $\boldsymbol{Y} = \{\mathbf{x}_1, \ldots, \mathbf{y}_i, \ldots, \mathbf{y}_j, \ldots, \mathbf{x}_N\}$ and it is accepted with probability $\min(1, r_c)$ according to the Metropolis rule. It is not difficult to see that the Metropolis-Hastings ratio is

$$r_c = \exp\left\{-\frac{H(\mathbf{y}_i) - H(\mathbf{x}_i)}{t_i} - \frac{H(\mathbf{y}_j) - H(\mathbf{x}_j)}{t_j}\right\} \frac{T(\boldsymbol{Y}, \boldsymbol{X})}{T(\boldsymbol{X}, \boldsymbol{Y})}. \tag{11.8}$$

Here, the proposal transition is

$$T(\boldsymbol{X}, \boldsymbol{Y}) = P[(\mathbf{x}_i, \mathbf{x}_j) \mid \boldsymbol{X}] P[(\mathbf{y}_i, \mathbf{y}_j) \mid (\mathbf{x}_i, \mathbf{x}_j)],$$

where $P[(\mathbf{x}_i, \mathbf{x}_j) \mid \boldsymbol{X}]$ denotes the selection probability of $(\mathbf{x}_i, \mathbf{x}_j)$ from the population $\boldsymbol{X}$ and $P[(\mathbf{y}_i, \mathbf{y}_j) \mid (\mathbf{x}_i, \mathbf{x}_j)]$ denotes the probability of generating $(\mathbf{y}_i, \mathbf{y}_j)$ from the parents $(\mathbf{x}_i, \mathbf{x}_j)$. If it is accepted, the current population $\boldsymbol{X}$ is replaced by $\boldsymbol{Y}$; otherwise, the population $\boldsymbol{X}$ is kept unchanged.

The selection procedure can be a weighted sampling as

$$P((\mathbf{x}_i, \mathbf{x}_j) \mid \boldsymbol{X}) \propto [\exp\{-H(\mathbf{x}_i)/t\} + \exp\{-H(\mathbf{x}_j)/t\}], \; \mathbf{x}_i \neq \mathbf{x}_j. \quad (11.9)$$

It is not unusual to choose $t = 1$. This sampling can be achieved by first selecting $\mathbf{x}_i$ with probability proportional to $\exp\{-H(\mathbf{x}_i)/t\}$, and then choosing $\mathbf{x}_j$ independent of $\mathbf{x}_i$, but with the same sampling probability. If $\mathbf{x}_i = \mathbf{x}_j$, then we discard them and repeat the sampling until we obtain a distinct pair.

The possible choices for the crossover operator include 1-point crossover, 2-point crossover, and uniform crossover. It is clear that all of the above crossover operators are symmetric (i.e., the generating probability of $\boldsymbol{Y}$ given $\boldsymbol{X}$ is equal to that of $\boldsymbol{X}$ given $\boldsymbol{Y}$) and the generating probability can be canceled in Equation (11.8). The ratio of the transition probabilities in (11.8) is reduced to that of the selection probabilities.

A new crossover operator, *adaptive crossover*, is also introduced (Liang and Wong 2000). In adaptive crossover, the two offsprings are generated as follows. If $\mathbf{x}_i$ and $\mathbf{x}_j$ takes the same value at their kth position of the binary string [i.e., $\mathbf{x}_i(k) = \mathbf{x}_j(k)$], then offsprings $\mathbf{y}_i$ and $\mathbf{y}_j$ copy that value and independently inherit that value with probability p_0 (and flip to the opposite with probability $1 - p_0$). If $\mathbf{x}_i$ and $\mathbf{x}_j$ have different values at position k, then $\mathbf{y}_i(k)$ inherits $\mathbf{x}_i(k)$ with probability p_1 and $\mathbf{y}_j(k)$ inherits $\mathbf{x}_j(k)$ with probability p_2, all independently. In this way, different genes in one chromosome have different probabilities to be inherited by the offsprings and the same gene also has a different inheritance probability in different generations. The adaptation is determined by the genes of the parent chromosomes. The adaptive crossover has a tendency to keep the better genes found in the early simulation.

Exchange. This operation is same as that introduced in parallel tempering/exchange Monte Carlo (Geyer 1991, Hukushima and Nemoto 1996). Given the current population $\boldsymbol{X}$, the proposed new configuration $\boldsymbol{Y}$ only differs from $\boldsymbol{X}$ by an exchange between $\mathbf{x}_i$ and $\mathbf{x}_j$. The new population is accepted or rejected according to the Metropolis rule (Section 10.4)

11.3.2 Evolutionary movements in continuous space

Suppose $\mathbf{x} = (x_1, \ldots, x_d)$ is defined in a d-dimensional Euclidean space, and the *tempered population* is still defined as in the previous section; that is, we define a system on $\boldsymbol{X} = (\mathbf{x}_1, \ldots, \mathbf{x}_N)$ with an augmented Boltzmann distribution as in (11.5). A mutation operation can still be any kind of Metropolis-Hastings move independently for each chain. A convenient one is perhaps the random-walk-type Metropolis move.

In the crossover step, one randomly picks a pair of "chromosomes," x_i and x_j, from the current population and mate them to produce a new pair, y_i and y_j. The new pair is then subjected to acceptance-rejection decision based on the usual Metropolis-Hastings rule. Liang and Wong (2001) discussed two choices of the crossover operator. In a *real crossover* operator, the pair x_i and x_j exchange their components (digits) in the same way as that for binary-coded space described in the previous section. As a consequence, every component of an offspring, say $y_i(k)$, has a certain probability to be $x_i(k)$ and the remaining probability to be $x_j(k)$. Another operator, the *snooker crossover*, is based on the algorithm described in Section 11.1 and can be implemented as follows.

- Select one chromosome, say x_i, at random from the current population X.

- Select another chromosome x_j from the remaining population with probability proportional to $w_j = \exp\{-H(x_j)/t_s\}$, its "Boltzmann weight," where t_s is called a selection temperature. The selected x_j is called the *anchor chromosome*.

- Let $e = (x_j - x_i)/\|x_j - x_i\|$, and $y_i = x_j + re$, where $r \in (-\infty, \infty)$ is a random variable sampled from the density

$$f(r) \propto |r|^{d-1} \pi(x_j + re). \tag{11.10}$$

- Construct a new population by replacing x_i with the "offspring" y_i.

Note that the sampling of $f(r)$ can be replaced by a number of Metropolis-Hastings moves or a multiple-try Metropolis move as discussed in Sections 5.5. The selected anchor chromosome can also be updated by some local optimization procedure, as in CGMC (Section 11.2). In summary, the evolutionary Monte Carlo algorithm works as follows:

1. Choose the population size m and the population temperatures $T = \{t_1, \ldots, t_m\}$. Initialize the population X at random.

2. Calculate the fitness of each chromosome.

3. Apply either a mutation or a crossover operator to the current population with probabilities q and $1 - q$.

4. Apply the exchange step. A chromosome x_i is chosen at random and is subjected to the proposal of a configuration swap with one of its neighbors (see Section 10.4).

5. The algorithm stops if a termination criteria is met. Otherwise, go to Step 3.

There are three user-set parameters in the algorithm, namely m, T, and q. The mutation rate q can be chosen to achieve a good trade-off between exploration and convergence of the algorithm. For a problem with a small population, q should be large in order to provide better chance to explore the sample space. For example, setting the mutation rate $q=1$ is equivalent to parallel tempering. However, for a problem with a large population, q can be set to a small value to get a fast convergence. Liang and Wong (2001) set q to around 0.2 for a problem with population size less than 50. The population size m and temperature ladder T can be set as in simulated tempering and parallel tempering (Marinari and Parisi 1992, Geyer 1991, Hukushima and Nemoto 1996). Roughly speaking, t_i should be set such that

$$\text{var}\{H(\mathbf{x}_i)\}\delta^2 = O(1), \qquad (11.11)$$

where $\delta = 1/t_{i+1} - 1/t_i$ and the variance is taken with respect to the target distribution. A key measurement is the acceptance ratio for the exchange operations, which should be kept at about 50%. This condition of δ is equivalent to requiring that the histograms of $H(\mathbf{x}_i)$ and $H(\mathbf{x}_{i+1})$ overlap substantially.

Clearly, all evolutionary Monte Carlo (EMC) operators including mutation, crossover, and exchange satisfy detailed balance. Hence, EMC sampler possesses the necessary properties of a MCMC sampler. Inference and estimation for each distribution $\pi_i(\mathbf{x}_i)$ can be made with the samples at the corresponding temperature level.

11.4 Some Further Thoughts

The EMC has been shown effective in many difficult problems (Liang and Wong 2000, Liang and Wong 2001). Careful examinations of these algorithms, however, still reveal some less satisfactory features. For example, the crossover operator is arguably the most important feature of the genetic algorithm that has not been previously incorporated into Monte Carlo computation. This operator still performs less satisfactorily in EMC because of the the reversibility constraint in the MCMC design. For example, in the current setting, the crossover proposal forces the offsprings produced after mating to completely replace the parental chromosomes. As a consequence, such a move is not easily accepted because the fitness of the new offsprings are usually not comparable to their parents. Occasionally, a good offspring (or a pair of good ones) is produced from the mating between a good parent and a not-so-good parent. Then, the acceptance of the new generation means that the good parental chromosome will perish, which is not a desirable feature in difficult simulation problems. Here, we describe a potentially useful idea to improve the performance of crossover operator.

In addition to the current temperature ladder $T = \{t_1, \ldots, t_m\}$, we can add a new "infinite temperature" level $t_0 = \infty$ at the bottom of the temperature ladder. The new (single) offspring obtained by a random mating of the current population is automatically "stored" at level t_0 (as $\mathbf{x}_0$) without subjecting it to any acceptance-rejection criterion. With the use of the exchange operator, the new configuration, if it is a good one, can then rise along the temperature ladder, as in the usual parallel tempering setting. In particular, when an exchange of $\mathbf{x}_0$ with $\mathbf{x}_1$ is proposed (note that $t_0 = \infty$), we accept the proposal with probability

$$\min\left\{1,\ \exp\left\{\frac{H(\mathbf{x}_1) - H(\mathbf{x}_0)}{t_1}\right\}\right\}.$$

Because the offspring produced by random mating is stored without being properly "adjusted," this procedure will not produce properly distributed Monte Carlo samples (with respect to the target distribution π). Instead, when using this new crossover procedure, the distribution of the samples at each temperature level shows a slight bias toward the "random mating" distribution (i.e., the distribution of $\mathbf{x}$ that is produced by randomly mating two members in a stationary population). The hope is that this bias is moderate and will have little effect in optimization problems. Note that this procedure would have been proper if the distribution of $\mathbf{x}_0$ was indeed uniform. Histograms of the distributions at all levels demonstrated this slight bias.

A way of alleviating the random mating bias is to employ the idea of histogram reweighting (i.e., the ideas in Chapter 10.5) as follows. Imagine that the system is indeed in equilibrium with respect to the proper target distributions [i.e., each $\mathbf{x}_i$ follows distribution $\pi_i(\mathbf{x}) \propto \exp\{-H(\mathbf{x})/t_i\}$]. Let a daughter chromosome $\mathbf{x}_0$ be produced by mating two randomly chosen members in the population. Then, $\mathbf{x}_0$ must have its distribution. We assume, without much theoretical basis, that this distribution is of the form

$$\pi_0(\mathbf{x}) \propto \exp\{-g(H(\mathbf{x}))\}. \tag{11.12}$$

Indeed, the accuracy of this method can be further improved if we assume a more refined form of π_0 [e.g., $\pi_0(\mathbf{x}) \propto \exp\{-g(H(\mathbf{x}), M(\mathbf{x}))\}$, where $M(\mathbf{x})$ is another "summary statistics," such as the total magnetization, of the system $\mathbf{x}$]. With these assumptions, we can iteratively estimate the unknown function $g(\ )$.

Now we describe an algorithm when form (11.12) is assumed. Let $W_0 = H(\mathbf{x}_0)$ and $W_1 = H(\mathbf{x}_1)$. Then, the distribution of W_0 is of the form

$$p_0(w) \propto N(w)\exp\{-g(w)\}$$

and W_1 is of the form

$$p_1(w) \propto N(w)e^{-w/t_1}.$$

Hence, we have

$$p_0(w)/p_1(w) \propto \exp\{-g(w) + w/t_1\}.$$

This implies that if we have good histogram estimates, $\hat{p}_1(w)$ and $\hat{p}_2(w)$, of $p_1(w)$ and $p_0(w)$, we can approximate the unknown function g well. Thus, when the exchange of $\mathbf{x}_0$ and $\mathbf{x}_1$ is proposed, we can accept the proposal with

$$p = \min\left\{1, \frac{\hat{p}_0(H(\mathbf{x}_1))\hat{p}_1(H(\mathbf{x}_0))}{\hat{p}_0(H(\mathbf{x}_0))\hat{p}_1(H(\mathbf{x}_1))}\right\}.$$

11.5 Numerical Examples

11.5.1 Simulating from a bimodal distribution

Consider a two-dimensional mixture Gaussian target distribution

$$0.34 \times N_2(\mathbf{0}, I_2) + 0.33 \times N_2\left\{\begin{pmatrix} -6 \\ -6 \end{pmatrix}, \begin{pmatrix} 1 & 0.9 \\ 0.9 & 1 \end{pmatrix}\right\}$$
$$+ 0.33 \times N_2\left\{\begin{pmatrix} 4 \\ 4 \end{pmatrix}, \begin{pmatrix} 1 & -0.9 \\ -0.9 & 1 \end{pmatrix}\right\}.$$

Here, the covariance matrices in the three components are identical to those in Gilks, Roberts and Sahu (1998),
but the mean vectors are separated by a larger distance in each dimension.

We started two independent Metropolis samplers with starting points drawn from Uniform$[-0.5, 0.5]^2$. A spherical proposal function was employed: A direction was generated uniformly and then the radius drawn from Uniform$[0, a]$, where a (which is equal to 4 in our case) is calibrated so that the Metropolis sampler had an acceptance rate of about 0.23 (Gelman, Roberts and Gilks 1995). A total of 200,000 iterations of the Metropolis step was conducted for each sampler, which took about 28 seconds of CPU time from a Sun Ultra 2 workstation. In Figure 11.1, we plotted the histograms and autocorrelations for one of the variables (left panels). It is seen that the Metropolis sampler moves very slowly due to the low-probability barriers between the modes, and the mixture proportions were very poorly estimated.

The CGMC method was applied to this problem with $m=2$ streams and 20,000 iterations for each. Each iteration consists of two Metropolis steps and one adaptation step. Thus, a total of 100,000 random draws from π, which took about 27 seconds of CPU time from the same computer, were produced as the program ended. The proposal function for the Metropolis step was the same spherical distribution as in the previous case but with

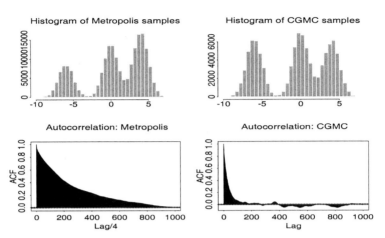

FIGURE 11.1. A comparison of results obtained by the Metropolis sampler and by the CGMC. The autocorrelation plot of the Metropolis samples has taken the computational cost into account.

a narrower range for the radius: $[0, 2.5]$ (corresponding to an acceptance rate of 0.37). For the CGMC, a small Metropolis step is beneficial for the purpose of exploring local features. The line sampling proposal was a univariate Gaussian with variance $= 10^2$ and the number of tries $k = 5$. This corresponds to an acceptance rate of 0.47. Our experience shows that an acceptance rate between 0.4 and 0.5 for the multiple-try step is appropriate. In Figure 11.1, we plotted the histograms and autocorrelations for one of the variables in one stream (right panels).

Using the heuristic of *integrated autocorrelation time* (IAT), which equals the sum of all lag autocorrelations, we can estimate that with the *same* amount of CPU time, the IAT for the Metropolis algorithm is about 249 after adjusting for the computational cost (4 to 1 ratio), whereas for each stream of the CGMC, the IAT is about 34. This translates to a seven-fold improvement.

11.5.2 Comparing algorithms for a multimodal example

To compare performances of different algorithms (e.g., EMC with real crossover, EMC with snooker crossover, and parallel tempering), Liang and Wong (2001) consider the simulation from a two-dimensional mixture Gaussian distribution:

$$\pi(\mathbf{x}) = \frac{1}{\sqrt{2\pi}\sigma} \sum_{i=1}^{20} w_i \exp\left\{-\frac{1}{2\sigma^2}(\mathbf{x} - \boldsymbol{\mu}_i)'(\mathbf{x} - \boldsymbol{\mu}_i)\right\},$$

$\sigma = 0.1$, $w_1 = \cdots = w_{20} = 0.05$. The mean vectors $\mu_1, \mu_2, \ldots, \mu_{20}$ are drawn uniformly from the rectangle $[0, 10] \times [0, 10]$. Among these centers, components 2, 4, and 15 are well separated from other components.

In their computer experiment, the technical settings were kept the same for all the algorithms: A population size of 20 was used; the highest temperature was 5.0, the lowest, 1.0, and the intermediate ones equally spaced in between; and all the chains were initiated with a vector uniformly drawn from $[0, 1] \times [0, 1]$. The EMC algorithm was run until 100,000 iterations were generated. It took slightly longer (30% more) for parallel tempering to generate the same amount of random draws. Figure 11.2 displays the scatterplots of Monte Carlo samples obtained from the first 100,000 iterations of both the real-coded EMC and parallel tempering. The EMC algorithm successfully sampled all 20 components with correct proportion of times, whereas parallel tempering completely missed the two corner clusters (northeast and southeast) represented by components 2, 4, and 15.

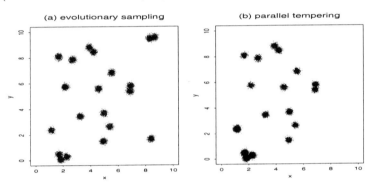

FIGURE 11.2. Scatterplots of the first 100,000 iterations of (a) the EMC algorithm and (b) parallel tempering.

Suppose one is interested in estimating the mean vector, (μ_1, u_2), and the covariance terms, σ_1^2, σ_2^2, and σ_{12}, of π. Then, the estimation accuracy of the three algorithms can also be compared. From Table 11.1, we observe that the snooker crossover made a significant improvement over the real crossover in accelerating the mixing of the system and that parallel tempering failed the test completely.

11.5.3 Variable selection with binary-coded EMC

Linear regression models have been used extensively in many application areas for fitting the data and for making predictions. Although linear models have been studied extensively in the past few decades, the variable selection problem (i.e., how to select a subset of good predictors from a set of potential explanatory variables) remains an intriguing and important topic in statistical research. For example, one may be interested in predicting

parameter	truth	EMC-SC		EMC-RC		PT	
		Estimate	S.E.	Estimate	S.E.	Estimate	S.E.
μ_1	4.478	4.481	0.0043	4.444	0.0259	3.781	0.0316
μ_2	4.905	4.909	0.0076	4.862	0.0230	4.337	0.0435
σ_1^2	5.552	5.549	0.0062	5.544	0.0507	3.656	0.1114
σ_2^2	9.861	9.841	0.0097	9.775	0.0481	8.546	0.0485
σ_{12}^2	2.605	2.591	0.0105	2.580	0.0434	1.294	0.0839

TABLE 11.1. Parameter estimation based on the samples over 20 independent runs. SD denotes the standard deviation of the estimate. EMC-SC: with the real crossover and the snooker crossover; EMC-RC: with only the real crossover; PT: parallel tempering.

the future performance of a company's stock price and is faced with many potential predictors such as the global economic indicator, currency exchange rate, industrial indicators, the company's size, its profit margin, its R&D budget, its stock price history, its trading volume history, its earning history, its insider trading records, who the president of the United States is, who won last year's Superbowl, whether there is a flood somewhere in China, etc. It is of interest to select among these alleged predictors the truly useful ones. A statistical formulation of the variable selection problem is as follows.

Let observations be of the form $(y_i; z_{1i}, \ldots, z_{Ki})$, $i = 1, \ldots, n$, where y_i is the response variable and $z_{1i}, \ldots, z_{Ki}$ are K potential predictors. A linear regression model is

$$y_i = \beta_0 + \beta_1 z_{1i} + \cdots + \beta_K z_{Ki} + \epsilon_i,$$

where $\epsilon_i \sim N(0, \sigma^2)$ and the β_j and σ^2 are unknown. We further assume that there is only a subset of the coefficients, $\beta_1, \ldots, \beta_K$, that are nonzero. Thus, the actual model can be reduced to

$$y = \beta_0 \sum_{i \in S} \beta_i z_i + \epsilon,$$

where S is a subset of $\{1, \ldots, K\}$. In order to select a good subset of predictors, Mallows (1973) proposed a C_p statistics defined as

$$C_p = \frac{\text{RSS}_p}{\hat{\sigma}^2} + 2p - n, \tag{11.13}$$

where p is the number of variables included in the prediction equation, RSS_p is the residual sum of squares of a submodel with p predictors, $\hat{\sigma}^2$ is the estimated error variance calculated from the full model, and n is the number of observations. Mallows (1973) suggested that good models have $C_p \cong p$. However, searching for a subset with good C_p is a *nondeterministic polynomial time* (NP)-hard problem.

We can reformulate the optimization problem as a simulation problem; that is, one can simulate from a *nominal* distribution defined on the space of subsets models

$$\pi(\mathbf{m}) \propto \exp\{-C_p(\mathbf{m})/t\}, \tag{11.14}$$

where $C_p(\mathbf{m})$ is the C_p value of model $\mathbf{m}$, $Z(t) = \sum_\mathbf{m} \exp\{-C_p(\mathbf{m})/t\}$. It can be shown that distribution (11.14) with $t = 2$ can be used to approximate a Bayesian posterior distribution of $\mathbf{m}$ (Liang, Truong and Wong 2001). Other distributions derived from a Bayesian formulation (Chen and Liu 1996, George and McCulloch 1997) can also be used in the place of (11.14). Since each model configuration $\mathbf{m}$ can be coded by a 0-1 string of length N, the binary-coded EMC algorithm can be applied to simulate from this model. A "1" at position k of the binary model vector indicates that the variable z_k is included in the model, and a "0" otherwise.

With population sizes of 5 and 20, respectively, and temperature range from $T = 1$ to $T = 5$ in each experiment, two comparison tests between the EMC and parallel tempering were conducted for the *highway* dataset of Weisberg (1985), where there are 10 potential variables. In this example, one can afford to use an exhaustive search method (a total of 2^{10} possible subset models) to get the ground truth.

Figures 11.3(c) and 11.3(d) compare the convergence performances of parallel tempering and the EMC for population sizes 5 and 20, respectively, where the y axis is the L^2 distance between the estimated frequency and the true distribution of C_p. In the two figures, EMC samples have been adjusted to have the same time scale as parallel tempering. Parallel tempering was also ran until it converged, defined by a distance measure below 0.01, and it took three times longer than that of the EMC in both cases. A few more difficult examples have also been examined by Liang and Wong (2000), and in every case, the EMC approach was shown to significantly outperform simulated tempering.

11.5.4 *Bayesian neural network training*

Suppose we observe $\mathbf{D} = \{(y_1, \mathbf{z}_1), (y_2, \mathbf{z}_2), \ldots, (y_n, \mathbf{z}_n)\}$ from a model

$$y_t = f(\mathbf{z}_t) + \epsilon_t, \tag{11.15}$$

where $y_t \in \mathbb{R}^1$, $\mathbf{z}_t \in \mathbb{R}^p$, $\epsilon_t \sim N(0, \sigma^2)$ for $t = 1, \ldots, n$. Our task is to estimate $f(\ )$ and to make predictions of y for a future explanatory variable $\mathbf{z}$. In the following, we assume that $\mathbf{z}$ has included the constant term $z_0 = 1$.

In *parametric* approaches (e.g., linear and logistic regressions), we assume that $f(\ )$ takes a parametric form known up to a finite number of tuning parameters (e.g., unknown coefficients in a linear regression model). Then, we use the observed data to estimate (or "train") the model. In *nonparametric* approaches, we do not not assume any particular form for $f(\ )$

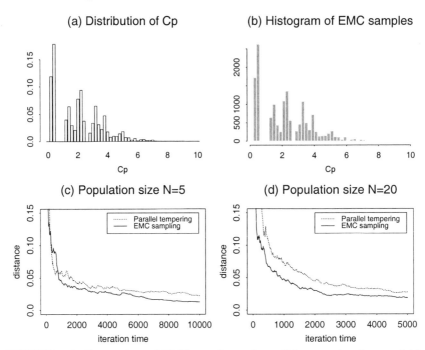

FIGURE 11.3. Comparison of EMC sampler and parallel tempering for a variable selection problem. (Courtesy of Professor Faming Liang.)

and use methods such as window smoothing or smoothing spline to estimate $f(\ )$ (Green and Silverman 1994). Nonparametric methods become ineffective when dimensionality of the explanatory variable $\mathbf{z}$ is high.

Here, we take a parametric approach, but with a more complex neural network (NN) model. Suppose $f(\ )$ can be approximated by a feed-forward neural network with one hidden-layer (Section 10.6.2):

$$\hat{f}(\mathbf{z}_t) = \sum_{j=1}^{M} \beta_j \psi(\mathbf{z}_t' \gamma_j), \qquad (11.16)$$

where M denotes the number of the hidden units, $\beta_j \in R$, the connection weights from the hidden unit j to the output unit, and $\gamma_j \in \mathbb{R}^p$, the connection weights from the input units to the hidden unit j. The activation function $\psi(\cdot)$ is a tanh function,

$$\psi(z) = \frac{\exp(z) - \exp(-z)}{\exp(z) + \exp(-z)}. \qquad (11.17)$$

A Bayesian method (Buntine and Weigend 1991, Neal 1996) is used to estimate unknown parameters in this model (or *train the network*).

Similar to the approach we took in Section 9.6.1, the NN model (11.16) is augmented by an error term and the NN training problem is viewed as a nonlinear regression of response y on the covariates $\mathbf{z}$:

$$y_t = \sum_{j=1}^{M} \beta_j \psi(\mathbf{z}_t' \gamma_j) + \epsilon_t, \tag{11.18}$$

where $\epsilon_t \sim N(0, \sigma^2)$ for $t = 1, \ldots, n$. The prior distributions for the parameters are, respectively, $\beta_j \sim N(0, \sigma_\beta^2)$, $\gamma_j \sim N(0, \sigma_\gamma^2 I)$ for $j = 1, \ldots, M$, and $\sigma^{-2} \sim \text{Gamma}(\nu, \delta)$. We can then derive the log-posterior of the parameters (up to an additive constant):

$$\log \pi(\beta, \gamma, \sigma^{-2} | D) = C - \left(\frac{n}{2} + \nu - 1\right) \log(\sigma^2) \tag{11.19}$$

$$- \frac{1}{2\sigma^2} \left\{ 2\delta + \sum_{t=1}^{n} \left[y_t - \sum_{j=1}^{M} \beta_j \psi(\mathbf{z}_t' \gamma_j) \right]^2 \right\}$$

$$- \sum_{j=1}^{M} \frac{\beta_j^2}{2\sigma_\beta^2} - \sum_{j=1}^{M} \sum_{i=0}^{p} \frac{\gamma_{ij}^2}{2\sigma_\gamma^2}. \tag{11.20}$$

This posterior distribution is difficult to simulate from for two main reasons: nonlinearity and multimodality. For example, the posterior distribution is invariant with respect to an arbitrary relabeling of the hidden units. Since $\psi(-x) = -\psi(x)$, the distribution is also invariant with respect to the simultaneous negation of both β_j and γ_j. To avoid this multimodality, one may impose a constraint on the parameter space [e.g., $0 < \gamma_{11} < \gamma_{21} < \cdots < \gamma_{M1}$; see Müller and Insua (1998)]. No such constraints were given, however, in the following study so as to fully examine the capability of EMC in the presence of multimodality and nonlinearity.

Under the Bayesian framework, the point prediction can be obtained by integrating out the nuisance parameters; that is,

$$\hat{y}_{\text{new}} = E(y_{\text{new}} | \mathbf{z}_{\text{new}}, \text{Data})$$

$$= \iiint \left(\sum_{j=1}^{M} \beta_j \psi(\mathbf{x}_t' \gamma_j) \right) \pi(\beta, \gamma, \sigma^{-2} | \text{Data}) d\beta d\gamma d\sigma^{-2}. \tag{11.21}$$

When $\mathbf{z}_{\text{new}}$ is not fully available (e.g., such as a time series with missing observations), one may have to impute it based on the model. Note that the posterior mean in (11.21) is not affected by the unidentifiability of the hidden units and the multimodality of the posterior density.

To test the EMC algorithm, Liang and Wong (2001) simulated $y_1, \ldots, y_{100}$ from Equation (11.18) with $M = 2$, $\gamma_1 = (\gamma_{10}, \gamma_{11}) = (2, -1)$, $\gamma_2 = (\gamma_{20}, \gamma_{21}) = (1, 1.5)$, $\beta = (20, 10)$, $\sigma = 0.1$, and the input variable $\mathbf{z}_t = (1, z_t)$, where $z_t = t * 0.1$ for $t = 1, 2, \ldots, 100$. The simulated data are shown in Figure 11.4. A NN model with the same structure is fit to the data

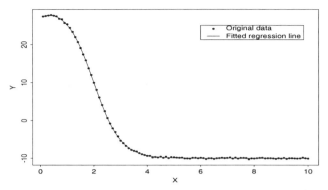

FIGURE 11.4. The original data overlayed by the MAP estimate (found by the EMC algorithm) of the unknown function.

by the Bayesian method outlined above [thus, the approximation model (11.16) has the identical structure as the "true" model]. The purpose of this exercise is to test EMC's ability in fitting highly nonlinear curves.

With the same amount of computing time, the EMC algorithm with both real and snooker crossovers outperformed parallel tempering significantly. All of the five independent EMC runs found and settled at the region of global maximum of the posterior, whereas all of the five parallel tempering runs got stuck in local maxima. Figure 11.5 depicts the comparison of two methods. One can also see from the figure that the mean square errors between fitted and observed values of the y_k converge to zero rather rapidly for the EMC.

11.6 Problems

1. Directly prove Lemma 11.1 by reparameterizing the state space.

2. Show that the CGMC algorithm is proper.

3. Implement both the snooker algorithm and the CGMC for the multimodal examples in Section 11.5.1.

4. Prove that the EMC algorithm for both continuous and binary-coded spaces satisfy the detailed balance.

5. Discuss why in CGMC the derivation of the anchor point can only depend on the current generation, but not on, say, $\mathcal{S}^{(t-1)}$.

6. Discuss why it is important to use tempered distributions for the population construction in EMC.

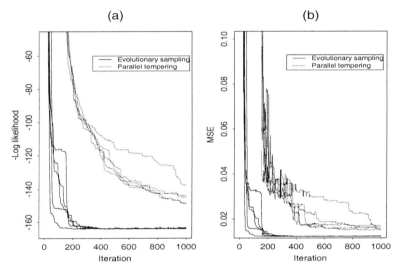

FIGURE 11.5. Comparison between EMC and parallel tempering. (a) The negative of log-posterior values of the samples versus computing time. (b) The mean square errors between the fitted and the observed values for each Monte Carlo sample of the NN parameters.

12
Markov Chains and Their Convergence

12.1 Basic Properties of a Markov Chain

When running a MCMC sampler, one is often fascinated by the fact that the sampler can produce desirable random samples from a target distribution by making a series of local changes to an arbitrary initial state. It is therefore a natural question to ask: What makes this operation work? Why can we obtain "typical samples" from a target distribution by conducting a series of local moves? A basic tool for studying theoretical properties of these Monte Carlo algorithms is the Markov chain theory.

Consider here a sequence of random variables $\mathbf{x}^{(0)}, \mathbf{x}^{(1)}, \ldots$, defined on a finite state space $\mathcal{X}$. This sequence is called a *Markov chain* if it satisfies the *Markov property*:

$$P(\mathbf{x}^{(t+1)} = \mathbf{y} \mid \mathbf{x}^{(t)} = \mathbf{x}, \ldots, \mathbf{x}^{(0)} = \mathbf{z}) = P(\mathbf{x}^{(t+1)} = \mathbf{y} \mid \mathbf{x}^{(t)} = \mathbf{x}); \quad (12.1)$$

that is, the value of $\mathbf{x}^{(t+1)}$ is dependent on its history only through its nearest past, $\mathbf{x}^{(t)}$. If the form of the transition probability $P(\mathbf{x}^{(t+1)} = \mathbf{y} \mid \mathbf{x}^{(t)} = \mathbf{x})$ is time homogeneous (i.e., it does not change with t), then it is often expressed as a *transition function*, $A(\mathbf{x}, \mathbf{y})$. Therefore, the simple property of *any* transition function is

$$\sum_{\mathbf{y}} A(\mathbf{x}, \mathbf{y}) = 1 \text{ for all } \mathbf{x}.$$

The first consequence of (12.1) is that for any time $s > 0$,

$$P(\mathbf{x}^{(t+s)} = \mathbf{y} \mid \mathbf{x}^{(t)} = \mathbf{x}, \ldots, \mathbf{x}^{(0)} = \mathbf{z}) = P(\mathbf{x}^{(t+s)} = \mathbf{y} \mid \mathbf{x}^{(t)} = \mathbf{x}). \quad (12.2)$$

12. Markov Chains and Their Convergence

We leave it to the reader to check the correctness of (12.2) (Hint: by induction). When the state space $\mathcal{X}$ is continuous (as in many examples of the previous chapters), the above transition probability function is replaced by the *transition density function* and summations replaced by integrations.

Example 1. Simple random walk on a line. Suppose the Z_n are i.i.d. Bernoulli random variables (coin tosses), with $P(Z_i = 1) = 1 - P(Z_i = -1) = p$. Let $S^{(0)} = 0$ and let $S^{(t)} = Z_1 + \cdots + Z_t$ (the deference between the number of heads and the number of tails of t coin tosses). Then, $S^{(t)}$, $t = 0, 1, \ldots$, forms a Markov chain. As $t \to \infty$, however, $S^{(t)}$ does not "converge" to a stable distribution and will drift to $\pm \infty$ (according to whether $p < 0.5$, $= 0.5$, or > 0.5).

Example 2. Let $\mathbf{x}^{(0)} = (-1, -1, \ldots, -1)$ be a vector of length d. We let $\mathbf{x}^{(t+1)}$ be generated recursively as follows: Randomly pick a coordinate of $\mathbf{x}^{(t)}$ and negate its current value. Then, the sequence $\mathbf{x}^{(0)}, \mathbf{x}^{(1)}, \ldots$, forms a Markov chain. If we write two consecutive states as $\mathbf{x} = (x_1, \ldots, x_d)$ and $\mathbf{y} = (y_1, \ldots, y_d)$, then the transition function for this chain is

$$A(\mathbf{x}, \mathbf{y}) = d^{-1} \text{ if } x_i = y_i \text{ for all but one component.}$$

This chain is often referred to as the *simple random walk* on a d-dimensional cube, $\mathbb{Z}_2^d$. For example, when $d = 3$, all the possible configurations this chain is allowed to visit correspond to the eight vertices of a three-dimensional cube). Many examples of Markov chains are constructed similarly: A new configuration $\mathbf{x}^{(t+1)}$ is generated from the previous one, $\mathbf{x}^{(t)}$, by a random local perturbation.

As $t \to \infty$, the simple-random-walk chain stabilizes to the *uniform* distribution. In other words, when t is large enough, the chance that you will guess correctly which vertex of the cube is occupied by $\mathbf{x}^{(t)}$ is roughly 2^{-d} (if I only tell you that t is greater than a certain large number but do not tell you whether t is even or odd). There is, however, one problem — the chain is *periodic*; that is, when t is odd, we are *sure* that $\mathbf{x}^{(t)}$ differs from $\mathbf{x}^{(0)}$ with only an odd number of +1's; whereas when t is even, the difference in the number of +1's has to be even. Thus, no matter how large t is, we still know something that relates $\mathbf{x}^{(t)}$ with $\mathbf{x}^{(0)}$.

To overcome this parity problem, we can incorporate a small "holding" probability in the Markov chain's transition rule. The modified rule is are follows: Pick a component $x_i^{(t)}$ of $\mathbf{x}^{(t)}$ at random; negate its value with probability $1 - \epsilon$ and leave it unchanged with probability ϵ. With this new transition rule, the theorem to be stated in Section 12.3 guarantees that the chain converges to a stable distribution, a uniform distribution in this case, and the realized value of $\mathbf{x}^{(t)}$ is becoming independent of the starting configuration $\mathbf{x}^{(0)}$.

Example 3. Random-to-top shuffling (Diaconis 1988). Suppose n cards are placed as a pile on a table. We consider the following slow card shuffling procedure: Pick a card from the deck at random and place it on the top of the deck. It is easier to describe the procedure than to implement it manually because most people have a tendency to pick a card more "inside" the deck instead of giving all the cards equal probabilities. It is easy to verify that this is a Markov chain. State configurations of the chain are the set of all permutations on $\{1, \ldots, n\}$, also called the *permutation group*.

Example 4. The Metropolis-type algorithms and Gibbs sampling algorithms described in the previous sections are all Markov chains.

12.1.1 Chapman-Kolmogorov equation

The most useful feature of a "good" Markov chain is its fast *forgetfulness* of its past; that is, after the Markov chain has evolved for a period of time, the realized value of the current state of the chain, $\mathbf{x}^{(t)}$, becomes nearly independent of starting state $\mathbf{x}^{(0)}$. This is precisely the reason why people shuffle the cards a number of times before they play another card game. Heuristically, this feature occurs in the same way as that in an analogous deterministic iterative system: If the system has a fixed point and if there is a "force" to drive the system toward that fixed point, then, with a sufficient number of iterations, the system will converge to (or be trapped in) that point regardless of what the starting configuration of the system is. In a stochastic system such as a Markov chain, the "fixed point" is a probability distribution, the so-called *invariant* or *equilibrium* distribution, and the "force" that drives the process toward this distribution is an appropriate transition rule. Several excellent books (Asmussen 1987, Ethier and Kurtz 1986, Nummelin 1984) explain this phenomenon rigorously.

Let $A^{(n)}(\mathbf{x}, \mathbf{y})$ denote the n-step transition function of a Markov chain:

$$A^{(n)}(\mathbf{x}, \mathbf{y}) = P(\mathbf{x}^{(n)} = \mathbf{y} \mid \mathbf{x}^{(0)} = \mathbf{x}).$$

Then, with $A^{(1)}(\mathbf{x}, \mathbf{y}) = A(\mathbf{x}, \mathbf{y})$, the *Chapman-Kolmogorov* equations take the following form:

$$A^{(n+m)}(\mathbf{x}, \mathbf{y}) = \int A^{(n)}(\mathbf{x}, \mathbf{z}) p^{(m)}(\mathbf{z}, \mathbf{y}) d\mathbf{x},$$

for any $n, m > 0$.

If the state space is countable, say, $\mathcal{X} = \{1, 2, \ldots, N\}$, then the transition matrix can be written as a matrix $A = (a_{ij})$, where a_{ij} is the transition probability from state i to state j. The Chapman-Kolmogorov equation takes the form

$$A^{(n+m)}(i, j) = \sum_{k=1}^{N} A^{(n)}(i, k) A^{(m)}(k, j).$$

248   12. Markov Chains and Their Convergence

This expression coincides perfectly with a matrix operation and is easy to verify from the Markov property (12.2):

$$P(\mathbf{x}^{(m+n)} = j \mid \mathbf{x}^{(0)} = i) = \sum_{k=1}^{N} P(\mathbf{x}^{(m+n)} = j, \mathbf{x}^{(n)} = k \mid \mathbf{x}^{(0)} = i)$$

$$= \sum_{k=1}^{N} A^{(n)}(i,k) A^{(m)}(k,j).$$

In summary, we have

$$A^{(m+n)} = A^{(m)} A^{(n)} = A^{m+n}.$$

In the later part of this book, we use $A^{(n)}$ and A^n interchangeably, with the former emphasizing on the conditional probability aspect and latter on the matrix aspect of the n-step transition function of A.

12.1.2 Convergence to stationarity

Let p_0 be the distribution of $\mathbf{x}^{(0)}$, which can be degenerate at a point mass $\mathbf{x}$. A nearly rigorous way to understand why a Markov chain converges is to analyze its transition matrix A directly. It is easy to see that because $\sum_{j=1}^{k} a_{ij} = 1$ for all i, A has an eigenvalue 1. Consider now an arbitrary function h_0 on the state space $\mathcal{X}$. Because of the discreteness of $\mathcal{X}$, this function can be written as a vector $\mathbf{h}_0$. Then, the new vector $\mathbf{h}_1 = A\mathbf{h}_0$ is simply the conditional expectation function

$$h_1(\mathbf{x}) = E[h(\mathbf{x}^{(1)}) \mid \mathbf{x}^{(0)} = \mathbf{x}]$$

for $\mathbf{x}$ running through $\mathcal{X}$. Because

$$\mathrm{var}\{E[h(\mathbf{x}^{(1)}) \mid \mathbf{x}^{(0)} = \mathbf{x}]\} \leq \mathrm{var}\{h(\mathbf{x}^{(1)})\},$$

we see that all of A's eigenvalues have to be smaller or equal to 1 in absolute value. Additionally, theory for positive matrices (Berman and Plemmons 1994) indicates that all of A's eigenvalues are real and A can be diagonalized; that is, we have the expression

$$A = B \Lambda B^{-1},$$

where $\Lambda = \mathrm{diag}(1, \lambda_2, \ldots, \lambda_k)$ with $1 \geq |\lambda_2| \geq \cdots \geq |\lambda_k|$. Hence, as $n \to \infty$,

$$A^n = B \begin{pmatrix} 1 & 0 & & 0 \\ 0 & \lambda_2^n & & \\ & & \ddots & \\ 0 & & & \lambda_k^n \end{pmatrix} B^{-1} \stackrel{n \to \infty}{\longrightarrow} B \begin{pmatrix} 1 & 0 & & 0 \\ 0 & 0 & & \\ & & \ddots & \\ 0 & & & 0 \end{pmatrix} B^{-1}$$

if and only if $|\lambda_2| < 1$. Because $\pi A^n = \pi$ and the limit of A^n is of rank 1, every row of the limiting matrix A^∞ must be the same as π.

The above argument loosely shows that the chain will converge to its unique invariant distribution if and only if matrix A's second largest eigenvalue in modular is strictly less than 1. It also shows that in the finite state-space case, the convergence rate is geometric (i.e., the "distance" between the distribution of $\mathbf{x}^{(t)}$ and the target distribution decreases geometrically).

The most commonly used distance measure between two probability measures P and Q is the *variation distance* or the L^1-*distance* defined as

$$\|P - Q\|_{\text{var}} \equiv \sup_{S \in \mathcal{X}} |P(S) - Q(S)|$$
$$= \frac{1}{2} \sum_{\mathbf{x} \in \mathcal{X}} |P(\mathbf{x}) - Q(\mathbf{x})| \equiv \frac{1}{2} \|P - Q\|_{L^1}. \qquad (12.3)$$

When the two distribution have density functions, $p(\mathbf{x})$ and $q(\mathbf{x})$, the above distance can be re-expressed as

$$\|P - Q\|_{\text{var}} = \frac{1}{2} \int |p(\mathbf{x}) - q(\mathbf{x})| d\mathbf{x}.$$

Another useful distance is the χ^2-distance defined as

$$\|P - Q\|_{\chi^2}^2 = \text{var}_P\{Q(\mathbf{x})/P(\mathbf{x})\} = \sum_{\mathbf{x} \in \mathcal{X}} |Q(\mathbf{x}) - P(\mathbf{x})|^2 / P(\mathbf{x}). \qquad (12.4)$$

It can be shown that the χ^2-distance is "stronger" than the previous two ones; that is, if the chain converges to the target distribution geometrically in χ^2-distance, it will also converge geometrically in variation distance. See Diaconis and Stroock (1991) for more details. The geometric convergence property of a Markov chain can be precisely expressed as follows: There exists a constant c and $|\lambda_2| < 1$ such that

$$\|A^{(n)}(\mathbf{x}^{(0)}, \cdot) - \pi(\cdot)\|_{L^k} \leq c|\lambda_2|^n.$$

For a finite state space, geometric convergence holds for both L^1- and χ^2-distances ($k=1$ or 2).

However, it is not so easy to show algebraically that A's second largest eigenvalue in modular is strictly less than 1. Classical Markov chain theory offers an attractive solution to this problem from a probabilistic argument. The following two basic properties are important for a Markov chain's asymptotic behavior.

Definition 12.1.1 *A state* $\mathbf{x} \in \mathcal{X}$ *is said to be irreducible if under the transition rule one has nonzero probability of moving from* $\mathbf{x}$ *to any other state and then coming back in a finite number of steps. A state* $\mathbf{x} \in \mathcal{X}$ *is said to be aperiodic if the greatest common divider of* $\{n : A^{(n)}(\mathbf{x}, \mathbf{x}) > 0\}$ *is 1.*

Clearly, if one state in $\mathcal{X}$ is irreducible, then all the states in $\mathcal{X}$ must also be. The aperiodicity basically means that there are no parity problems. Clearly, if a state **x** is aperiodic and the chain is irreducible, then every state in $\mathcal{X}$ must be aperiodic. If a finite-state Markov chain is irreducible and aperiodic, then it converges to the stationary distribution exponentially fast — this is a classical result which will be proved in Section 12.3. Before we get into the serious theorem-proof mode, we digress to illustrate the "coupling" method, one of the most useful and elegant probabilistic tools, through two simple examples.

12.2 Coupling Method for Card Shuffling

12.2.1 Random-to-top shuffling

The coupling method can be applied elegantly to analyze the card shuffling scheme described in Example 3 of Section 12.1. Imagine that we have two decks of n playing cards. The **x** deck starts with a special configuration $\mathbf{x}^{(0)}$ (all ordered, say), whereas the **y** deck has been thoroughly shuffled (i.e., it is a random sample from the population of $n!$ permutations). We can implement the following procedure to "couple" the two decks of cards. Suppose at the tth step, a random card in the **x** deck, say, the six of hearts, is chosen and placed at the top of the deck. We can also choose the same card, the six of hearts, in the **y** deck and put it also at the top of the deck. Note that although the physical process of choosing this card in the **y** deck is a bit different from that in the **x** deck, this six of hearts is indeed a card selected at random (i.e., having equal probability of being any of the n cards). Thus, this process does not disturb the stationarity of the **y** deck; that is, the **y** deck is still well shuffled after this move. Now, all the cards that have been touched up to time t are in identical positions in both decks. Therefore, time T when *all* the cards in the **x** deck have been touched according to the random-to-top process is the time the **x** deck and the **y** deck become identical in order.

Since the **y** deck starts from the equilibrium distribution and each move does not disturb its stationarity, the **y** deck still follows the stationary distribution (i.e., the uniform distribution in this case) at and after the "coupling time" T. Because of the coupling, the **x** deck also follows its stationary distribution at and after time T. Now, the problem is reduced to the *coupon collector's problem* in elementary probability: How many times do we have to draw the card in order to touch all of them? It is easy to see that

$$E(T) = n\left\{\frac{1}{n} + \frac{1}{n-1} + \cdots + 1\right\} \approx n\log n,$$

and var$(T) \approx 0.72n$. So the answer is: it will take about $n(\log n + c)$ random-to-top shuffles to completely mix a deck of n cards.

12.2.2 Riffle shuffling

How many ordinary riffle shuffles are needed to bring a deck of cards close to random? The answer is 7 times for the regular deck (52) of playing cards. This result of Diaconis and co-workers (Diaconis 1988) is one of the most elegant convergence results in modern Markov chain theory. Here, we describe an easier argument of Reeds (1981), which gives us a slightly worse result (about 11 times).

To prove mathematical results for riffle shuffling, we need a statistical model to describe the process. One reasonable model assumes that a deck of n cards is first cut into two piles according to the Bionom$(n, 1/2)$ distribution. Suppose k cards are cut off and held in the left hand and the remaining $n - k$ cards are held in the right hand. Drop cards with probability proportional to packet size. For example, if a deck of 52 cards is cut to $k = 20$ and $n - k = 32$. The probability that the first card drops from the left hand is 20/52. As shown by a data analysis carried out by Diaconis and co-workers, this statistical model seems to be a reasonable approximation to real-life card shuffling, at least for ordinary people (shuffles of professional dealers are "neater").

It can be shown that the foregoing model is, in fact, mathematically equivalent to another model, the *inverse shuffling* model: Label the back of each card by 0 or 1 according to an independent fair coin toss; remove all cards labeled 0 and place them on top of the deck, keeping other relative orders unchanged.

The inverse shuffling model lends us some more insight into the analysis. After one inverse shuffle, all the cards labeled 0 must be above all those labeled 1. We can imagine that the effect of the second shuffle is to add a random 0 or 1 label to the *left* of the previous label on each card. Thus, all the cards with label 00 must be above those with 01, and then above those with 10, and then 11. After t shuffles, every card is associated with a random binary sequence of length t. At time T when all such randomly generated binary sequences of length T are *distinct*, the deck becomes completely random (i.e., all the cards are sorted according to a randomly generated order). It is easy to see that $P(T \leq t)$ is the same as the probability that n balls are dropped into 2^t boxes and there are not two or more balls in one box. Thus,

$$P(T > t) = 1 - \prod_{i=1}^{n}\left(1 - \frac{i}{2^t}\right). \tag{12.5}$$

A stopping T is called *strong uniform time* by Aldous and Diaconis (1987) if

$$P(\mathbf{x}_t \in S \mid T = t) = U(S), \quad \forall \, t \text{ and } S, \tag{12.6}$$

where U is the uniform probability measure on $\mathcal{X}$. It is easy to show from (12.6) and the Markov property that $P(\mathbf{x}_{t+s} \in S \mid T = t) = U(S)$ for all $s \geq 1$. The coupling time T for random-to-top shuffling is clearly a strong uniform time.

Let Q_t be the distribution of the card ordering after t shuffles. Then, for a strong uniform time T, we have

$$\begin{aligned}
\|Q_t - U\|_{\text{var}} &= \sup_S |P(\mathbf{x}_t \in S \mid T < t) P(T < t) \\
&\quad + P(\{\mathbf{x}_t \in S\} \cap \{T > t\} - U(S)| \\
&= \sum_S |P(\mathbf{x}_t \in S \mid T > t) P(T > t) - \pi(S) P(T > t)| \\
&\leq P(T > t). \tag{12.7}
\end{aligned}$$

Combining (12.5) and (12.7) gives us a bound for how fast the card shuffling process convergences.

12.3 Convergence Theorem for Finite-State Markov Chains

The following theorem is one of the most well-known classical results, for which we provide a proof based on the coupling method.

Theorem 12.3.1 *Suppose the state space $\mathcal{X}$ of a Markov chain is finite. The transition function of this chain is irreducible and aperiodic; then, $A^{(n)}(\mathbf{x}_0, \mathbf{y}) = P(\mathbf{x}^{(n)} = \mathbf{y} \mid \mathbf{x}^{(0)} = \mathbf{x}_0)$ as a probability measure on $\mathbf{y}$ converges to its invariant distribution $\pi(\mathbf{y})$ geometrically in variation distance; that is, there exists a $0 < r < 1$ and $c > 0$ such such*

$$\|A^{(n)}(x, \cdot) - \pi\|_{\text{var}} \leq c r^n$$

Proof: First, it can be seen that the irreducibility and aperiodicity conditions guarantee the existence of an integer n_0 such that $A^{(n_0)}(\mathbf{x}, \mathbf{y}) > 0$ for all $\mathbf{x}, \mathbf{y} \in \mathcal{X}$; that is, we are able to find a large enough n_0 so that it is possible to travel from any state to any other one in n_0 steps. Therefore, we can imagine having another Markov chain whose one-step transition is the same as the n_0-step transition in the original chain. If we can prove the convergence result for the new chain, we also have a similar convergence

12.3 Convergence Theorem for Finite-State Markov Chains

result for the original chain. Thus, it is sufficient for us to prove that the convergence result holds for a Markov chain with a transition matrix A that has *all* nonzero entries. From now on, we simply assume that $A(\mathbf{x}, \mathbf{y}) > 0$ for all $\mathbf{x}, \mathbf{y} \in \mathcal{X}$.

Because the state space is finite, we can find a small enough number $\epsilon > 0$ such that

$$A(\mathbf{x}, \mathbf{y}) \geq \epsilon \pi(\mathbf{y})$$

for all $\mathbf{x}, \mathbf{y} \in \mathcal{X}$. Let Π be a $|\mathcal{X}| \times |\mathcal{X}|$ matrix with each of its row vectors being π. Then, Π can be regarded as the transition matrix that generates independent draws from distribution π [i.e., $\Pi(\mathbf{x}, \mathbf{y}) \equiv \pi(\mathbf{y})$]. We then has the following simple identity:

$$A = (1 - \epsilon)A^* + \epsilon\Pi, \tag{12.8}$$

where $A^* = (A - \epsilon\Pi)/(1 - \epsilon)$ is clearly another transition function (i.e., all of its entries are non-negative and every of its rows adds up to 1). An "operational" interpretation of (12.8) is as follows: We can implement the transition rule by first tossing an ϵ-coin (i.e., the one that shows head with probability ϵ and tail with probability $1 - \epsilon$); if the coin shows head, the next state is generated by Π, whereas if the coin shows tail, the next state is generated by A^*. This probabilistic mixture of transitions is clearly equivalent to A.

Now, our basic setups are mostly in place and we can use the *coupling argument* to prove convergence. Imagine that we start two Markov chains $\mathbf{x}^{(0)}, \mathbf{x}^{(1)}, \ldots$ and $\mathbf{y}^{(0)}, \mathbf{y}^{(1)}, \ldots$, both evolving according to the transition function A. The first chain is started with $\mathbf{x}^{(0)} = \mathbf{x}_0$, and the second chain is started from the *equilibrium* distribution (i.e., $\mathbf{y}^{(0)} \sim \pi$). Thus, because of the Markov property, the $\mathbf{y}$ chain is *always* in stationarity (i.e., $\mathbf{y}^{(t)} \sim \pi$ for all t). Therefore, if we can link the evolvement of the two processes and can find a "coupling time" when the two chains become identical, then the $\mathbf{x}$ chain becomes stationary after the coupling time.

In order to couple the two chains, we evolve them jointly by using the mixture interpretation of (12.8). At every step t we first toss an ϵ-coin. If a tail shows up, $\mathbf{x}^{(t)}$ and $\mathbf{y}^{(t)}$ are evolved separately according to A^* to their corresponding next configurations. If a head shows up, we update both the $\mathbf{x}$ chain and the $\mathbf{y}$ chain by a common $\mathbf{z}$ drawn from π. In this latter case, the two chains are "coupled" and they will be always coupled thereafter. Let T denote this coupling time. Since we start the $\mathbf{y}$ chain from the stationary distribution, we know that the $\mathbf{x}$ chain is also in stationarity once the coupling event occurs; that is, for any $S \subseteq \mathcal{X}$,

$$P(\mathbf{x}^{(T)} \in S) = \sum_{t=1}^{\infty} P(\mathbf{y}^{(t)} \in S \mid T = t) P(T = t) = \pi(S).$$

The last equality follows because the occurrence of the coupling event (based on a coin toss) is independent of the state variable.

Furthermore, we have

$$P(T > t) = \epsilon\{(1-\epsilon)^t + (1-\epsilon)^{t+1} + \cdots\} = (1-\epsilon)^t.$$

Thus, for any set $S \subseteq \mathcal{X}$, we have

$$\begin{aligned}|P(\mathbf{x}^{(t)} \in S) - \pi(S)| &= |P(\mathbf{x}^{(t)} \in S) - P(\mathbf{y}^{(t)} \in S)| \\ &\leq P(\mathbf{x}^{(t)} \neq \mathbf{y}^{(t)}) = P(T > t) = (1-\epsilon)^t.\end{aligned}$$

This means that the variation distance between the distribution of $\mathbf{x}^{(t)}$ and the stationarity distribution converges to zero geometrically. ◇

It needs a bit of thought to convince oneself that the foregoing coupling prescription is a valid operation; that is, the operation does not violate the transition rule for both the $\mathbf{x}$ chain and the $\mathbf{y}$ chain.

Once we have shown the convergence of a finite-state Markov chain to its unique stationary distribution, we can now go back to use the algebraic method to see that the convergence rate is actually governed by $|\lambda_2|$. The coupling argument similar to the above is also very useful for analyzing the infinite-state case and has been used extensively for studying Markov chains with a continuous state space (Nummelin 1984).

12.4 Coupling Method for General Markov Chain

As demonstrated by a number of researchers (Athreya and Ney 1978, Asmussen 1987, Mykland, Tierney and Yu 1995, Nummelin 1984, Nummelin and Tweedie 1978, Rosenthal 1995, Roberts and Rosenthal 1998, Murdoch and Green 1998) in the past decade, a very useful concept in studying an infinite-state Markov chain is that of the *small set*, or *minorisation condition*. This condition requires the existence of a set $C \subseteq \mathcal{X}$ of which all the states have a common component. As with the previous sections, we let $A^{(n)}(\mathbf{x}, \mathbf{y})$ be the n-step transition function:

$$A^{(n)}(\mathbf{x}, \mathbf{y}) = \int\int\int A(\mathbf{x}, \mathbf{x}^{(1)}) \cdots A(\mathbf{x}^{(n-1)}, \mathbf{y}) d\mathbf{x}^{(1)} \cdots d\mathbf{x}^{(n-1)}.$$

Clearly, $A^{(n)}(\mathbf{x}, \cdot)$ is a probability measure (density or distribution function) on $\mathcal{X}$. A main goal of all the convergence analyses is to provide a quantitative or qualitative guide on how fast this probability measure converges to the target distribution π. The following two definitions are crucial in the later analysis.

Definition 12.4.1 *A subset $C \subseteq \mathcal{X}$ is small if there exists an integer $n_0 > 0$, $\epsilon > 0$, and a probability measure $\nu(\cdot)$ on $\mathcal{X}$ such that*

$$A^{(n_0)}(\mathbf{x}, \cdot) \geq \epsilon\nu(\cdot), \quad \forall \mathbf{x} \in C.$$

12.4 Coupling Method for General Markov Chain

It is worthwhile to note here that when $\mathcal{X}$ is finite and the chain is irreducible, the whole space $\mathcal{X}$ is a *small* set. The coupling approach used in the proof of Theorem 12.3.1 relies heavily on the irreducibility condition which guarantees the existence of an n_0 such that $A^{(n_0)}(\mathbf{x}, \cdot)$ has a common component for all $\mathbf{x} \in \mathcal{X}$. With the small set, we can now contemplate a similar coupling argument for the general Markov chain.

Consider initiating two Markov chains: $\mathbf{x}^{(0)}$ is drawn from an arbitrary initial distribution and $\mathbf{y}^{(0)}$ is drawn from the equilibrium distribution π. Without loss of much generality, we suppose $n_0 = 1$ for the small set C. Then, we can couple the two chains as follows. (A) If at time t, both $\mathbf{x}^{(t)}$ and $\mathbf{y}^{(t)}$ are in C, we toss an ϵ-coin. If it shows head, then we draw $\mathbf{x} \sim \nu(\cdot)$ and let $\mathbf{x}^{(t+1)} = \mathbf{y}^{(t+1)} = \mathbf{x}$; if it shows tail, we update $\mathbf{x}^{(t+1)}$ and $\mathbf{y}^{(t+1)}$ independently according to the "complementary" measures, $\{A(\mathbf{x}^{(t)}, \cdot) - \epsilon\nu(\cdot)\}/(1-\epsilon)$ and $(A(\mathbf{y}^{(t)}, \cdot) - \epsilon\nu(\cdot))/(1-\epsilon)$, respectively. (B) If $\mathbf{x}^{(t)}$ and $\mathbf{y}^{(t)}$ are not all in C, we simply update them independently according to $A(\mathbf{x}^{(t)}, \cdot)$ and $A(\mathbf{y}^{(t)}, \cdot)$, respectively. Once the two chains are coupled, they will be identical thereafter. Since the $\mathbf{y}$ chain starts from equilibrium, we have

$$\|A^{(t)}(\mathbf{x}^{(0)}, \cdot) - \pi(\cdot)\|_{\mathrm{var}} \leq P(\mathbf{x}^{(t)} \neq \mathbf{y}^{(t)}) \leq P(T > t),$$

where T is the coupling time.

Each time both $\mathbf{x}^{(t)}$ and $\mathbf{y}^{(t)}$ are in C together, there is a probability of ϵ of coupling. Let N_t be the number of times both $\mathbf{x}^{(s)}$ and $\mathbf{y}^{(s)}$ are in C for $1 \leq s \leq t$. Then,

$$\begin{aligned} P(T > t) &= P(T > t \text{ and } N_t \geq j) + P(T > t \text{ and } N_t < j) \\ &\leq (1-\epsilon)^j + P(N_t < j). \end{aligned}$$

We now need only to address the following question: How frequently will the $\mathbf{x}$ chain and the $\mathbf{y}$ chain visit C together? Clearly, if the small set C is physically large (e.g., $C = \mathcal{X}$), then N_t is large and $P(N_t < j)$ is small for $j \sim O(t)$. When the common component is also "big" (i.e., ϵ big), the chain converges to its stationary distribution quickly. In many cases, one can directly studies the behavior of the chain to get a bound on quantity $P(N_t < j)$, which reflects the "attractiveness" of the small set C.

In order to obtain a convergence bound for a general Markov chain, Meyn and Tweedie (1994) introduced the following drift condition.

Definition 12.4.2 *The chain is said to satisfy a geometric drift condition for the small set C if there is a π-almost everywhere finite function $V(\mathbf{x}) \geq 1$, constant $\lambda < 1$, and constant $b < \infty$, such that*

$$E[V(\mathbf{x}^{(1)}) \mid \mathbf{x}^{(0)} = \mathbf{x}] \equiv \int V(\mathbf{y}) A(\mathbf{x}, \mathbf{y}) d\mathbf{y} \leq \lambda V(\mathbf{x}) + b \mathbf{1}_C(\mathbf{x}),$$

where $\mathbf{1}_C(\mathbf{x})$ is equal to 1 when $\mathbf{x} \in C$ and 0 otherwise.

With this condition, Meyn and Tweedie (1994) showed that

$$\|A^n - \pi\|_{L^1} \leq \rho^{n+1}/(\rho - \vartheta), \quad \forall \rho > \vartheta,$$

where $\vartheta = 1 - M_\alpha^{-1}$,

$$M_\alpha = \frac{1}{(1-\lambda)^2}\left[1 - \lambda + b + b^2 + \zeta_\alpha(b(1-\lambda) + b^2)\right],$$

and

$$\zeta_\alpha = \sup_{|z| \leq 1} \left|\sum_{m=0}^{\infty} [A^m(\alpha,\alpha) - A^{m-1}(\alpha,\alpha)]z^n\right|.$$

Under a slightly different drift condition: $\exists\, V(\cdot) \geq 0$, $\lambda < 1$, $b < \infty$, $d > 0$, and the small set $C = \{\mathbf{x} : V(\mathbf{x}) \leq d\}$, such that

$$E[V(\mathbf{x}^{(1)}) \mid \mathbf{x}^{(0)} = \mathbf{x}] \leq \lambda V(\mathbf{x}) + b,$$

Rosenthal (1995) showed that for *any* $0 < r < 1$ and $M > 0$,

$$\|A^n(\mathbf{x}^{(0)}, \cdot) - \pi(\cdot)\|_{L^1} \leq (1-\epsilon)^{\lfloor \frac{rn}{n_0} \rfloor} + C_0(M)(\alpha A)^{-1}\left(\alpha^{-(1-rn_0)} A^r\right)^n,$$

where

$$\alpha^{-1} = \frac{1 + 2Mb + \lambda Md}{1 + Md}, \quad A = 1 + 2(\lambda d + b)M,$$

and

$$C_0(M) = 1 + \frac{Mb}{1-\lambda} + MV(\mathbf{x}^{(0)}).$$

In practice, finding an appropriate test function $V(\cdot)$ is often difficult.

12.5 Geometric Inequalities

Over the past two decades, probabilists and computer theorists have found that a number of elementary inequalities in differential geometry are very useful for bounding eigenvalues of the transition matrix of a Markov chain. In this section, we follow Diaconis and Stroock (1991) to introduce a few techniques such as the Poincaré inequalities and Cheeger's inequality for analyzing a finite-state Markov chain. The central idea of these techniques is to relate the Markov chain convergence rate to a measure of "bottlenecks" in the Markov transition graph.

12.5.1 Basic setup

Let $A(\mathbf{x}, \mathbf{y})$ be an irreducible Markov transition function on a finite state space $\mathcal{X}$. We assume in this section that $A(\mathbf{x}, \mathbf{y})$ satisfies the detailed balance with respect to its invariant probability measure π and define

$$Q(\mathbf{x}, \mathbf{y}) = \pi(\mathbf{x})A(\mathbf{x}, \mathbf{y}) = \pi(\mathbf{y})A(\mathbf{y}, \mathbf{x}).$$

Such a Markov chain is often called *reversible* in the literature.

As we have discussed earlier, A has a set of eigenvalues

$$1 = \beta_0 > \beta_1 \geq \cdots \geq \beta_{m-1} \geq -1, \quad \text{where } m = |\mathcal{X}|.$$

It is sometimes more convenient to consider a related matrix, the *Laplacian* $L = I - A$, because its eigenvalues, $\lambda_i = 1 - \beta_i$, $i = 0, \ldots, m-1$, are easier to characterize by a minimax argument; that is, we have the relationship (Horn and Johnson 1985)

$$\lambda_1 = \inf\left\{ \frac{\mathcal{E}(\phi, \phi)}{\text{var}(\phi)} \: : \: \text{var}(\phi) > 0 \right\}, \tag{12.9}$$

where $\text{var}(\phi) = E_\pi[\phi(\mathbf{x})^2] - E_\pi^2[\phi(\mathbf{x})]$ is the variance with respect to π and

$$\mathcal{E}(\phi, \phi) = \frac{1}{2} \sum_{\mathbf{x}, \mathbf{y}} [\phi(y) - \phi(x)]^2 Q(\mathbf{x}, \mathbf{y}). \tag{12.10}$$

For historical reasons, this is also called the *quadratic form* of Laplacian L.

A finite-state Markov chain naturally induces a "connection" graph $G = (\mathcal{X}, E)$, where the set of vertices is the whole space and the set of edges consists of all the pairs of states with nonzero transition probabilities:

$$E = \{(\mathbf{x}, \mathbf{y}) \: : \: \text{if and only if } Q(\mathbf{x}, \mathbf{y}) > 0\}$$

For each distinct pair of states $\mathbf{x}$ and $\mathbf{y}$ in $\mathcal{X}$, we can choose one path γ_{xy} that leads from $\mathbf{x}$ to $\mathbf{y}$. (At least one such path exists because the chain is irreducible.) Let Γ be the collection of all the paths we have chosen (one for each ordered pair $\mathbf{x}$ and $\mathbf{y}$). We define the "length" of the path γ_{xy} as

$$\|\gamma_{xy}\| = \sum_{e \in \gamma_{xy}} Q(e)^{-1},$$

where the sum is over all the edges in the path and the edge weight $Q(e)$ is equal to $Q(e^+, e^-)$, where e^+ and e^- are the two vertices connected by edge e.

12.5.2 Poincaré inequality

An important geometric quantity that measures how well all the states in $\mathcal{X}$ are "connected" is defined as

$$\kappa = \kappa(\Gamma) = \max_e \sum_{\gamma_{xy} \ni e} \|\gamma_{xy}\| \pi(\mathbf{x}) \pi(\mathbf{y}), \tag{12.11}$$

where the maximum is taken over all the directed edges in the graph and the sum is over all paths which traverse e. Clearly, κ will be very large if there is an edge e which has a small $Q(e)$ but is used heavily by many paths in the graph (like the usual cause of a traffic jam). With these notations, we have the following theorem.

Theorem 12.5.1 (Poincaré inequality) *The second largest eigenvalue of an irreducible Markov chain A satisfies*

$$\beta_1 \leq 1 - \kappa^{-1},$$

where κ is defined by (12.11).

Proof: Because of (12.9), we need only to show that $\text{var}(\phi) \leq \gamma \mathcal{E}(\phi, \phi)$. Define for any edge $\phi(e) = \phi(\mathbf{w}) - \phi(\mathbf{v})$, where e is the directed edge connecting from $\mathbf{v}$ to $\mathbf{w}$. Then,

$$\text{var}(\phi) = \frac{1}{2} \sum_{\mathbf{x},\mathbf{y} \in \mathcal{X}} [\phi(\mathbf{x}) - \phi(\mathbf{y})]^2 \pi(\mathbf{x})\pi(\mathbf{y})$$

$$= \frac{1}{2} \sum_{\mathbf{x},\mathbf{y} \in \mathcal{X}} \left[\sum_{e \in \gamma_{xy}} \left(\frac{Q(e)}{Q(e)}\right)^{1/2} \phi(e) \right]^2 \pi(\mathbf{x})\pi(\mathbf{y})$$

$$\leq \frac{1}{2} \sum_{\mathbf{x},\mathbf{y} \in \mathcal{X}} \left[\left\{ \sum_{e \in \gamma_{xy}} Q(e)\phi(e)^2 \right\} \left\{ \sum_{e \in \gamma_{xy}} Q(e)^{-1} \right\} \pi(\mathbf{x})\pi(\mathbf{y}) \right]$$

$$= \frac{1}{2} \sum_{\mathbf{x},\mathbf{y} \in \mathcal{X}} \|\gamma_{xy}\| \pi(\mathbf{x})\pi(\mathbf{y}) \sum_{e \in \gamma_{xy}} Q(e)\phi(e)^2$$

$$= \frac{1}{2} \sum_{e} Q(e)\phi(e)^2 \sum_{\gamma_{xy} \ni e} \|\gamma_{xy}\| \pi(\mathbf{x})\pi(\mathbf{y}) \leq \kappa \mathcal{E}(\phi, \phi).$$

The inequality follows from the Cauchy-Schwartz; the last equality is due to the exchange of the two summation signs. ◇

Instead of using the inequality in the foregoing proof, we can apply the Cauchy-Schwartz inequality slightly differently to obtain another Poincaré inequality:

$$\begin{aligned}
\text{var}(\phi) &= \frac{1}{2} \sum_{\mathbf{x},\mathbf{y}\in\mathcal{X}} \left[\sum_{e\in\gamma_{xy}} 1\cdot\phi(e)\right]^2 \pi(\mathbf{x})\pi(\mathbf{y}) \\
&\leq \frac{1}{2} \sum_{\mathbf{x},\mathbf{y}\in\mathcal{X}} \left\{\sum_{e\in\gamma_{xy}} \phi(e)^2\right\} \left\{\sum_{e\in\gamma_{xy}} 1\right\} \pi(\mathbf{x})\pi(\mathbf{y}) \\
&= \frac{1}{2} \sum_{\mathbf{x},\mathbf{y}\in\mathcal{X}} \left\{\sum_{e\in\gamma_{xy}} Q(e)\phi(e)^2\right\} Q(e)^{-1}|\gamma_{xy}|\pi(\mathbf{x})\pi(\mathbf{y}) \\
&= \sum_e Q(e)\phi(e)^2 \sum_{\gamma_{xy}\ni e} Q(e)^{-1}|\gamma_{xy}|\pi(\mathbf{x})\pi(\mathbf{y}),
\end{aligned}$$

where $|\gamma_{xy}|$ is the number of edges in γ_{xy}. Therefore, we have

$$\text{var}(\phi) \leq K\mathcal{E}(\phi,\phi),$$

where

$$K = \max_e \left\{ Q(e)^{-1} \sum_{\gamma_{xy}\ni e} |\gamma_{xy}|\pi(\mathbf{x})\pi(\mathbf{y}) \right\}. \tag{12.12}$$

Sometimes this newer version of the Poincaré inequality is better than the one stated in the theorem and sometimes vice versa (Diaconis and Stroock 1991, Fill 1991).

A similar technique can also be used to bound the smallest eigenvalue and is omitted here (Diaconis and Stroock 1991). It can be seen from the definition that $\|\gamma_{xy}\|$ can be very large, which gives rise to a very large κ and a bad convergence bound, when the target distribution π is very nonuniform (i.e., having a very small probability of visiting certain states). Cheeger's inequality to be discussed later is useful to get around this technical difficulty.

12.5.3 Example: Simple random walk on a graph

Let $G = (V, E)$ be a undirected graph where V is the vertex set and E is the edge set. A nearest-neighbor random walk on G can be described as follows: Suppose the walker is at a vertex $\mathbf{x}$ at time t. Then, at time $t+1$, he chooses at random to visit one of its nearest neighbors (i.e., the set of all vertices that are connected with $\mathbf{x}$ by an edge). The *degree* of a vertex, $d(\mathbf{x})$, is just the number of its nearest neighbors:

$$d(\mathbf{x}) = \#\{\mathbf{y} \in \mathcal{X} : \{\mathbf{x},\mathbf{y}\} \in E\}.$$

Thus, the transition function $A(\mathbf{x}, \mathbf{y})$ is equal to $1/d(\mathbf{x})$ if $\mathbf{y}$ is a neighbor of $\mathbf{x}$ and is zero otherwise. It is not difficult to see that the invariant distribution of this chain is

$$\pi(\mathbf{x}) = d(\mathbf{x})/2|E|,$$

where $|E|$ is the total number of edges, and $Q(\mathbf{x},\mathbf{y}) = 1/2|E|$ if $(\mathbf{x},\mathbf{y}) \in E$ and 0 otherwise. For any given path, we have

$$\|\gamma_{xy}\| = |\gamma_{xy}|/2|E| \leq \gamma_*/2|E|,$$

where $|\gamma|$ is the number of edges in a path γ and γ_* is the length of the longest path.

It is trivial to verify that

$$\kappa(\Gamma) \leq \left(\frac{d_*}{2|E|}\right)^2 2|E|\gamma_* b,$$

where $d_* = \max d(\mathbf{x})$ and b measures how busy the "busiest" street is:

$$b = \max_e \#\{\gamma \in \Gamma : \gamma \ni e\}.$$

A direct application of the Poincaré inequality gives rise to the following result.

Corollary 12.5.1 *For the simple random walk on a connected graph $G = (V, E)$, the second largest eigenvalue is bounded by*

$$\beta_1 \leq 1 - \frac{2|E|}{d_*^2 \gamma_* b},$$

where d_*, γ_*, and b are as defined earlier.

Simple random walk on a cube. A cube on an d-dimensional space (denoted as $\mathbb{Z}_2^d$) can be coded by the set of all 2^d binary strings of length d, $\mathbf{x} = (x_1, \ldots, x_d)$, where $x_j = 0$ or 1. A simple random walk on it can be implemented as follows: Randomly pick a coordinate x_j and change its current value to $1 - x_j$. The graph for this Markov chain has 2^d vertices and $d2^{d-1}$ edges. Every vertex has the same degree, d. The length of the longest path is also d. For any pair $\mathbf{x}, \mathbf{y} \in \mathbb{Z}_2^d$, we can define a path from $\mathbf{x}$ to $\mathbf{y}$ by changing every coordinate of $\mathbf{x}$ to the corresponding coordinate of $\mathbf{y}$, from left to right. For this choice of Γ, $b = 2^{d-1}$. Thus, we get a bound

$$\beta_1 \leq 1 - 2/d^2.$$

This is a very reasonable bound, but is not exactly the "correct answer." From other analyses, it is in fact known that $\beta_1 = 1 - 2/d$ (Diaconis 1988).

Thus, the bound we obtained is off by a factor of d. Diaconis and Stroock (1991) noticed that the correct answer can be obtained if we project the corresponding random walk on $\mathbb{Z}_2^d$ to the Ehrenfest chain [i.e., a Markov chain that counts the "distance" from origin (the number of 1's in the vertex)]. For this chain, the corresponding states are $\{0, 1, \ldots, d\}$, the transition probability is $A(j, j+1) = A(j, j-1) = 1 - j/d$, and the stationary distribution is $\pi(j) = \binom{d}{j}/2^d$. The reader can verify that the worst edge is the middle edge, which gives

$$\kappa = \frac{1}{4} d \log d (1 + o(1)) \text{ and } K = d/2,$$

where κ and K are as defined in (12.11) and (12.12). Thus, the correct bound can be obtained by using the second version of the Poincaré inequality.

12.5.4 Cheeger's inequality

In late 1980s, researchers (Jerrum and Sinclair 1989, Holley and Stroock 1988) discovered that the Cheeger's inequality can be used effectively to bound a Markov chain's convergence rate. A successful example is Jerrum and Sinclair's analyses of a Markov chain Monte Carlo algorithm for computing the permanent of a 0-1 matrix.

For any subset $S \in \mathcal{X}$, we define $S^c = \mathcal{X} \setminus S$ and

$$Q(S, S^c) = \sum_{\mathbf{x} \in S} \sum_{\mathbf{y} \in S^c} Q(\mathbf{x}, \mathbf{y}).$$

The conductance of the chain is defined as

$$h = \min_{S:\ \pi(S) \leq 1/2} \frac{Q(S, S^c)}{\pi(S)}.$$

The quantity $Q(S, S^c)/\pi(S)$ measures the relative flow out of S when the chain is in stationarity. If this is large for all S, then there are no bottlenecks and the chain should converge fast.

Theorem 12.5.2 (Cheeger's inequality) *Let β_1 be the second largest eigenvalue of the Markov chain transition matrix A. Then,*

$$1 - 2h \leq \beta_1 \leq 1 - \frac{h^2}{2}.$$

Since the proof of the theorem is a bit complicated, we refer the interested reader to Diaconis and Stroock (1991) for more details. We give here a few examples for using the inequality to bound Markov chain convergence.

Simple random walk on a cube. As discussed in the previous subsection, the state space of this Markov chain has $|\mathcal{X}| = 2^d$ vertices and the stationary distribution is $\pi(\mathbf{x}) = 2^{-d}$. For any subset $S \in \mathcal{X}$, $\pi(S) = |S|/2^d$, where $|S|$ denotes the size of set S. Moreover, it is not difficult to see that $Q(S, S^c)/\pi(S) = |\partial S| \times \frac{1}{d|S|}$, where $|\partial S|$ is the number of "boundary links," or "outward connections," S has. An induction argument (on dimensionality) shows that $|\partial S|/|S| \geq 1$ for $|S| \leq 2^{d-1}$ and the equality is achieved if we let $S = \{\mathbf{x} : x_1 = 0\}$ (corresponding to one-half of a cube). Thus, we have $h = 1/d$, which gives us the bounds

$$1 - \frac{2}{d} \leq \beta_1 \leq 1 - \frac{1}{2d^2}.$$

Note that the true answer is $\beta_1 = 1 - 2/d$. Thus, the Cheeger's lower bound is tight, whereas the upper bound is slightly worse than that obtained by the Poincaré inequality.

To show that $h = 1/d$ for $\mathbb{Z}_2^d$, we first verify that when $d = 1$, the lower bound of $|\partial S|/|S|$ is one. Suppose the lower bound of $|\partial S|/|S|$ is also one for $d = l - 1$. For $d = l$, we assume that the lower bound for $|\partial S|/|S|$ is achieved by a set S^l. We can further divide S^l into $S_0^l = \{\mathbf{x} \in S^l, x_1 = 0\}$ and $S_1^l = S^l \setminus S_0^l$. Because $|S^l| \leq 2^{l-1}$, we assume, without loss of any generality, that $|S_0^l| \leq 2^{l-2}$. By the induction assumption, set S_0^l has at least $|S^l(0)|$ "horizontal outward links" on the half-cube $\{\mathbf{x} : x_1 = 0\}$. If $|S_1^l| \leq 2^{l-2}$ as well, we can easily show that

$$|\partial S^l| = |\partial(S_0^l \cup S_1^l)| \geq |S_0^l| + |S_1^l| = |S^l|.$$

On the other hand, if $|S_1^l| > 2^{l-2}$, then S_1^l has at least $2^{l-1} - |S_1^l|$ horizontal outward links and $|S_1^l| - |S_0^l|$ vertical ones. Hence, we have

$$|\partial S^l| \geq |S_0^l| + (|S_1^l| - |S_0^l|) + (2^{l-1} - |S_1^l|) = 2^{l-1}.$$

Hence, we showed that $|\partial S^l| \geq |S^l|$.

It seems rather difficult to find the exact value of h for a general problem. Jerrum and Sinclair (1989) introduces an elegant geometric argument to give a bound on h. First, they noticed that for any subset $S \in \mathcal{X}$ with $\pi(S) \leq 1/2$,

$$W = \pi(S)\pi(S^c) \geq \pi(S)/2.$$

Second, they introduce a "canonical path" γ_{xy} for every pair of vertices $\mathbf{x}, \mathbf{y} \in \mathcal{X}$. Moreover,

$$W = \sum_{\mathbf{x} \in S} \sum_{\mathbf{y} \in S^c} \pi(\mathbf{x})\pi(\mathbf{y}) \leq \sum_{e \in \partial S} \sum_{\gamma_{xy} \ni e} \pi(\mathbf{x})\pi(\mathbf{y}),$$

where the first summation after the inequality sign is over all the cutting edges between S and S^c. Let

$$\eta = \max_e Q(e)^{-1} \sum_{\gamma_{xy} \ni e} \pi(\mathbf{x})\pi(\mathbf{y}). \quad (12.13)$$

We can easily show that

$$W \leq \eta \sum_{e \in \partial S} Q(e) = \eta Q(S, S^c).$$

Hence, we have $h \geq 1/2\eta$ and the following theorem.

Theorem 12.5.3 (Jerrum and Sinclair) *The second largest eigenvalue of a finite-state reversible and irreducible Markov chain satisfies*

$$\beta_1 \leq 1 - 1/8\eta^2, \tag{12.14}$$

where η is defined in (12.13).

For the simple random walk on $\mathbb{Z}_2^d$, we can define a path from $\mathbf{x} = (x_1, \ldots, x_d)$ to $\mathbf{y} = (y_1, \ldots, y_d)$ as the one that changes one coordinate of $\mathbf{x}$ a time from left to right:

$$(x_1, x_2, \ldots, x_d) \to (y_1, x_2, \ldots, x_d) \to (y_1, y_2, x_3, \ldots, x_d) \to \cdots.$$

For any edge in the graph, there are 2^{d-1} such paths that use the edge. Thus,

$$\eta = \left(\frac{1}{d2^d}\right)^{-1} 2^{d-1} \frac{1}{2^d} \times \frac{1}{2^d} = \frac{d}{2}.$$

By using (12.14), we obtain the same upper bound for β_1 as the one that uses the exact value of h.

Both the Poincaré and Cheeger inequalities are very powerful tools and, if used properly, can provide us a great deal of insights into the convergence issue of a Markov chain. However, so far, most of the success stories of these geometric techniques are for random-walk-type Markov chains where the target distribution is by and large a uniform distribution. Although the path arguments have been extended to deal with some infinite state spaces and special nonuniform distributions (Rosenthal 1996), their usefulness in analyzing general Monte Carlo Markov chains still needs further demonstrations from interested researchers.

12.6 Functional Analysis for Markov Chains

Suppose $\mathbf{x}^{(0)}, \mathbf{x}^{(1)}, \ldots$ are consecutive states from a *stationary* Markov chain with equilibrium distribution $\pi(\mathbf{x})$ and transition function $A(\mathbf{x}, \mathbf{y})$. This section introduces a few basic concepts in functional analysis that have been used to analyze a Markov chain.

12.6.1 Forward and backward operators

Definition 12.6.1 *Let the space of all mean zero and finite variance functions (with respect to π) be denoted as*

$$L_0^2(\pi) = \left\{ h(\mathbf{x}) : \int h^2(\mathbf{x})\pi(\mathbf{x})d\mathbf{x} < \infty \text{ and } \int h(\mathbf{x})\pi(\mathbf{x})d\mathbf{x} = 0 \right\}.$$

This is a Hilbert space when equipped with an inner product

$$\langle h(\mathbf{x}), g(\mathbf{x}) \rangle = E_\pi \{ h(\mathbf{x}) \cdot s(\mathbf{x}) \}. \tag{12.15}$$

Two operators F and B, where F stands for "forward" and B stands for "backward," that map $L_0^2(\pi)$ to itself can be defined as:

$$F h(\mathbf{x}) \stackrel{\text{def}}{=} \int h(\mathbf{y}) A(\mathbf{x}, \mathbf{y}) d\mathbf{y} = E\{t(\mathbf{x}^{(1)}) \mid \mathbf{x}^{(0)} = \mathbf{x}\}, \tag{12.16}$$

$$B h(\mathbf{y}) \stackrel{\text{def}}{=} \int h(\mathbf{x}) \frac{A(\mathbf{x}, \mathbf{y})\pi(\mathbf{x})}{\pi(\mathbf{y})} d\mathbf{x} = E\{h(\mathbf{x}^{(0)}) \mid \mathbf{x}^{(1)} = \mathbf{y}\}. \tag{12.17}$$

The norm of an operator F is defined as

$$\|F\| = \max_{\|h\|=1} \|Fh\|, \quad \text{where } \|h\|^2 = \langle h, h \rangle,$$

and the spectral radius of F is

$$r_F = \lim_{n \to \infty} \|F^n\|^{1/n}.$$

By the Markov property we see that F and B are adjoint to each other:

$$\langle Fh, g \rangle = \langle h, Bg \rangle,$$

and

$$\begin{aligned} F^n h(\mathbf{x}) &= E\{h(\mathbf{x}^{(n)}) \mid \mathbf{x}^{(0)} = \mathbf{x}\}, \\ B^n h(\mathbf{y}) &= E\{h(\mathbf{x}^{(0)}) \mid \mathbf{x}^{(n)} = \mathbf{y}\}. \end{aligned}$$

Using inequality

$$\text{var}[E\{h(\mathbf{x}^{(1)}) \mid \mathbf{x}^{(0)}\}] \leq \text{var}\{h(\mathbf{x}^{(1)})\},$$

we easily show that the norms of the two operators are all bounded above by 1. When the Markov chain is defined on a finite state space, F is just a matrix and r_F is simply the largest eigenvalue in norm of F. A useful link between the covariance structure of a Markov chain with the two operators is proved as follows.

12.6 Functional Analysis for Markov Chains

Lemma 12.6.1 *Suppose* $\mathbf{x}^{(0)} \sim \pi$. *For any* $h, g \in L_0^2(\pi)$, *we have*

$$\text{cov}(h(\mathbf{x}^{(n)}), g(\mathbf{x}^{(0)})) = \text{cov}_\pi\{F^k h(\mathbf{x}), B^{n-k} g(\mathbf{x})\} \qquad (12.18)$$

for any $0 \leq k \leq n$.

Proof: Since $E_\pi h(\mathbf{x}) = E_\pi g(\mathbf{x}) = 0$, we have

$$\begin{aligned}
\text{cov}(h(\mathbf{x}^{(n)}), g(\mathbf{x}^{(0)})) &= E\{h(\mathbf{x}^{(n)})g(\mathbf{x}^{(0)})\} \\
&= E[E\{h(\mathbf{x}^{(n)})g(\mathbf{x}^{(0)})|\mathbf{x}^{(n-1)}\}] \\
&= E[E\{h(\mathbf{x}^{(n)})|\mathbf{x}^{(n-1)}\} \cdot E\{g(\mathbf{x}^{(0)})|\mathbf{x}^{(n-1)}\}] \\
&= E\{Fh(\mathbf{x}^{(n-1)}) \cdot B^{n-1} g(\mathbf{x}^{(n-1)})\} \\
&= \text{cov}_\pi\{Fh(\mathbf{x}), B^{n-1} g(\mathbf{x})\}.
\end{aligned}$$

The last equality holds because $\mathbf{x}^{(n-1)}$ follows the stationary distribution π. An induction argument leads to the conclusion of the lemma. Letting $n = 1$ in the lemma shows that F and B are self-adjoint operators. ◇

The following lemma proves another equivalent relationship.

Lemma 12.6.2 *If the Markov chain is reversible, then* $F = B$.

Proof: Since the reversibility of a Markov chain implies the detailed balance condition, $\pi(\mathbf{x}) A(\mathbf{x}, \mathbf{y}) = \pi(\mathbf{y}) A(\mathbf{y}, \mathbf{x})$, we have

$$\begin{aligned}
Bh(\mathbf{y}) &= \int h(\mathbf{x}) \frac{\pi(\mathbf{x}) A(\mathbf{x}, \mathbf{y})}{\pi(\mathbf{y})} d\mathbf{x} \\
&= \int h(\mathbf{x}) A(\mathbf{y}, \mathbf{x}) d\mathbf{x} = Fh(\mathbf{y}).
\end{aligned}$$

Hence, operators F and B are equal and self-adjoint. ◇

Theorem 12.6.1 *For a reversible Markov chain, if* $\mathbf{x}^{(0)} \sim \pi$, *then* $\forall g \in L_0^2(\pi)$,

$$\begin{aligned}
\text{cov}\{g(\mathbf{x}^{(0)}), g(\mathbf{x}^{(2m)})\} &= E_\pi[\{F^m g(\mathbf{x})\}^2] = E_\pi[(B^m g(\mathbf{x}))^2] \\
&= \text{var}[E[\cdots E[E\{g(\mathbf{x}^{(0)}) \mid \mathbf{x}^{(1)}\} \mid \mathbf{x}^{(2)}] \mid \cdots \mid \mathbf{x}^{(m)}]].
\end{aligned}$$

Hence it is a non-negative monotone decreasing function of m. *Furthermore,*

$$|\text{cov}\{g(\mathbf{x}^{(0)}), g(\mathbf{x}^{(2m+1)})\}| \leq \text{cov}\{g(\mathbf{x}^{(0)}), t(\mathbf{x}^{(2m)})\}.$$

Proof: The first conclusion follows from lemma 12.6.1 with $k = m$ and lemma 12.6.2. The monotonicity of even-lag autocorrelations follows from inequality

$$\text{var}[E\{g(\mathbf{x}) \mid \mathbf{y}\}] \leq \text{var}\{g(\mathbf{x})\}.$$

The last inequality follows from the Hölder inequality:

$$\begin{aligned}|\text{cov}\{g(\mathbf{x}^{(0)}), g(\mathbf{x}^{(2m)})\}| &= |\text{cov}(F^m g(\mathbf{x}), F^{m+1} g(\mathbf{x}))| \\ &\leq \sqrt{\text{var}\{F^m g(\mathbf{x})\} \cdot \text{var}\{F^{m+1} g(\mathbf{x})\}} \\ &\leq \text{var}\{F^m g(\mathbf{x})\} = \text{cov}\{g(\mathbf{x}^{(0)}), g(\mathbf{x}^{(2m)})\}).\end{aligned}$$

◇

12.6.2 Convergence rate of Markov chains

The concept of χ^2-distance (12.4) can be linked with the norm and spectral radius of a forward operator. Let $E_n\{h(\mathbf{x})\}$ denote the expectation taken under the measure $P^{(n)}(d\mathbf{x})$ of the nth-step evolution from an initial distribution $P^{(0)}(d\mathbf{x})$, then

$$\begin{aligned}|E_n\{h(\mathbf{x})\} - E_\pi\{h(\mathbf{x})\}|^2 &\leq \int h(\mathbf{x}) \left|\frac{P^{(n)}(d\mathbf{x})}{\pi(d\mathbf{x})} - 1\right| \pi(d\mathbf{x}) \\ &\leq \|h\|^2 \|P^{(n)} - \pi\|_{\chi^2}^2.\end{aligned}$$

Taking $h = I_S - \pi(S)$ for $\pi(S) \leq 1/2$, the indicator function of a measurable set A, we see that

$$|P^{(n)}(S) - \pi(S)| \leq \frac{1}{2}\|P^{(n)} - \pi\|_{\chi^2},$$

implying that the variation distribution is bounded above by χ^2 distance. Suppose the starting distribution must have finite χ^2-distance from π:

Condition (A). The density $P^{(0)}(d\mathbf{x})$ satisfies the condition

$$c_0^2 = \int \frac{P_0^2(d\mathbf{x})}{\pi(d\mathbf{x})} - 1 < \infty.$$

We have the following simple result.

Lemma 12.6.3 *The spectral radius of operators F and B are equal. If this radius is less than 1, then the chain converges to its stationary distribution geometrically in χ^2-distance, provided that the starting density $P_0(d\mathbf{x})$ has a finite χ^2-distance from the stationary distribution π.*

Proof: It is well known that the adjoint operators in a Hilbert space have the same norms and spectral radii (i.e., equality $\|F^n\| = \|B^n\|$ holds for all n). Hence, we have the equality between the two spectral radii:

$$r_F = \lim_{n\to\infty} \|F^n\|^{\frac{1}{n}} = \lim_{n\to\infty} \|B^n\|^{\frac{1}{n}} = r_B. \tag{12.19}$$

Furthermore, there exists a n_0 such that $\| F^{n_0} \| < 1$. By using Lemma 12.6.1 and writing $g_0(\mathbf{x}) = P^{(0)}(d\mathbf{x})/\pi(d\mathbf{x})$, we can convert the convergence rate problem into a covariance problem:

$$\begin{aligned}
|E_n h(\mathbf{x}) - E_\pi h(\mathbf{x})| &= \left| \int h(\mathbf{y}) A^n(\mathbf{x}, d\mathbf{y}) \left[\frac{P^{(0)}(d\mathbf{x})}{\pi(d\mathbf{x})} - 1 \right] \pi(d\mathbf{x}) \right| \\
&= |\text{cov}_\pi \{ h(\mathbf{x}^{(n)}), g_0(\mathbf{x}^{(0)}) \}| \\
&= |\text{cov}_\pi \{ F^n h(\mathbf{x}), g_0(\mathbf{x}) \}| \\
&\le c_0 \| F^n \| \cdot \| h \| \cdot \| g_0 \|.
\end{aligned}$$

Combining this with (12.19) and the condition that $\|g_0\| < \infty$, we proved the result. $\diamond$

12.6.3 Maximal correlation

As indicated in Lemma 12.6.3, the convergence rate of a Markov chain is closely related to how *correlated* the two states of the chain, $\mathbf{x}^{(0)}$ and $\mathbf{x}^{(n)}$ are. To explore this connection, we define a general notion of *maximal correlation* between two random variables.

Definition 12.6.2 *Suppose random variables* $\mathbf{x}$ *and* $\mathbf{y}$ *have the joint distribution* $\pi(\mathbf{x}, \mathbf{y})$. *Let* $L_x^2(\pi) = \{f(\mathbf{x}) : \text{var}_\pi \{f(\mathbf{x})\} < \infty\}$, $L_y^2(\pi) = \{g(\mathbf{y}) : \text{var}\{g(\mathbf{y})\} < \infty\}$; *the maximal correlation* γ *between* $\mathbf{x}$ *and* $\mathbf{y}$ *is defined as*

$$\gamma = \sup_{f \in L_x^2(\pi), g \in L_y^2(\pi)} \text{corr}\{f(\mathbf{x}), g(\mathbf{y})\}. \tag{12.20}$$

The concept of maximal correlation has been developed since the 1930s. For $(\mathbf{x}, \mathbf{y})$ to be bivariate normal, several pioneer researchers (Maung 1941, Lancaster 1958) proved that the maximal correlation is exactly the same as the absolute value of its ordinary correlation. Some works on general bivariate random variables were done by Lancaster (1958), Csáki and Fischer (1960), Renyi (1959), and Sarmanov (1958). Breiman and Friedman (1985) also make use of the maximal correlation concept in proving properties for their ACE algorithm. It is not difficult to verify that the maximal correlation can be expressed another way (Liu 1991):

$$\gamma = \sum_{E(h)=0, \text{var}(h)=1} \text{var}[E\{h(\mathbf{x}) \mid \mathbf{y}\}]. \tag{12.21}$$

This implies that γ is the *largest eigenvalue* of the conditional expectation operator defined on the space of mean zero functions.

With Lemma 12.6.3, we can set up a duality between convergence rate and correlation structure. More precisely, it follows from Lemma 12.6.3 that the following result is true.

268   12. Markov Chains and Their Convergence

Lemma 12.6.4

$$\| F^n \| = \| B^n \| = \gamma_n,$$

where γ_n is the lag-n maximal correlation:

$$\gamma_n = \sup_{g,h \in L^2(\pi)} \text{corr}\{g(\mathbf{x}^{(0)}), h(\mathbf{x}^{(n)})\}, \qquad (12.22)$$

where $\mathbf{x}^{(0)}$ and $\mathbf{x}^{(n)}$ are n-lag apart in a stationary Markov chain.

If the mean-zero function $h(\mathbf{x})$ is an eigenfunction of operator F corresponding to eigenvalue λ, then from Lemma 12.6.4 we have

$$|\lambda^n| \|h\| = |\langle F^n h, h \rangle| \leq \|F^n\| \cdot \|h\|,$$

which implies that $|\lambda^n| \leq \gamma_n$. On the other hand, we also know that $\lim_{n \to \infty} (\gamma_n)^{1/n}$ is equal to the spectral radius of F.

Suppose F for a reversible Markov chain is a compact operator (for those who are not familiar with functional analysis, just treat F as a finite symmetric matrix). Thus, F has a set of discrete eigenvalues $|\lambda_1| \geq |\lambda_2| \geq \cdots$, with the corresponding orthonormal eigenfunctions $\alpha_1(\mathbf{x}), \alpha_2(\mathbf{x}), \ldots,$. These eigenfunctions form a basis of the space. Since F is self-adjoint in this case, we have the equality

$$\gamma_n = \gamma_1^n = |\lambda_1|^n.$$

For any $h(\mathbf{x}) \in L_0^2(\pi)$, we can express it in a *spectral decomposition* form:

$$h(\mathbf{x}) = c_1 \alpha_1(\mathbf{x}) + c_2 \alpha_2(\mathbf{x}) + \cdots.$$

Hence, for any n,

$$F^n h(\mathbf{x}) = c_1 \lambda_1^n \alpha_1(\mathbf{x}) + c_2 \lambda_2^n \alpha_2(\mathbf{x}) + \cdots.$$

It is seen that

$$\rho_n(h) \equiv \text{corr}\{F^n h(\mathbf{x}), h(\mathbf{x})\} = \frac{\langle F^n h(\mathbf{x}), h(\mathbf{x}) \rangle}{\langle h(\mathbf{x}), h(\mathbf{x}) \rangle} = \frac{c_1^2 \lambda_1^n + c_2^2 \lambda_2^n + \cdots}{c_1^2 + c_2^2 + \cdots}.$$

Provided that $c_1 \neq 0$, we have

$$\lim_{n \to \infty} \{|\rho_n(h)|\}^{1/n} = |\lambda_1|.$$

This ties in with the autocorrelation analysis in Section 5.8, suggesting that monitoring autocorrelation curves in a MCMC sampler are a useful diagnostics on the sampler's speed of convergence.

12.7 Behavior of the Averages

After all the dust settles, we need to go back to our original purpose of running the Monte Carlo Markov chain: We want to estimate certain "averge" of the system. Suppose $\mathbf{x}^{(0)}, \mathbf{x}^{(1)}, \ldots, \mathbf{x}^{(m)}$ is the realization of a single long run of the Markov chain. Most practitioners use

$$\bar{h}_m = \frac{1}{m}\sum_{j=1}^{m} h(\mathbf{x}^{(j)})$$

to estimate the expectation $\mu = E_\pi[h(\mathbf{x})]$. We know that the laws of large numbers and the central limit theorem (see the Appendix) hold for $\bar{h}_m$ if the $\mathbf{x}^{(j)}$ are i.i.d. random variables with a finite second moment. A natural question is, then, whether the Markovian relationship among the $\mathbf{x}^{(j)}$ changes the overall behavior of such an average. Fortunately, the answer is no for most real problems.

In order to make the argument rigorous in an infinite state-space, one has to introduce a number of subtle concepts, definitions, and lemmas that are beyond the scope of this book. The interested reader is referred to Ethier and Kurtz (1986), Nummelin (1984), and Tierney (1994). Here, we only state two theorems for the finite-state Markov chain.

Theorem 12.7.1 *Suppose the finite-state Markov chain* $\mathbf{x}^{(0)}, \mathbf{x}^{(1)}, \ldots$ *is irreducible and aperiodic, then* $\bar{h}_m \to E_\pi(h)$ *almost surely for any initial distribution on* $\mathbf{x}^{(0)}$.

Theorem 12.7.2 *Under the same conditions as in Theorem 12.7.1,*

$$\sqrt{m}[\bar{h}_m - E_\pi(h)] \to N[0, \sigma(h)^2]$$

weakly (i.e., in distribution) for any initial distribution on $\mathbf{x}^{(0)}$.

In practice, the variance term $\sigma(h)^2$ needs to be estimated from the Monte Carlo samples based on (5.13) of page 125; that is

$$\sigma(h)^2 = \sigma^2\left[1 + 2\sum_{j=1}^{\infty}\rho_j\right] \equiv 2\tau_{\text{int}}(h)\sigma^2,$$

where $\sigma^2 = \text{var}_\pi[h(\mathbf{x})]$, $\rho_j = \text{corr}\{h(\mathbf{x}^{(1)}), h(\mathbf{x}^{(j+1)})\}$, and $\tau_{\text{int}}(h)$ is the integrated autocorrelation time.

13
Selected Theoretical Topics

Topics in this chapter include the covariance analysis of iterative conditional sampling; the comparison of Metropolis algorithms based on Peskun's theorem; the eigen-analysis of the independence sampler; perfect simulation, convergence diagnostics, and a theory for dynamic weighting. The interested reader is encouraged to read the related literature for more detailed analyses.

13.1 MCMC Convergence and Convergence Diagnostics

When the state space of the MCMC sampler is finite, theorems in Chapter 12 can be used to judge their convergence. Two key concepts are the irreducibility and aperiodicity. These two concepts can be generalized to continuous state spaces and be used to prove convergence. One of such theorem is described in (Tierney 1994), who used the techniques developed in Nummelin (1984) for the proof.

Theorem 13.1.1 (Tierney) *Suppose A is π-irreducible and $\pi A = \pi$. Then A is positive recurrent and π is the unique invariant distribution of A. If A is also aperiodic, then, for all x but a subset whose measure under π is zero (i.e., π-almost all $\mathbf{x}$),*

$$\|A^{(n)}(\mathbf{x},\cdot) - \pi\|_{\text{var}} \to 0, \qquad (13.1)$$

where $\|\cdot\|_{\text{var}}$ denote the total variation distance.

272   13. Selected Theoretical Topics

Clearly, all the MCMC algorithms covered in this book have an invariant measure π. In most cases they are also π-irreducible and aperiodic. Hence, their convergence can be "assured," except that we still have no idea how fast they converge.

Our view on the convergence diagnosis issue concurs with that of Cowles and Carlin (1996): A combination of Gelman and Rubin (1992) and Geyer (1992) can usually provide an effective, yet simple, method for monitoring convergence in MCMC sampling. Many other approaches, which typically consume a few times more computing resources, can only provide marginal improvements. The *coupling from the past* (CFTP) algorithm described in Section 13.5 is an exciting theoretical breakthrough and its value in assessing convergence of MCMC schemes has been noticed. However, the method is still not quite ready for a routine use in MCMC computation. Interested reader may find Mengersen, Robert and Cuihenneuc-Jouyaux (1999) a useful reference, which provided an extensive study on convergence diagnostics.

Based on a normal theory approximation to the target distribution $\pi(\mathbf{x})$, Gelman and Rubin (1992) propose a method that involves the following few steps:

1. Before sampling begins, obtain a simple "trial" distribution $f(\mathbf{x})$ which is overdispersed relatively to the target distribution π. Generate m (say, 10) i.i.d. samples from $f(\mathbf{x})$. A side note is that in high-dimensional problems, it is often not so easy to find a suitable over-dispersed starting distribution.

2. Start m independent samplers with their respective initial states being the ones obtained in Step 1. Run each chain for $2n$ iterations.

3. For a scalar quantity of interest (after appropriate transformation to approximate normality), say $\theta = \theta(\mathbf{x})$, we use the sample from the last n iterations to compute W, the average of m *within-chain* variances, and B, the variance between the means θ from the m parallel chains.

4. Compute the "shrink factor"

$$\sqrt{\hat{R}} = \sqrt{\left(\frac{n-1}{n} + \frac{m+1}{mn}\frac{B}{W}\right)\frac{\text{df}}{\text{df}-2}}$$

Here df refers to the degree of freedom in a t-distribution approximation to the empirical distribution of θ.

Gelman and Rubin (1992) suggested using $\theta = \log \pi(\mathbf{x})$ as a general diagnosis benchmark. Other choices of θ have been reviewed in Cowles and Carlin (1996). Geyer's (1992) main criticism to Gelman and Rubin's approach is that for difficult MCMC computation, one should concentrate all the resources to a single chain iteration: The latter 9000 samples from a single run of 10,000 iterations are much more likely to come from the target distribution π than those samples from 10 parallel runs of 1000 iterations. In addition, good convergence criterion such as the integrated autocorrelation time (Section 5.8) used in physics can be produced with a single chain.

Concerning the generic use of MCMC methods, we advocate a variety of diagnostic tools rather than any single plot or statistic. In our own work, we often run a few (three to five) parallel chains with relatively scattered starting states. Then, we inspect these chains by comparing many of their aspects, such as the histogram of some parameters, autocorrelation plots, and Gelman and Rubin's $\hat{R}$.

13.2 Iterative Conditional Sampling

13.2.1 Data augmentation

It has been shown in Sections 6.4 and 6.6.1 that *data augmentation* (DA) can be viewed as a two-component Gibbs sampler with a *deterministic scan*. The transition function from $\mathbf{x}^{(0)}$ to $\mathbf{x}^{(1)}$ in DA is, then,

$$A(\mathbf{x}^{(0)}, \mathbf{x}^{(1)}) = \pi(x_1^{(1)} \mid x_2^{(0)})\pi(x_2^{(1)} \mid x_1^{(1)}), \tag{13.2}$$

where $\mathbf{x}^{(t)} = (x_1^{(t)}, x_2^{(t)})$. An important result proved in Section 6.6.1 is the expression (6.6) for the one-lag autocovariance of DA. Here we present more details about the theory developed in Liu, Wong, and Kong (1994, 1995).

From Figure 6.2 and the proof of Theorem 6.6.1, we make the following two observations:

(i) The *marginal chains*, $\{x_1^{(t)}, t = 1, 2, \ldots,\}$ and $\{x_2^{(t)}, t = 1, 2, \ldots,\}$, are all reversible Markov chains. In particular, the transition function for, say, the first chain is

$$A_1(x_1^{(0)}, x_1^{(1)}) = \int \pi(x_1^{(1)} \mid x_2^{(0)})\pi(x_2^{(0)} \mid x_1^{(0)})dx_2^{(0)}. \tag{13.3}$$

(ii) The two marginal chains have the *interleaving Markov property* defined below.

Definition 13.2.1 *A stationary Markov chain $\{x^{(t)}, t = 1, 2, \ldots,\}$ is said to have the interleaving Markov property if there exists a conjugate Markov chain $\{y^{(t)}, t = 1, 2, \ldots,\}$ such that*

(a) $x^{(t)}$ and $x^{(t+1)}$ are conditionally independent given $y^{(t)}$, $\forall t$,

(b) $y^{(t)}$ and $y^{(t+1)}$ are conditionally independent given $x^{(t+1)}$, $\forall k$,

(c) $(y^{(t-1)}, x^{(t)})$, $(x^{(t)}, y^{(t)})$ and $(y^{(t)}, x^{(t+1)})$ are identically distributed.

The two chains are said to be mutually interleaving. The interleaving property implies the reversibility of both chains.

Lemma 13.2.1 *The marginal chains $\{x_1^{(t)}\}$ and $\{x_2^{(t)}\}$ constructed in data augmentation are mutually interleaving.*

Consider the forward operator for the marginal chain $\{x_1^{(t)}\}$. From (13.3), we have

$$F_1 h(x_1) = \int h(x_1^{(1)}) A_1(x_1, x_1^{(1)}) dx_1^{(1)}$$
$$= E_\pi[E_\pi\{h(x_1)|x_2\}|x_1]$$

If we let γ_0 be the *maximal correlation* between x_1 and x_2 under π, then the norm $\|F_1\|$ is γ_0^2. Using expressing (12.21), we can write

$$\|F_1\|^{1/2} = \gamma_0 = \sup_{h \in L_0^2(\pi)} \operatorname{var}[E\{h(x_1) \mid x_2\}].$$

Similarly, the norm of F_2 for the companion chain is also γ_0^2. Since both marginal chains are reversible, the two forward operators are self-adjoint. Thus, their norms are equal to their spectral radii and are equal to each other. This means that the two chains *converge* at the same speed. On the other hand, the joint forward operator F has the property

$$F h(\mathbf{x}^{(0)}) = \int\int h(x_1^{(1)}, x_2^{(1)}) \pi(x_2^{(1)} \mid x_1^{(1)}) \pi(x_1^{(1)} \mid x_2^{(0)}) dx_1^{(1)} dx_2^{(1)},$$

whose norm is shown to be $\|F\| = \gamma_0$. It is not difficult to show that the spectral radius of F is also equal to γ_0^2, as expected.

The above discussion has two implications: (a) One need only to establish convergence properties of one of the marginal chains in order for the joint chain to converge; (b) the convergence rate of data augmentation is completely determined by the maximal correlation between the two components. In statistical missing problems, one component often corresponds to missing data, $\mathbf{y}_{\text{mis}}$, and the other to the parameter θ. The maximal correlation between these two components reflects the "maximal fraction of missing information" (Little and Rubin 1987, Liu 1994b), defined as

$$r = \max_{0 < \operatorname{var}_\pi(h) < \infty} \frac{\operatorname{var}[E\{h(\theta) \mid \mathbf{y}_{\text{mis}}\}]}{\operatorname{var}_\pi\{h(\theta)\}}.$$

Hence, the more the fraction of missing information, the larger the maximal correlation between θ and $\mathbf{y}_{\text{mis}}$ and the slower the corresponding data augmentation scheme. On the other hand, because of this duality, we can also estimate the *maximal faction* of missing information in a missing data problem from the output of its data augmentation scheme (Liu 1994b).

Furthermore, because

$$1 = \frac{E[\text{var}\{h(\theta) \mid \mathbf{y}_{\text{mis}}\}]}{\text{var}_\pi\{h(\theta)\}} + \frac{\text{var}[E\{h(\theta) \mid \mathbf{y}_{\text{mis}}\}]}{\text{var}_\pi\{h(\theta)\}},$$

a larger fraction of $E[\text{var}\{h(\theta) \mid \mathbf{y}_{\text{mis}}\}]$ in the total variance means that x_1 can move more freely conditional on x_2 and, hence, a faster scheme. Thus, when giving two data augmentation schemes with the same target distribution for θ, we prefer the scheme that gives a larger conditional variance for $h(\theta)$, for all functions $h(\ )$ (Liu and Wu 1999, Meng and van Dyk 1999).

13.2.2 Random-scan Gibbs sampler

As shown in Lemma 6.6.1 of Section 6.6.2, the random-scan Gibbs sampler (RSGS) has a similar expression for its first-order autocorrelation to that of data augmentation. Thus, the Markov chain produced by the RSGS also has the interleaving property, with its conjugate process $(\mathbf{i}^{(t)}, \mathbf{x}^{(t)}_{[-\mathbf{i}^{(t)}]})$, for $t = 1, 2, \ldots$. Here, $\mathbf{i}^{(t)}$ is the random variable that indicates which of the d components is updated at the t-th iteration.

The geometric convergence property of the RSGS process is not very difficult to prove and the interested reader is referred to Liu, Wong and Kong (1995) and Schervish and Carlin (1992). What is a little new here is an expression to bound the convergence rate of the RSGS. Based on the theory discussed in the previous subsection, we have, for any $\|h\| = 1$,

$$\begin{aligned}
\|Fh\|^2 &= 1 - E[\text{var}\{h(\mathbf{x}) \mid \mathbf{i}, \mathbf{x}_{[-\mathbf{i}]}\}] \\
&= 1 - \sum_{i=1}^{d} \alpha_i E[\text{var}\{h(\mathbf{x}) \mid \mathbf{x}_{[-i]}\}] \\
&= \sum_{i=1}^{d} \alpha_i \text{var}[E\{h(\mathbf{x}) \mid \mathbf{x}_{[-i]}\}].
\end{aligned} \quad (13.4)$$

Suppose we can find a pair of functions $h_i(x_i)$, and $g_i(\mathbf{x}_{[-i]})$ with unit variance such that

$$\begin{aligned}
E\{h_i(x_i) \mid \mathbf{x}_{[-i]}\} &= \gamma_i g_i(\mathbf{x}_{[-i]}), \\
E\{g_i(\mathbf{x}_{[-i]}) \mid x_i\} &= \gamma_i h_i(x_i).
\end{aligned}$$

Then, using (13.4) with h replaced by h_i, we have

$$\|F\|^2 \geq \langle Fh_i, h_i \rangle = 1 - \alpha_i(1 - \gamma_i^2).$$

Letting $i = 1, \ldots, d$, we have

$$\|F\|^2 \geq \max_i \{1 - \alpha_i(1 - \gamma_i^2)\}.$$

Heuristically, we may want to find the vector $(\alpha_1, \ldots, \alpha_d)$ so that this lower bound for the convergence rate is minimized. This is equivalent to finding

$$\max_{\boldsymbol{\alpha}} \left[\min_i \{\alpha_i(1-\gamma_i^2)\}\right].$$

It is easy to see that the solution is $\alpha_i \propto (1-\gamma_i^2)^{-1}$. This result is rather intuitive: In order to achieve a better convergence rate, one should spend more resources (number of updates roughly proportional to the inverse of the spectral gap) on those "stickier" components. On the other hand, if one is interested in estimating a particular expectation, $Ef(\mathbf{x})$, after the chain becomes stationary, one should allocate resources differently.

We can also use (13.4) to show that *grouping* and *collapsing* (Section 6.7) in a random scan are always preferable (a stronger result than that for the deterministic scans).

Theorem 13.2.1 *Suppose we can either group the first two components together or integrate out the first component so as to result in a RSGS with $d-1$ components. We also assume that the scheduling probability α_i remains unchanged for $i = 3, \ldots, d$. Then, the collapsing sampler converges faster than the grouping sampler, and the grouping sampler faster than the original RSGS.*

Proof: We directly compare (13.4) for the three samplers. For grouping,

$$\|F_g h\|^2 = (\alpha_1 + \alpha_2)\mathrm{var}[E\{h(\mathbf{x}) \mid \mathbf{x}_{[-1,-2]}\}]$$
$$+ \sum_{i=3}^{d} \alpha_i \mathrm{var}[E\{h(\mathbf{x}) \mid \mathbf{x}_{[-i]}\}];$$

for collapsing,

$$\|F_c h\|^2 = (\alpha_1 + \alpha_2)\mathrm{var}[E\{h(\mathbf{x}) \mid \mathbf{x}_{[-1,-2]}\}]$$
$$+ \sum_{i=3}^{d} \alpha_i \mathrm{var}[E\{h(\mathbf{x}) \mid \mathbf{x}_{[-1,-i]}\}].$$

For notational simplicity, here we take the test function for the collapsed sampler, which only has $d-1$ components, as $E[g(\mathbf{x}) \mid \mathbf{x}_{[-1]}]$. Because

$$E\{h(\mathbf{x}) \mid \mathbf{x}_{[-1,-i]}\} = E[E\{h(\mathbf{x}) \mid \mathbf{x}_{[-i]}\} \mid \mathbf{x}_{[-1]}],$$

it follows that

$$\mathrm{var}[E\{h(\mathbf{x}) \mid \mathbf{x}_{[-1,-i]}\}] \leq \mathrm{var}[E\{h(\mathbf{x}) \mid \mathbf{x}_{[-i]}\}],$$

for $i = 2, \ldots, d$. Hence, the theorem is proved. $\diamond$

13.3 Comparison of Metropolis-Type Algorithms

13.3.1 Peskun's ordering

As an alternative to the acceptance-rejection criterion of Metropolis et al. (1953), Barker (1965) proposes a more "continuous" acceptance function

$$r_B(\mathbf{x}, \mathbf{y}) = \frac{\pi(\mathbf{y})T(\mathbf{x}, \mathbf{y})}{\pi(\mathbf{x})T(\mathbf{x}, \mathbf{y}) + \pi(\mathbf{y})T(\mathbf{y}, \mathbf{x})};$$

that is, one accepts the proposed $\mathbf{y}$ with probability r_B and rejects the proposal with probability $1 - r_B$ (Section 5.2). To understand whether Barker's proposal has any advantage, Peskun (1973) introduced a *partial ordering* among all finite-state reversible Markov transition matrices that have the same stationary distribution.

Definition 13.3.1 *Suppose two reversible transition kernels A_1 and A_2 have the same stationary distribution π. Matrix A_1 is said to dominate A_2 (i.e., $A_1 \succeq A_2$) if all the off-diagonal elements of A_1 is greater than or equal to the corresponding elements in A_2. This definition is generalized by Tierney (1998) as follows:*

$$P_1\left[\mathbf{x}^{(1)} \in S \setminus \{\mathbf{x}\} \mid \mathbf{x}^{(0)} = \mathbf{x}\right] \geq P_2\left[\mathbf{x}^{(1)} \in S \setminus \{\mathbf{x}\} \mid \mathbf{x}^{(0)} = \mathbf{x}\right]$$

for all measurable subset $S \subseteq \mathcal{X}$.

This dominance condition easily leads to a comparison between the lag-1 autocorrelations of two Markov chains.

Lemma 13.3.1 (Tierney) *Suppose A_1 and A_2 have the same invariant distribution π and satisfy $A_1 \succeq A_2$. Then, the corresponding forward operators, F_1 and F_2, satisfy*

$$\langle (F_2 - F_1)f, f \rangle \geq 0$$

for all $f \in L_0^2(\pi)$.

Proof: Here, we only prove the case when the state space is finite. Please refer to Tierney (1998) for a more general proof. Suppose the total number of states is N. In this case, we can express the target distribution as the vector $\pi = (\pi_1, \ldots, \pi_N)$ and the transition function as matrices. Then, any function $f \in L_0^2(\pi)$ can be expressed as a column vector of length N, $f = (f_1, \ldots, f_N)^T$, and $F_1 f$ is simply equal to the matrix product $A_1 f$. We define an $N \times N$ matrix as

$$H = \Delta(I + A_1 - A_2),$$

where I is the identity matrix and $\Delta = \mathrm{diag}(\pi_1, \ldots, \pi_N)$. Because A_1 dominates A_2, it is easy to check that H is a probability measure on the product space (all entries h_{ij} are non-negative and they sum to 1) and the both marginals of H are equal to π. Hence,

$$\begin{aligned}\langle (A_2 - A_1)f, f\rangle &= |!f^T \Delta (A_2 - A_1)f \\ &= f^T \{\Delta - H\} f \\ &= \sum_i f_i^2 \pi_i - \sum_i \sum_j f_i f_j h_{ij}, \\ &= \frac{1}{2} \left[2\sum_i \sum_j f_i^2 h_{ij} - 2 \sum_i \sum_j f_i f_j h_{ij} \right] \\ &= \frac{1}{2} \sum_i \sum_j (f_i - f_j)^2 h_{ij} \geq 0. \end{aligned}$$

◊

Suppose that we are interested in estimating $E_\pi f$ by using a MCMC sampler whose transition kernel is $A(\mathbf{x}, \mathbf{y})$. We can define the sampler's *asymptotic efficiency* as

$$v(f, A) = \lim_{n \to \infty} \frac{1}{n} \text{var} \left\{ \sum_{t=1}^n f(\mathbf{x}^{(t)}) \right\}, \qquad (13.5)$$

where $\mathbf{x}^{(0)}, \mathbf{x}^{(1)}, \ldots$ are stationary samples obtained from this sampler. Since we are ultimately interested in using a MCMC sampler to compute quantities of interest, this asymptotic efficiency measure seems to be a sensible criterion in comparing different schemes. Peskun (1973) proved that if $A_1 \succeq A_2$, then the first chain will be asymptotically more efficient.

Theorem 13.3.1 (Peskun) *Suppose A_1 and A_2 are reversible transition kernels with the same invariant distribution and $A_1 \succeq A_2$. Then, for all $f \in L_0^2(\pi)$ (i.e., mean zero functions), we have $v(f, A_1) \leq v(f, A_2)$.*

Proof: It is easy to see that for any transition matrix A,

$$v(f, A) = \langle f, \{I + (I-A)^{-1}\} f \rangle.$$

Note that operator $(I - A)$ is invertible only in the restricted space $L_0^2(\pi)$, but not in the unrestricted space $L^2(\pi)$. Define $A(\beta) = (1 - \beta)A_1 + \beta A_2$; then,

$$\begin{aligned} \frac{\partial v(f, A(\beta))}{\partial \lambda} &= \langle f, (I - A(\beta))^{-1}(A_2 - A_1)(I - A(\beta))^{-1} f \rangle \\ &= \langle (I - A(\beta))^{-1} \} f, (A_2 - A_1)(I - A(\beta))^{-1} f \rangle \geq 0 \end{aligned}$$

The second equality follows from the fact that if A is a self-adjoint operation, then all powers of A, and thus, $(I - A)^{-1}$, are also self-adjoint operators. The last inequality follows from Lemma 13.3.1. Hence, $v(f, A(\beta))$ is a monotone nondecreasing function in β and attains its minimum at $\beta = 0$ and maximum at $\beta = 1$. ◇

13.3.2 Comparing schemes using Peskun's ordering

The ordering among transition functions introduced by Peskun is very useful for comparing different schemes. Based on Theorem 13.3.1, for example, Peskun (1973) proved the following theorem.

Theorem 13.3.2 *For the same proposal transition $T(\mathbf{x}, \mathbf{y})$, the acceptance function suggested by Metropolis et al. dominates that proposed by Barker in terms of asymptotic efficiency (13.5).*

Proof: The transition function of the Metropolis algorithm is

$$A_M(\mathbf{x}, \mathbf{y}) = T(\mathbf{x}, \mathbf{y}) \min\{1, r(\mathbf{x}, \mathbf{y})\} \quad \text{for } \mathbf{x} \neq \mathbf{y},$$

where

$$r(\mathbf{x}, \mathbf{y}) = \frac{\pi(\mathbf{y}) T(\mathbf{y}, \mathbf{x})}{\pi(\mathbf{x}) T(\mathbf{x}, \mathbf{y})}.$$

On the other hand, the transition function for Barker's scheme is

$$A_B(\mathbf{x}, \mathbf{y}) = T(\mathbf{x}, \mathbf{y}) \frac{r(\mathbf{x}, \mathbf{y})}{1 + r(\mathbf{x}, \mathbf{y})} \quad \text{for } \mathbf{x} \neq \mathbf{y}.$$

It is easy to show that

$$\min\{1, r(\mathbf{x}, \mathbf{y})\} \geq \frac{r(\mathbf{x}, \mathbf{y})}{1 + r(\mathbf{x}, \mathbf{y})}.$$

Hence, $A_M \succeq A_B$. ◇

As a generalization of Peskun's result, Liu (1996a) showed that the "Metropolization" of the Gibbs sampler for a finite state space as described in Section 6.3.2 dominates the usual random-scan Gibbs sampler.

Theorem 13.3.3 *Suppose that $\mathbf{x} = (x_1, \ldots, x_d)$, where x_i takes $m_i < \infty$ possible values, and that $\pi(\mathbf{x})$ is the distribution of interest. Then, the Metropolized Gibbs sampler defined in Section 6.3.2 for discrete random variables is statistically more efficient than the random-scan Gibbs sampler.*

Proof: Suppose that in the random-scan Gibbs sampler, we choose each component with probability α_i. Then, all the nonzero elements of the transition matrix P_1 of the random scan Gibbs sampler are of the form

$$P_1(\mathbf{x}, \mathbf{y}) = \alpha_i \pi(y_i \mid \mathbf{x}_{[-i]}),$$

where $\mathbf{y} = \mathbf{x}$ except that y_i replaces x_i. In contrast, those nonzero off-diagonal elements in the transition matrix P_2 of the modified sampler are

$$P_2(\mathbf{x}, \mathbf{y}) = \alpha_i \min\left\{\frac{\pi(y_i \mid \mathbf{x}_{[-i]})}{1 - \pi(x_i \mid \mathbf{x}_{[-i]})}, \frac{\pi(y_i \mid \mathbf{x}_{[-i]})}{1 - \pi(y_i \mid \mathbf{x}_{[-i]})}\right\}.$$

Clearly, $P_2 \succeq P_1$. ◇

Although Theorem 13.3.3 does not even require m_i to be finite, the modification is likely to be most useful for components with m_i rather small. It is easily shown from inequality $v(f, P_1) \geq v(f, P_2), \forall f \in L^2(\pi)$, that the second largest eigenvalue of P_1 is greater than or equal to that of P_2. Frigessi et al. (1993) proved that for the binary Ising model, Metropolis converges faster than Gibbs for strong interaction and more slowly for weak interaction. This does not conflict with our result, which concerns statistical efficiency in equilibrium, rather than rate of convergence. Whereas the eigenvalues of the Gibbs sampler are necessarily non-negative (Liu, Wong and Kong 1995), slow Metropolis convergence under weak interaction is the product of a large negative eigenvalue.

Using the same technique, Besag et al. (1995) and Tierney (1998) proved another interesting result regarding the use of a mixture proposal in a Metropolis sampler. Let T_i be a sequence of proposal kernels and let $\alpha_i > 0$ with $\sum_i \alpha_i = 1$. Let A_i be the corresponding Metropolis transition kernel:

$$A_i(\mathbf{x}, \mathbf{y}) = T_i(\mathbf{x}, \mathbf{y}) \min\left\{1, \frac{\pi(\mathbf{y}) T_i(\mathbf{y}, \mathbf{x})}{\pi(\mathbf{x}) T_i(\mathbf{x}, \mathbf{y})}\right\}.$$

Furthermore, we define a mixture proposal

$$T^*(\mathbf{x}, \mathbf{y}) = \sum_i \alpha_i T_i(\mathbf{x}, \mathbf{y})$$

and its corresponding Metropolis transition function $A^*(\mathbf{x}, \mathbf{y})$.

Theorem 13.3.4 *The Metropolis transition with a mixture proposal dominates the corresponding mixture of Metropolis transitions; that is,*

$$A^* \succeq \sum_i \alpha_i A_i$$

Proof: Because of the simple inequality

$$\min(A_1, B_1) + \min(A_2, B_2) \leq \min(A_1 + A_2, B_1 + B_2),$$

we have that

$$\sum_i \alpha_i A_i(\mathbf{x}, \mathbf{y}) = \sum_i \min\left\{\alpha_i T_i(\mathbf{x}, \mathbf{y}), \frac{\pi(\mathbf{y})}{\pi(\mathbf{x})} \alpha_i T_i(\mathbf{y}, \mathbf{x})\right\}$$
$$\leq \min\left\{\sum_i \alpha_i T_i(\mathbf{x}, \mathbf{y}), \frac{\pi(\mathbf{y})}{\pi(\mathbf{x})} \sum_i \alpha_i T_i(\mathbf{y}, \mathbf{x})\right\}$$
$$= \min\left\{T^*(\mathbf{x}, \mathbf{y}), \frac{\pi(\mathbf{y})}{\pi(\mathbf{x})} T^*(\mathbf{y}, \mathbf{x})\right\} = A^*(\mathbf{x}, \mathbf{y})$$

Hence, the theorem is proved. ◇

This theorem may also shed some light on the issue of whether the multi-point Metropolis method is superior to the ordinary Metropolis algorithm (with a comparable number of proposals).

13.4 Eigenvalue Analysis for the Independence Sampler

In the special case that the proposal is an *independent* transition function (Section 5.4.2), we have a rather clean result on the analysis of all the eigenvalues of the Metropolis-Hastings transition matrix (Liu 1996a). In this section, we assume that the state space $\mathcal{X}$ is finite. Without loss of generality, we let $\mathcal{X} = \{1, \ldots, m\}$. Two probability measures $\pi(\cdot)$ and $p(\cdot)$ are then abbreviated as $\pi_i = \pi(i)$, and $p_i = p(i)$, $i = 1, \ldots, m$. We introduce the following four notations: $F_\pi(k) = \pi_1 + \cdots + \pi_k$, $S_\pi(k) = 1 - F_\pi(k-1) = \pi_k + \cdots + \pi_m$, $F_p(k) = p_1 + \cdots + p_k$, and $S_p(k) = 1 - F_p(k-1)$.

For any $i, j \in \mathcal{X}$, we can write down the transition probability from i to j for the Metropolized independence sampler

$$A(i,j) = \begin{cases} p_j \min\{1, w_j/w_i\} & \text{if } j \neq i \\ p_i + \sum_k p_k \max\{0, 1 - w_k/w_i\} & \text{if } j = i, \end{cases}$$

where $w_i = \pi_i/p_i$ is the *importance ratio*. Without loss of generality, we further assume that the states are sorted according to the magnitudes of their importance ratios; that is, the elements in $\mathcal{X}$ are labeled so that

$$w_1 \geq w_2 \geq \cdots \geq w_m.$$

The transition matrix can then be written as

$$A = \begin{pmatrix} p_1 + \lambda_1 & \pi_2/w_1 & \pi_3/w_1 & \cdots & \pi_{m-1}/w_1 & \pi_m/w_1 \\ p_1 & p_2 + \lambda_2 & \pi_3/w_2 & \cdots & \pi_{m-1}/w_2 & \pi_m/w_2 \\ \vdots & \vdots & \vdots & \ddots & \vdots & \vdots \\ p_1 & p_2 & p_3 & \cdots & p_{m-1} + \lambda_{m-1} & \pi_m/w_{m-1} \\ p_1 & p_2 & p_3 & \cdots & p_{m-1} & p_m \end{pmatrix},$$

where

$$\lambda_k = \sum_{i=k}^{m}(p_i - \pi_i/w_k) = S_p(k) - S_\pi(k)/w_k, \quad (13.6)$$

which is just the probability of being rejected in the next step if the chain is currently at state k.

For any function $f(x)$ defined on $\mathcal{X}$, we denote

$$f^+(x) = \begin{cases} f(x) & \text{if } f(x) > 0; \\ 0 & \text{if } f(x) \leq 0. \end{cases}$$

It is noted that λ_k has another expression:

$$\lambda_k = \sum_{i \geq k}(\pi_i/w_i - \pi_i/w_k) = E_\pi \left\{ \frac{1}{w(X)} - \frac{1}{w_k} \right\}^+,$$

where the expectation is taken with respect to $X \sim \pi$. Apparently, if two states i and $i+1$ have equal importance ratios, then $\lambda_i = \lambda_{i+1}$. Let $\mathbf{p} = (p_1, \ldots, p_m)^T$ denote the column vector of the trial distribution and let $\mathbf{e} = (1, \ldots, 1)^T$. Then, A can be expressed as

$$A = G + \mathbf{e}\mathbf{p}^T,$$

where G is an upper triangular matrix of the form

$$G = \begin{pmatrix} \lambda_1 & \frac{p_2(w_2 - w_1)}{w_1} & \cdots & \frac{p_{m-1}(w_{m-1} - w_1)}{w_1} & \frac{p_m(w_m - w_1)}{w_1} \\ \vdots & \vdots & \ddots & \vdots & \vdots \\ 0 & 0 & \cdots & \lambda_{m-1} & \frac{p_m(w_m - w_{m-1})}{w_{m-1}} \\ 0 & 0 & \cdots & 0 & 0 \end{pmatrix}$$

Note that $\mathbf{e}$ is a common right eigenvector for both A and $A - G$, corresponding to the largest eigenvalue 1. Since $A - G$ is of rank 1, the rest of the eigenvalues of A and G have to be the same. Hence, the eigenvalues for A are $1 > \lambda_1 \geq \lambda_2 \geq \cdots \geq \lambda_{m-1}$.

When m is fixed and the number of iterations goes to infinity, the mixing rate of this Metropolis Markov chain is asymptotically dominated by the second largest eigenvalue λ_1, which is equal to $1 - 1/w_1$. All the eigenvectors of G can also be found explicitly. We first note that the vector $\tilde{\mathbf{v}}_1 = (1, 0, \ldots, 0)^T$ is a right eigenvector corresponding to λ_1. Checking one more step, we find that $\tilde{\mathbf{v}}_2 = (\pi_2, 1 - \pi_1, 0, \ldots, 0)^T$ is a right eigenvector of λ_2. Generalizing the result, we obtain the following result.

Lemma 13.4.1 *The eigenvectors and eigenvalues of G are λ_k, and $\tilde{\mathbf{v}}_k = (\pi_k, \ldots, \pi_k, S_\pi(k), 0, \ldots, 0)^T$, for $k = 1, \ldots, m - 1$, where there are k nonzero entries in $\tilde{\mathbf{v}}_k$.*

13.4 Eigenvalue Analysis for the Independence Sampler

Theorem 13.4.1 *For the Metropolized independence sampler, all the eigenvalues of its transition matrix are $1 > \lambda_1 \geq \lambda_2 \geq \cdots \geq \lambda_{m-1} \geq 0$, where $\lambda_k = \sum_{i=k}^{m}(p_i - \pi_i/w_k) = E_\pi\{1/w(X) - 1/w_k\}^+$. The right eigenvector $\mathbf{v}_k$ corresponding to λ_k is*

$$\mathbf{v}_k \propto (0,\ldots 0, S_\pi(k+1), -\pi_k, \ldots, -\pi_k)^T,$$

where there are $k-1$ zero entries.

Proof: Since $A = G + \mathbf{e}\mathbf{p}^T$, $A\tilde{\mathbf{v}}_k = G\tilde{\mathbf{v}}_k + \mathbf{e}(\mathbf{p}^T\tilde{\mathbf{v}}_k)$. It is further noted that

$$\mathbf{p}^T\tilde{\mathbf{v}}_k = S_\pi(k)\pi_k + p_k S_p(k) = \pi_k(1-\lambda_k).$$

Hence, $A\tilde{\mathbf{v}}_k = \lambda_k \tilde{\mathbf{v}}_k + \pi_k(1-\lambda_k)\mathbf{e}$. Since $\mathbf{e}$ is a right eigenvector of A with eigenvalue 1, we have, for any t,

$$A(\tilde{\mathbf{v}}_k - t\mathbf{e}) = \lambda_k\left\{\tilde{\mathbf{v}}_k - \frac{t - \pi_k(1-\lambda_k)}{\lambda_k}\mathbf{e}\right\}.$$

Solving $t = \{t - \pi_k(1-\lambda_k)\}/\lambda_k$, we find that $\mathbf{v}_k = \tilde{\mathbf{v}}_k - \pi_k\mathbf{e}$ is a right eigenvector of A corresponding to λ_k. ◇

The coupling method can also be used to bound the convergence rate for this sampler and the argument does not require that the state space of the chain is discrete. Suppose two independence sampler chains $\{\mathbf{x}^{(0)}, \mathbf{x}^{(1)}, \ldots\}$ and $\{\mathbf{y}^{(0)}, \mathbf{y}^{(1)}, \ldots\}$ are simulated, of which the $\mathbf{x}$ chain starts from a fixed point $\mathbf{x}^{(0)} = \mathbf{x}_0$ (or a distribution) and the $\mathbf{y}$ chain starts from the equilibrium distribution π. The two chains can be "coupled" in the following way. Suppose $\mathbf{x}^{(t)} = \mathbf{x}$ and $\mathbf{y}^{(t)} = \mathbf{y}$. At step $t+1$, a new state $\mathbf{z}$ is drawn according to distribution $p(\cdot)$, its associated importance ratio $w_z \equiv \pi(\mathbf{z})/p(\mathbf{z})$ is computed and a uniform random variable u is generated independently. There are three scenarios: (i) If $u \leq \min\{w_z/w_x, w_z/w_y\}$, then both chains accept $\mathbf{x}$ as their next states (i.e., $\mathbf{x}^{(t+1)} = \mathbf{y}^{(t+1)} = \mathbf{z}$); (ii) if $u \geq \max\{w_z/w_x, w_z/w_y\}$, then both chains reject so that $\mathbf{x}^{(t+1)} = \mathbf{x}$ and $\mathbf{y}^{(t+1)} = \mathbf{y}$; and (iii) if u lies between w_z/w_x and w_z/w_y, then the chain with larger ratio accepts and the chain with smaller ratio rejects. It is clear that the first time when scenario (i) occurs is the *coupling time*, the time at and after which the realizations of the two chains become identical. The probability of the occurrence of (i) can be bounded from below:

$$P(\text{accept} \mid X_k = x, Y_k = y) = \sum_{i=1}^m p_i \min\left\{1, \frac{w_i}{w_x}, \frac{w_i}{w_y}\right\}$$

$$= \sum_{i=1}^m \pi_i \min\left\{\frac{1}{w_i}, \frac{1}{w_x}, \frac{1}{w_y}\right\} \geq \frac{1}{w_1},$$

where w_1 is the largest importance ratio. Hence, from the Markov property, the number of steps for the chains to be coupled is bounded by a geometric distribution

$$P(N > n) \le (1 - w_1^{-1})^n.$$

Consequently, for any measurable subset $S \subset \mathcal{X}$,

$$\begin{aligned}|p^{(n)}(S) - \pi(S)| &= |P(X_n \in S) - P(Y_n \in S)| \\ &\le P(X_n \ne Y_n) = P(N > n) \le (1 - w_1^{-1})^n.\end{aligned}$$

13.5 Perfect Simulation

When running a MCMC sampler, we have always to wait for a period of "burn-in" time (or called the time for *equilibration*). Samples obtained after this period of time can be regarded as approximately following the target distribution π and be used in Monte Carlo estimation. In practice, however, one is never sure how long the "burn-in" period should be and it is always a distracting question for researchers to know when to declare "convergence" of the chain. A surprising discovery recently made by Propp and Wilson (1996) is that *perfect* random samples can be obtained, in finite (but stochastic) time, from many Markov chain samplers. Their algorithm is also called *coupling from the past* (CFTP).

Under mild conditions (irreducibility, aperiodicity, and a drift condition) which we have assumed throughout of the book, the Markov chain underlying a MCMC sampler would have been in its stationary distribution had it been iterated for infinite steps. Thus, if the chain had been started from $t = -\infty$, the infinite past, then at time $t = 0$, the chain would have been in equilibrium and a sample produced at $t = 0$ would have been an exact sample from the target distribution π. This fact has already been known to all the probabilists a long time ago. What Propp and Wilson discovered is that one can figure out what the *current sample* is without actually tracing back to the infinite past. The strategy they took was also known to probabilists a long time ago: *coupling* and *coalescence*.

Suppose the Markov chain under consideration is defined on a finite space $\mathcal{X} = \{1, \ldots, |\mathcal{X}|\}$. Let the transition matrix be $A(\mathbf{x}, \mathbf{y})$ and the equilibrium distribution be π. Consider all possible ways from $\mathbf{x}^{(-1)} \to \mathbf{x}^{(0)}$. The transition function tells us that

$$\Pr(\mathbf{x}^{(0)} = j \mid \mathbf{x}^{(-1)} = i) = A(i, j).$$

If we want to simulate this step on a computer, we will first compute the *cumulative transition probabilities*:

$$G(i, j) = \sum_{k=1}^{j} A(i, k) \equiv \Pr(\mathbf{x}^{(0)} \le j \mid \mathbf{x}^{(-1)} = i).$$

Then, we generate a uniform random number [i.e., $u_0 \sim \text{Uniform}(0,1)$]. Finally, we let $\mathbf{x}^{(0)} = j$ if $G(i, j-1) < u_0 \leq G(i,j)$. In a usual computer algorithm for realizing a forward Markov transition, one generates a random number whenever needed and they may be different had $\mathbf{x}^{(-1)}$ been $i' \neq i$. However, there is no reason why we cannot use the *same* random number u_0 generated beforehand and use it for all possible states $i \in \mathcal{X}$ at time -1. More abstractly, we can think of the above sampling step as a mapping

$$\mathbf{x}^{(0)} = \phi(u_0, \mathbf{x}^{(-1)}). \tag{13.7}$$

A distinctive feature of (13.7) is that the chains starting from all possible states are *coupled* by the same random number u_0.

If it so happens that our random number u_0 makes all the chain "coupled;" that is,

$$\phi(u_0, i) \equiv j_0 \quad \text{for all } i, \tag{13.8}$$

then $\mathbf{x}^{(0)} = j_0$ must be a perfect sample from the target distribution π. To see this point, imagine that the Markov chain has been run from $t = -\infty$ and entered into time $t = -1$. Then, it must be in stationarity at time $t = -1$. Because of the construction of ϕ, the next step $\mathbf{x}^{(0)}$ must still be in stationarity. Because of (13.8), $\mathbf{x}^{(0)}$ has to take value j_0 no matter what state $\mathbf{x}^{(-1)}$ takes.

Of course, the chance that the chains are all coupled in one step is too small. If they are not all coupled, we can iterate (13.7) backward. Since

$$\mathbf{x}^{(-n)} = \phi(u_{-n}, \mathbf{x}^{(-n-1)}) \tag{13.9}$$

for all n, we have

$$\mathbf{x}^{(0)} = \phi(u_0, \phi(u_{-1}, \ldots, \phi(u_{-N+1}, \mathbf{x}^{(-N)}) \cdots)).$$

In fact, we can even imagine that the sequence of uniform random numbers, $\ldots, u_{-N}, \ldots, u_{-1}, u_0$, has been given in advance, and we realize a stationary Markov chain by composing (13.9) from the infinite past.

Now, consider starting $|\mathcal{X}|$ parallel Markov chains at time $t = -N$, each with a different starting state [i.e., $\mathbf{x}^{(-N,j)} = j$]. Then, after one iteration of the ϕ function, we have, for all the chains,

$$\mathbf{x}^{(-N+1,j)} = \phi(u_{-N+1}, j) \quad \text{for } j = 1, \ldots, |\mathcal{X}|. \tag{13.10}$$

Hence, the $\mathbf{x}^{(-N+1,j)}$ have fewer distinct values than that of $\mathbf{x}^{(-N,j)}$. This means that each iteration of the ϕ function will "coalesce" some chains. If N is large enough, then all the chains starting at $t = -N$ will coalescence and produce a single random sample, $\mathbf{x}_0$, at time $t = 0$. Since a Markov chain that comes from the infinite past has to get into time $-N$ and then passes through recursion (13.10), the sample obtained at time $t = 0$ has to be

$\mathbf{x}^{(0)}$ (if the same set of uniform random numbers has been used from time $-N$ to 0). Thus, this $\mathbf{x}^{(0)}$ is an exact draw from the stationary distribution π. If N is not large enough, we will need to move K steps backward to time $-N-K$ and try again, reusing all the previously generated random numbers.

The CFTP algorithm can be implemented, at least conceptually, as follows:

1. Generate $u_0 \sim \text{Uniform}(0,1)$ and compute $f_{-1}(i) = \phi(u_0, i)$, for $i = 1, \ldots, |\mathcal{X}|$.

   (a) If the $f_{-1}(i)$ are all equal, then the common value $f_{-1}(i)$ is retained as a random sample from π and the algorithm is stopped.

   (b) If not all the $f_{-1}(i)$ are the same, set $n=2$ and go to Step 2.

2. At time $-n$, we generate $u_{-n+1} \sim \text{Uniform}(0,1)$ and update

$$f_{-n}(i) = f_{-n+1}\{\phi(u_{-n+1}, i)\} \quad \text{for } i = 1, \ldots, |\mathcal{X}|. \tag{13.11}$$

   (a) If all the $f_{-n}(i)$ are the same, return the common value $f_{-n}(i)$ as a sample from π and stop.

   (b) If not all the $f_{-n}(i)$ are the same, set $n \leftarrow n+1$ and return to Step 2.

It is important to notice the difference between the forward coupling expression (13.9) and the backward coupling formula (13.11). More explicitly, $f_{-n}(i)$, for all n, refers to a possible state for $\mathbf{x}^{(0)}$ at time 0 instead of that for $\mathbf{x}^{(-n)}$ at time $-n$.

It is often too slow to move one-step backward a time. A preferable approach is to modify Step 2(b) in the foregoing CFTP algorithm by setting $n \leftarrow 2n$; that is, one doubles the backward steps if not all the chains coalesce at time 0 in n steps. A main difficulty in applying the CFTP algorithm in interesting cases is that it is often impossible to monitor simultaneously all the chains starting from *all* possible states. For example, an Ising model on a 64×64 lattice has 2^{64^2} possible states, which are impossible to follow. A useful method (Propp and Wilson 1996) is to establish an *ordering* "$\preceq$" among all the states, so that this ordering is maintained after the one-step coupled Markov transition:

$$\mathbf{x} \preceq \mathbf{y} \quad \rightarrow \quad \phi(u, \mathbf{x}) \preceq \phi(u, \mathbf{y}),$$

for all $0 < u < 1$. Suppose a "maximal state" and a "minimal state" under this ordering exist. Then, one needs only to monitor two chains on the computer: one started from the maximal state and the other from the minimal state. When these two chains are coupled, then chains from all other states must be coupled to the same state.

The work of Propp and Wilson (1996) has stimulated a lot of interest from computer scientists, probabilists, and statisticians. Many new tricks have been developed to tackle various situations. One of the main concerns regarding the CFTP algorithm is the "user impatience" bias; that is, the user may stop the algorithm when it takes too long to find an appropriate past time $-N$ or the algorithm is stopped before schedule because of emergency (an electricity outage, say). Both cases will create a bias in the produced samples. In a sense, the CFTP cannot be *interrupted*. Fill (1998) recently proposed an interruptible algorithm that alleviates this concern. See Green and Murdoch (1998), Fill (1998), and Propp and Wilson (1998).

13.6 A Theory for Dynamic Weighting

13.6.1 Definitions

Suppose the configuration state $\mathcal{X}$ is augmented by a one-dimensional weight space so that the *current* state in a dynamic weighting Monte Carlo scheme is $(\mathbf{x}, w)$. Most of the analysis presented in this section are adapted from Liu et al. (2001).

Let constant $\theta \geq 0$ be given in advance. The Q-type and the R-type moves that we will study in this section are defined as follows.

Q-type Move:

- Propose the next state $\mathbf{y}$ from the proposal $T(\mathbf{x}, \cdot)$ and compute the *Metropolis ratio*

$$r(\mathbf{x}, \mathbf{y}) = \frac{\pi(\mathbf{y})T(\mathbf{y}, \mathbf{x})}{\pi(\mathbf{x})T(\mathbf{x}, \mathbf{y})}. \qquad (13.12)$$

- Draw $U \sim \text{Uniform}(0, 1)$. Update $(\mathbf{x}, w)$ to $(\mathbf{x}', w')$ as

$$(\mathbf{x}', w') = \begin{cases} (\mathbf{y}, \max\{\theta, wr(\mathbf{x}, \mathbf{y})\}) & \text{if } U \leq \min\{1, wr(\mathbf{x}, \mathbf{y})/\theta\} \\ (\mathbf{x}, aw) & \text{otherwise.} \end{cases} \qquad (13.13)$$

where $a > 1$ can be either a constant or a random variable independent of all other variables.

R-type Move

- Propose $\mathbf{y}$ and compute $r(\mathbf{x}, b\mathbf{y})$ as in the Q-type move.

- Draw $U \sim \text{Uniform}(0,1)$. Update $(\mathbf{x}, w)$ to $(\mathbf{x}', w')$ as

$$(\mathbf{x}', w') = \begin{cases} (\mathbf{y}, wr(\mathbf{x},\mathbf{y}) + \theta), & \text{if } U \leq \dfrac{wr(\mathbf{x},\mathbf{y})}{wr(\mathbf{x},\mathbf{y}) + \theta}; \\ (\mathbf{x}, w(wr(\mathbf{x},\mathbf{y}) + \theta)/\theta), & \text{Otherwise.} \end{cases} \tag{13.14}$$

The design of the algorithms was motivated by the following consideration.

Definition 13.6.1 *Random variable* $\mathbf{x}$ *is said correctly weighted by* w *with respect to* π *if* $\sum_w w f(\mathbf{x}, w) \propto \pi(\mathbf{x})$, *where* $f(\mathbf{x}, w)$ *is the joint distribution of* $(\mathbf{x}, w)$. *A transition rule is said to be invariant with respect to importance weighting (IWIW) if it maintains the correctly weightedness of* $(\mathbf{x}, w)$.

It is easy to verify that the R-type move satisfies IWIW property, whereas the Q-type move does not. Although the R-type move satisfies IWIW, it can be shown that with $\theta = 1$ the stationary distribution of the weight process, if it exists at all, will necessarily has an infinite expectation. It is also possible that the weight process does not have a stationary distribution, which renders the scheme useless.

Although θ is an adjustable parameter that can depend on previous value of $(\mathbf{x}, w)$, we focus only on the case when $\theta \equiv$ constant. Since any nonzero constant leads to the same weight behavior, we need only to discuss two cases: $\theta \equiv 1$ and $\theta \equiv 0$. One may think that using $\theta = 0$ would be a sensible choice. But we will show that this is not very interesting: Either the scheme becomes the usual importance sampling or the weight converges to zero. Consequently, in practice, we choose $\theta = 1$ and use a stratified truncation method to deal with the weight process whose expectation is infinite.

13.6.2 Weight behavior under different scenarios

Case (i): $\theta = 0$ and $T(\mathbf{x}, \mathbf{y})$ is reversible.

In this case the Q- and R-type moves are identical, and both can be viewed as a generalization of the importance sampling. More precisely, every proposed move will be accepted and the weight is updated as

$$w' = w\, r(\mathbf{x}, \mathbf{x}').$$

Suppose $g(\mathbf{x})$ is the invariant distribution for $T(\mathbf{x}, \mathbf{y})$ and we let $g(\mathbf{x}, \mathbf{y}) = g(\mathbf{x}) T(\mathbf{x}, \mathbf{y})$. Then, $g(\ )$ is a symmetric function. Let $u(\mathbf{x}) = \pi(\mathbf{x})/g(\mathbf{x})$ be the importance ratio function. The update formula for the weight process can be written as

$$w' = w \frac{u(\mathbf{x}') \, g(\mathbf{x}', \mathbf{x})}{u(\mathbf{x}) \, g(\mathbf{x}, \mathbf{x}')} = \frac{u(\mathbf{x}')}{u(\mathbf{x})}. \tag{13.15}$$

13.6 A Theory for Dynamic Weighting

Hence, if we start with $\mathbf{x}^{(0)}$ and $w^{(0)} = c\, u(\mathbf{x}^{(0)})$, then for any $t > 0$, $w^{(t)} = c\, u(\mathbf{x}^{(t)})$. These weights are the same as those from the standard importance sampling using the trial distribution g.

Case (ii): $\theta = 1$ and $T(\mathbf{x}, \mathbf{y})$ is reversible.

If the proposal chain is reversible, the Q-type sampler converges to a a joint state of $(\mathbf{x}, w)$ that is similar to case (i). To see this, let $u_0 = \min_{\mathbf{x}}\{\pi(\mathbf{x})/g(\mathbf{x})\}$. Once the pair $(\mathbf{x}, w)$ satisfies $w = c_0 u(\mathbf{x})$ with $c_0 u_0 \geq 1$, the proposed transition $\mathbf{y}$ will *always* be accepted according to (13.13), and the new weight will be $c_0 u(\mathbf{y})$ because of (13.15). Thus, the weight will be stabilized at $w(\mathbf{x}) = c_0 u(\mathbf{x})$ once $c_0 u_0 \geq 1$, and the equilibrium distribution of $\mathbf{x}$ will be $g(\mathbf{x})$. Therefore, for any starting value of w, the weight process will climb until it is large enough. After that, $(\mathbf{x}, w)$ behaves like a pair in an importance sampler.

The behavior of the weight process of an R-type move is less than satisfactory: It does *not* have a stable distribution. For example, let T be symmetric and let π be uniform on $\mathcal{X} = \{1, 2, 3\}$. It is easy to see that

$$w' = \begin{cases} w+1 & \text{if } U \leq \dfrac{w}{w+1} \\ w(w+1) & \text{otherwise.} \end{cases}$$

Therefore, the sequence of w is nondecreasing and diverges to infinity with probability 1. A simple way to fix this problem is to modify the weight update (13.14) by a random multiplier:

$$w' = \begin{cases} V(wr(\mathbf{x}, \mathbf{y}) + 1) & \text{if accepted} \\ Vw(wr(\mathbf{x}, \mathbf{y}) + 1) & \text{if rejected,} \end{cases} \quad (13.16)$$

where $V \sim \text{Uniform}(1-\delta, 1+\delta)$ is drawn independent of the $\mathbf{x}$. It is easy to see that this modified R-type move still satisfies IWIW. The parameter δ needs to be chosen properly so that $E(\log V)$ is not too small.

Case (iii): $\theta = 0$ and $T(\mathbf{x}, \mathbf{y})$ is nonreversible.

Dynamic weighting is most often used in combination with the regular Metropolis-Hastings's moves. Such a combination typically results in a nonreversible proposal transition. Thus, this is the case of most interest to us. When $\theta = 0$, the Q-type and the R-type moves are identical and have the following properties.

From the first equality in (13.15), we have

$$\log \frac{w'(\mathbf{x}')}{u(\mathbf{x}')} - \log \frac{w(\mathbf{x})}{u(\mathbf{x})} = \log \frac{g(\mathbf{x}', \mathbf{x})}{g(\mathbf{x}, \mathbf{x}')}.$$

Hence,

$$\log \frac{w^{(t)}}{u(\mathbf{x}^{(t)})} - \log \frac{w^{(0)}}{u(\mathbf{x}^{(0)})} = \sum_{s=1}^{t-1} \log \frac{g(\mathbf{x}^{(s+1)}, \mathbf{x}^{(s)})}{g(\mathbf{x}^{(s)}, \mathbf{x}^{(s+1)})}.$$

However, the following lemma shows that each term on the right-hand side has a negative expectation. Hence, the log-weight process is a cumulative sum of identically distributed (but correlated) random variables with a negative expectation, implying that the weight process converges to zero at an exponential rate.

Lemma 13.6.1 *Let $g(\ )$ be the invariant distribution of T and let $g(\mathbf{x},\mathbf{y}) = g(\mathbf{x})T(\mathbf{x},\mathbf{y})$. Then,*

$$e_0 = E_g\left\{\log\frac{g(\mathbf{y},\mathbf{x})}{g(\mathbf{x},\mathbf{y})}\right\} E_g\left\{\log\frac{T(\mathbf{y},\mathbf{x})}{T(\mathbf{x},\mathbf{y})}\right\} \leq 0. \qquad (13.17)$$

The equality holds only when T induces a reversible Markov chain.

Proof: By definition, we have

$$e_0 = E_g\left\{\log\frac{g(\mathbf{y},\mathbf{x})}{g(\mathbf{x},\mathbf{y})}\right\} \leq \log E_g\left\{\frac{g(\mathbf{y},\mathbf{x})}{g(\mathbf{x},\mathbf{y})}\right\} = 0.$$

We used the Jensen's inequality in which the equality holds only when $g(x_0, x_1) = g(x_1, x_0)$. ◇

Case (iv): $\theta = 1$ and $T(\mathbf{x},\mathbf{y})$ is nonreversible.
Consider the behavior of the log-weight process in Q-type moves:

$$\log w^{(t+1)} = \begin{cases} \max\{0, \log\left[w^{(t)}r(\mathbf{x}^{(t)}, \mathbf{x}^{(t+1)})\right]\}, & \text{if accept;} \\ \log w^{(t)} + \log a, & \text{if reject,} \end{cases} \qquad (13.18)$$

where the acceptance-rejection decision depends on θ. Thus, a nonzero θ prevents the weight process from converging to zero. In other words, zero works like a reflecting boundary. When $w^{(t)}$ becomes too small, θ ensures that some rejections will occur and the weight process will be forced to go up. On the other hand, when w becomes too large and no rejection occurs, the nonreversibility of T will produce a negative drift (Lemma 13.6.1) that prevents the process from drifting to infinity. To summarize, we state the following theorem of Liu et al. (2001) without giving a detailed proof.

Theorem 13.6.1 *Suppose the sample space $\mathcal{X}$ of $\mathbf{x}$ is finite and the proposal transition $T(\mathbf{x},\mathbf{y})$ is nonreversible. Then, the process $(\mathbf{x}^{(t)}, \log w^{(t)})$ induced by the Q-type move is positive recurrent and has a unique equilibrium distribution.*

Case (v): Mixing different types of moves ($\theta = 1$).
Suppose in each iteration that we make a Q-type move with probability α and a Metropolis move with probability $1-\alpha$. When w is sufficiently large,

there will be no rejection for the Q-type moves, thus, the actual transition for the Markov chain is approximately

$$A(\mathbf{x}, \mathbf{y}) = \alpha A_1(\mathbf{x}, \mathbf{y}) + (1 - \alpha) A_2(\mathbf{x}, \mathbf{y}),$$

where $A_1(\mathbf{x}, \mathbf{y})$ is just the proposal transition for the Q-type move and $A_2(\mathbf{x}, \mathbf{y})$ satisfies the detailed balance with respect to π. Let $g(\mathbf{x})$ be the invariant distribution of $A(\mathbf{x}, \mathbf{y})$ and let $\delta = 1$ or 2 be an indicator variable. When a move $\mathbf{x} \to \mathbf{y}$ occurs, the weight is updated as

$$w(\mathbf{y}) = w(\mathbf{x}) \frac{\pi(\mathbf{y}) A_\delta(\mathbf{y}, \mathbf{x})}{\pi(\mathbf{x}) A_\delta(\mathbf{x}, \mathbf{y})},$$

where $\delta = 1$ with probability α and $\delta = 2$ with $1 - \alpha$. It is easy to show that

$$E \log \frac{A_\delta(\mathbf{y}, \mathbf{x})}{A_\delta(\mathbf{x}, \mathbf{y})} = E \log \frac{g(\mathbf{y}) A_\delta(\mathbf{y}, \mathbf{x})}{g(\mathbf{x}) A_\delta(\mathbf{x}, \mathbf{y})} \leq \log E \frac{g(\mathbf{y}) A_\delta(\mathbf{y}, \mathbf{x})}{g(\mathbf{x}) A_\delta(\mathbf{x}, \mathbf{y})} = 0.$$

Hence, the log-weight process is a cumulative sum of terms with negative expectations. A similar argument as in Theorem 13.6.1 applies to show that the dynamic weighting process has a stable distribution.

13.6.3 Estimation with weighted samples

Suppose a set of weighted samples, $(\mathbf{x}^{(1)}, w^{(1)}), \ldots, (\mathbf{x}^{(m)}, w^{(m)})$, is obtained by running either a Q-type or an R-type scheme. We are interested in estimating $\mu = E_\pi h(\mathbf{x})$. The standard importance sampling estimate of μ is

$$\hat{\mu} = \frac{w^{(1)} h(\mathbf{x}^{(1)}) + \cdots + w^{(m)} h(\mathbf{x}^{(m)})}{w^{(1)} + \cdots + w^{(m)}}. \quad (13.19)$$

Since the weights derived from the Q- or R-type moves may have infinite expectations, it is not clear whether estimate (13.19) is still valid.

By a general weak law of large numbers (Chung 1974), Liu et al. (2001) show that this estimate still converges, although very slowly. Here we describe a stratified truncation method to improve the estimation and a brief justification for why it works.

Stratified Truncation for Weighted Estimate: Suppose of interest is the estimation of $\mu = E_\pi h(\mathbf{x})$. First, the samples $(\mathbf{x}^{(t)}, w^{(t)})$ are stratified according to the value of $h(\mathbf{x})$. Within each stratum, values of the $h(\mathbf{x}^{(t)})$ should be as close to constant as possible and the sizes of the strata are comparable. The highest $k\%$ (usually $k = 1$ or 2) of the weights within each stratum are then trimmed down to the value of the $(100 - k)$th percentile of the weights within the stratum. See Section 10.6 for some illustrations.

It is observed that the log-weight process has an exponential tail; that is, using a result of Kesten (1974), one can show that for the Q-type move on a finite state space satisfies

$$\lim_{c \to \infty} P(\log w^{(t)} > c \mid \mathbf{x}^{(t)} = \mathbf{x}) = K'u(\mathbf{x}),$$

where $u(\ )$ is as defined in Section 13.6.2. This implies that the upper k-percentile, $q_k(\mathbf{x})$, of those weights associated with $\mathbf{x}$ satisfies

$$q_k(\mathbf{x}) \propto u(\mathbf{x}) = \pi(\mathbf{x})/g(\mathbf{x}). \tag{13.20}$$

Since no rejection can occur in a Q-type move when w is large enough, those $\mathbf{x}$'s that are accompanied with large weights follow (approximately) distribution $g(\ )$. Hence, a weighted average of $h(\mathbf{x})$ using the upper percentiles of the weights provides us with an approximately correct estimate. Section 10.6 showed that the truncation method can provide good estimate for the Ising model simulation.

13.6.4 A simulation study

A simulation is designed to verify several predictions of our theory: (i) The tail distribution of the log-weight in a Q-type move is exponential with decay rate 1 (and a similar result for the R-type move); (ii) upper percentiles of the stratified weights are approximately proportional to $u(\mathbf{x})$; (iii) the plain estimate (13.19) converges slowly, but to the correct mean; and (iv) the estimation with stratified truncation gives us an approximately correct answer.

In our experiment, the state space of $\mathbf{x}$ is $\{1, 2, 3, 4, 5\}$ and a random 5×5 transition matrix was simulated (each row is independently drawn from a Dirichlet (1,1,1,1,1)):

$$T = \begin{pmatrix} 0.00370 & 0.15436 & 0.55588 & 0.15998 & 0.12608 \\ 0.18506 & 0.34190 & 0.17511 & 0.14471 & 0.15322 \\ 0.27798 & 0.26276 & 0.16575 & 0.21687 & 0.07664 \\ 0.29265 & 0.28028 & 0.22982 & 0.15994 & 0.03731 \\ 0.25206 & 0.23105 & 0.02426 & 0.22976 & 0.26287 \end{pmatrix}.$$

It is easy to check that the the transition matrix T is nonreversible and its invariant distribution is $g = (0.1987, 0.2611, 0.2398, 0.1782, 0.1222)$. We took the target distribution $\pi = (0.25, 0.1, 0.2, 0.4, 0.05)$.

With $a = 2$, a Q-type process was started with $w^{(0)} = 1$ and $\mathbf{x}^{(0)} \sim g$. A total of 200,000 iterations were carried out. Figure 13.1(a) shows the percentiles of weights stratified according to the state space. The percentages range from 70% to 99%. Figure 13.1(b) shows the histogram of the weight for $X = 3$, and Figure 13.1(c) shows the q-q plot of the tail of the weights versus the Exp(1) distribution. Estimating π by using stratified truncation at $k\% = 1\%, 2.5\%, 5\%$ [i.e., the upper k% of the weights in each stratum are trimmed down to be equal to the stratum's $(100 - k)$th percentile] gives us $\hat{\pi} = (0.2453, 0.0984, 0.2001, 0.4071, 0.0491), (0.2450, 0.1004, 0.2020, 0.4038, 0.0488)$, and $(0.2449, 0.1023, 0.1994, 0.4049, 0.0485)$, respectively. To show the slow convergence of the raw estimate, we run 2^{30} iterations, and at every 2^k epoch, we estimate π by using the raw estimate (13.19). Figure 13.1(d) shows the plot of the χ^2-distances between π and these estimates [i.e., $(\sum_{i=1}^{5}(\pi_i - \hat{\pi})^2/\pi_i)^{1/2}$] versus the logarithm of the number of iterations.

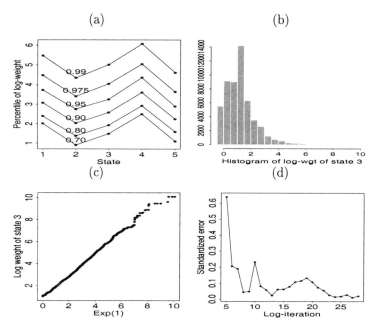

FIGURE 13.1. Results for the simulation study with Q-type moves. (a) The conditional percentiles of the weights. The parallel-ness of these percentiles is predicted by (13.20) and is the basis for the stratified truncation method; (b) the histogram of the log-weights; (c) the q-q plot of the upper tail of the log-weights versus the Exp(1) distribution; (d) the convergence of the raw weighted estimates.

We also applied the R-type moves to the same problem. The results showed that the weights resulting from R-type moves are appreciably greater than those from the Q-type, and the tail distribution of the weights seems to still be exponential but with a changing rate. Our predictions were also verified in more complicated examples, such as the Ising model simulation and a model selection problem (Liu et al. 2001).

Appendix A
Basics in Probability and Statistics

A.1 Basic Probability Theory

A.1.1 *Experiments, events, and probability*

Probability theory is a mathematical tool for modeling natural phenomena in which outcomes occur *nondeterministically*. Random phenomena considered in the probability theory are often referred to as *experiments*, although many of which are observational (i.e., experimental conditions cannot be controlled or intervened by the investigator) instead of experimental. Probability can be seen either as the *long-run frequency* of the occurrence of each potential outcome when the identical experiment is repeated indefinitely or as a numerical measure of subjective uncertainty on the experimental result. These two views underly two main different approaches to statistical inference: the frequentist school and the Bayesian school.

The set of all possible outcomes of an experiment is called a *sample space*, denoted as $\mathcal{X}$. An *event A* is a collection of individual outcomes that satisfy certain criterion. For example, in a coin toss experiment, event A can be "more than 8 heads in 10 tosses of a coin." Mathematically, A is a *measurable* subset of $\mathcal{X}$. Hence, $P(A \cup B \cup \cdots)$ is the probability that at least one of the events $A, B, \ldots$ occurs and $P(A \cap B \cap \cdots)$ is the probability that all the events $A, B, \ldots$ occur. The following simple axioms completely determine how probabilities operate:

(A1) $P(\emptyset) = 0$ and $P(\mathcal{X}) = 1$.

(A2) If $A \subset \mathcal{X}$, then $P(A) \geq 0$.

(A3) If $A_1, A_2, \ldots$ is are disjoint events (i.e., $A_i \cap A_j = \emptyset$ for any $i \neq j$), then,

$$P\left(\bigcup_{i=1}^{\infty} A_i\right) = \sum_{i=1}^{\infty} P(A_i).$$

From these axioms, one can derive some other useful properties such as $P(A^c) = 1 - P(A)$ and $P(A \cap B) = P(A) + P(B) - P(A \cup B)$. Here, $A^c \equiv \mathcal{X} \setminus A$ and reads as "A compliment." Furthermore, we define $P(A \mid B) = P(A \cap B)/P(B)$ as the *conditional probability* of A given B. The interpretation of this concept is that if we are told that event B has already occurred, the consideration of whether A occurs should be confined to the smaller universe — characterized by the occurrence of B. Therefore, we say that two events A and B are *mutually independent* if and only if $P(A \cap B) = P(A)P(B)$.

A.1.2 Univariate random variables and their properties

A one-dimensional random variable X is a mapping from the outcome space $\mathcal{X}$ to the real line $\mathbb{R}^1$. For example, in a coin toss experiment, the total number of heads observed in the first 10 tosses is a random variable. The *cumulative distribution function* (cdf) of random variable X is defined as $F(x) = P(X \leq x)$. Clearly, $F(-\infty) = 0$, $F(\infty) = 1$, and F is monotone nondecreasing. The *mathematical expectation* of a function of X, say $h(X)$, is defined as

$$E[h(X)] = \int h(x) dF(x),$$

where notation $\int$ should be understood as a Stieltjes integral. In most applications, the following two cases are sufficient: (a) F is differentiable, in which case X is called a *continuous* random variable, $f(x) = dF(x)/dx$ is called the *density function* of X, and $E[h(X)] = \int h(x)f(x)dx$; (b) F is a step-function with countably many jumps of sizes $p_1, p_2, \ldots$ at $x_1, x_2, \ldots$, respectively, in which case X is called a *discrete* random variable, and $E[h(X)] = \sum_{i=1}^{\infty} p_i h(x_i)$. The *mean* of the random variable X is defined as $E(X)$, the expectation of X, and the *variance* of X is $\text{var}(X) = E(X^2) - E(X)^2$. Heuristically, $E(X)$ is the center of probabilistic "gravity" and $\sqrt{\text{var}(X)}$ reflects the range of fluctuation of X.

Several important distributions are worth mentioning. The first one is the *normal distribution* (also called the Gaussian distribution), denoted as $N(\mu, \sigma^2)$, which has a density function

$$f(x) = \frac{1}{\sqrt{2\pi}\sigma} \exp\left\{-\frac{(x-\mu)^2}{2\sigma^2}\right\}.$$

This distribution has a mean μ and variance σ^2. When $\mu = 0$ and $\sigma = 1$, the distribution is referred to as the *standard* normal (or Gaussian) distribution. Due to the celebrated *central limit theorem* (Section A.1.4), the normal distribution is most frequently used in applied statistics to model continuous "errors" resulting from the aggregation of many sources of variations.

The next one is the *exponential distribution*, $\text{Exp}(\lambda)$, which has a density function

$$f(x) = \lambda \exp(-\lambda x), \quad \text{when } x \geq 0,$$

and $f(x) = 0$ when $x < 0$. The mean of this density is λ^{-1} and the variance is λ^{-2}. This distribution is often used to model the waiting time for the occurrence of rare events (e.g., car accidents in a given short segment of road or the emission of α-particles from a radioactive material).

A related distribution is the Gamma(α, λ) distribution with density

$$f(x) = \frac{\lambda^\alpha}{\Gamma(\alpha)} x^{\alpha-1} e^{-\lambda x}, \quad x > 0,$$

where $\alpha > 0$ and $\lambda > 0$ are adjustable parameters. Another most frequently used distribution is the *uniform* distribution, Uniform$[a, b]$, which has a density function

$$f(x) = (b-a)^{-1}, \quad \text{if } x \in [a, b],$$

and $f(x) = 0$ otherwise. In this case $E(X) = (a+b)/2$ and $\text{var}(X) = (b-a)^2/12$. All computer-based random number generators produce (pseudo-) random variables from Uniform[0,1].

If a random variable only takes value in the interval $[0, 1]$, a popular distribution for it is the Beta(α, β) distribution with density

$$f(x) = \frac{\Gamma(\alpha+\beta)}{\Gamma(\alpha)\Gamma(\beta)} x^{\alpha-1}(1-x)^{\beta-1}.$$

The mean and variance of this distribution are $\alpha/(\alpha+\beta)$ and $\alpha\beta/[(\alpha+\beta)^2(\alpha+\beta+1)]$, respectively. More interestingly, if $X \sim \text{Gamma}(\alpha, \lambda)$, $Y \sim \text{Gamma}(\beta, \lambda)$, and they are independent, then $X/(X+Y)$ is distributed as Beta(α, β). This property is often used in Monte Carlo simulation of a Beta random variable.

There are several useful discrete distributions. One is the *Binomial distribution*, denoted as Binom(n, θ), which records the probability of seeing x heads in n tosses of a biased coin (with θ probability of showing head). It has a distribution function

$$P(X = x) = \binom{n}{x} \theta^x (1-\theta)^{n-x}, \quad x = 0, 1, \ldots, n.$$

Clearly, $E(X) = n\theta$ and $\text{var}(X) = n\theta(1-\theta)$. A related distribution is the *geometric distribution*, $\text{Geo}(\theta)$, which models the number of tosses it takes to see the first head. Its distribution function has the form

$$P(X = k) = (1-\theta)^{k-1}\theta, \quad k = 1, 2, \ldots$$

The mean of this distribution is θ^{-1}, and the variance is $(1-\theta)/\theta^2$.

A generalization of the binomial distribution is the *multinomial distribution*, which is used to model the outcomes of tossing a loaded k-sided die. In particular, the multinomial random variable is a k-dimensional vector, $X = (x_1, \ldots, x_k)$, with $x_1 + \cdots + x_k = n$, where x_j records the number of occurrence of face j in n tosses of the die. Let $\boldsymbol{\theta} = (\theta_1, \ldots, \theta_k)$ be the probability vector that characterizes the "loadedness" of the die Then the multinomial distribution is

$$P(X = (x_1, \ldots, x_k)) = \frac{n!}{x_1! \cdots x_k!} \theta_1^{x_1} \cdots \theta_k^{x_k}.$$

Another famous discrete distribution is the *Poisson distribution*, $\text{Poi}(\lambda)$, who has a distribution function

$$P(X = x) = \frac{\lambda^x}{x!} e^{-\lambda}, \quad x = 0, 1, 2, \ldots$$

Here, the distribution is supported on all non-negative integers. The mean and the variance of this distribution are all equal to λ. The Poisson distribution can be viewed as the limit of a $\text{Binom}(n, p)$ distribution in which $n \to \infty$ and $np \to \lambda$. It is also related to the exponential distribution and is often used in modeling the number of occurrences of certain rare events in a region or a period of time (Pitman 1993)

A.1.3 Multivariate random variable

We can extend the distribution concept for univariate random variables to the multivariate case. Let us consider a random vector $\boldsymbol{X} = (X_1, \ldots, X_d)$ taking values in $\mathbb{R}^d$. Its cdf is defined as $F(\mathbf{x}) = P(\cap_{i=1}^d \{X_i \leq x_i\})$ $\mathbf{x} \in \mathbb{R}^d$. If $\boldsymbol{X}$ is a continuous random variable, its density is

$$f(x_1, \ldots, x_d) = \frac{\partial^d F(x_1, \ldots, x_d)}{\partial x_1 \cdots \partial x_d}.$$

For any function $h(\mathbf{x})$, we can compute its expectation as

$$E[h(\boldsymbol{X})] = \int h(\mathbf{x}) dF(\mathbf{x}) = \int h(\mathbf{x}) f(\mathbf{x}) dx_1 \cdots dx_d.$$

For example, a *multivariate normal distribution* has a density form

$$f(\mathbf{x}; \boldsymbol{\mu}, \Sigma) = |2\pi\Sigma|^{-1/2} \exp\left\{-\frac{1}{2}(\mathbf{x} - \boldsymbol{\mu})\Sigma^{-1}(\mathbf{x} - \boldsymbol{\mu})^T\right\}$$

in which **x** and **μ** are d-dimensional row vectors and Σ is a $d \times d$ positive-definite matrix.

Naturally, we say that random variables $X_1, \ldots, X_d$ are mutually *independent* if and only if their joint cdf is the product of the individual marginal cdf's. In the continuous random variable case, this really means that the joint density can be factored:

$$f(\mathbf{x}) = f_1(x_1) \times \cdots \times f_d(x_d).$$

An important concept in dealing with the multivariate case is that the variables often *vary* together; that is, a change in one component will usually influence others. A quantitative measure of this characteristic is the *covariance* between any pair of random variables, X_i and X_j:

$$\text{cov}(X_i, X_j) = E(X_i X_j) - E(X_i)E(X_j).$$

If X_i and X_j are mutually independent, then $\text{cov}(X_1, X_2) = 0$ (the reverse argument is not true). Using the definition of covariance, we can show that

$$\text{var}(X_1 + \cdots + X_d) = \sum_{i=1}^{d} \text{var}(X_i) + 2 \sum_{i<j} \text{cov}(X_i, X_j).$$

If the random variables $X_1, \ldots, X_n$ are mutually independent and have the same variance, then we have

$$\text{var}\left(\frac{X_1 + \cdots + X_n}{n}\right) = \frac{\text{var}(X_1) + \cdots + \text{var}(X_n)}{n^2} = \frac{\text{var}(X_1)}{n}, \quad (A.1)$$

implying that the "variability" of $\bar{X}$ decreases linearly as we average over more and more independent random variables with the *same* variance.

We have encountered a multivariate random variable — the multinomial distribution — in the previous section. This distribution is most frequently used in studying discrete categorical data and is the most basic probability distribution in all biological sequence analysis methods (Liu and Lawrence 1999). A closely related distribution is called the Dirichlet distribution of which the Beta distribution is a special case. This distribution is used for modeling the uncertainty in probability vector. For example, if we are given a loaded die that we know nothing about, it is convenient to assume that the long-run frequencies of all the faces follow a Dirichlet distribution. The density of a Dirichlet random variable has the form

$$f_0(\boldsymbol{\theta}) \propto \prod_{j=1}^{k} \theta_j^{\alpha_j - 1} \quad (A.2)$$

where $\alpha_j > 0$ for all j. Section A.3 gives more details on this distribution.

A.1.4 Convergence of random variables

It is often useful to think about how a sequence of random variables become more and more "stable" (i.e., converge to another random variable). Let $Y_1, Y_2, \ldots$ be a sequence of random variables and let $F_1(y), F_2(y), \ldots$ be the corresponding sequence of cumulative distribution functions. If their probability distributions look more and more similar to the distribution of a common random variable Y as $n \to \infty$, we say that they converge to Y *in distribution*. Mathematically, this means that

$$\lim_{n \to \infty} F_n(y) = F(y), \quad \forall y,$$

where $F(y)$ is the cdf of Y.

We say that Y_n converges to Y *in probability* if $\forall\, \epsilon > 0$,

$$\lim_{n \to \infty} P(|Y_n - Y| > \epsilon) = 0.$$

This means that the part of the sample space where X_n differs from X is getting smaller and smaller. Furthermore, we say that Y_n converges to Y *almost surely* (abbreviated as a.s.) if

$$P(\lim_{n \to \infty} |Y_n - Y| = 0) = 1.$$

In thinking about the latter two modes of convergence, it is useful to think of each random variable Y_n as the function of the outcomes in the sample space. Hence, almost sure convergence implies that for almost all s in $\mathcal{X}$ we have $Y_n(s)$ getting closer and closer to $Y(s)$. It can be shown that almost sure convergence implies the convergence in probability, which then implies convergence in distribution. The following three fundamental results for convergence of random variables are most widely used in practice.

The *weak law of large numbers* (WLLN) states that if $X_1, \ldots, X_n, \ldots$ are i.i.d. random variables with finite mean μ, then

$$\frac{X_1 + \cdots + X_n}{n} \to \mu \text{ in probability.}$$

If X_1 also has a finite variance (a weaker condition is also possible), then we have the *strong law of large numbers* (SLLN):

$$\frac{X_1 + \cdots + X_n}{n} \to \mu \text{ a.s.}$$

The *central limit theorem* (Chung 1974) says that for mutually independent (or weakly correlated) random variables $X_1, X_2, \ldots$ with mean μ and variance σ^2,

$$\frac{\sqrt{n}(\bar{X} - \mu)}{\sigma} \to N(0,1) \text{ in distribution.}$$

This means that the distributional shape of $\bar{X}$ is more and more like that of a Gaussian random variable as n increases. It gives one a sense of how fast $\bar{X}$ approaches to μ as n increases. This result is also an informal basis for the widespread use of the normal distributions in diverse application fields.

A.2 Statistical Modeling and Inference

A.2.1 Parametric statistical modeling

Statistical modeling and analysis, including the collection of data, the construction of a probabilistic model, the quantification and incorporation of expert opinions, the interpretation of the model and the results, and the prediction from the data, form an essential part of the scientific method in diverse fields. The key focus of statistics is on making inferences, where the word *inference* follows the dictionary definition as "the process of deriving a conclusion from fact and/or premise." In statistics, the facts are the observed data, the premise is represented by a probabilistic model of the system of interest, and the conclusions concern unobserved quantities. Statistical inference distinguishes itself from other forms of inferences by explicitly quantifying uncertainties involved in the premise and the conclusions.

In *nonparametric* statistical inference, one does not assume any specific distributional form for the probability law of the observed data, but only imposes on the data a dependence (or independence) structure. For example, an often imposed assumption in nonparametric analyses is that the observations are *independent and identically distributed* (i.i.d.). When the observed data are continuous quantities, what one has to infer for this nonparametric model is the whole density curve — an infinite-dimensional parameter. A main advantage of nonparametric methods is that the resulting inferential statements are relatively more robust than those from parametric methods. A main disadvantage of the nonparametric approach is, however, that it is difficult, and sometimes impossible, to build into the model more sophisticated structures based on our scientific knowledge (i.e., it does not facilitate researchers to "learn").

Indeed, it would be ideal and preferable if we could derive what we want without having to assume anything. On the other hand, however, the process of using simple models (with a small number of adjustable parameters) to describe natural phenomena and then improving upon them (e.g., Newton's law of motion versus Einstein's theory of relativity) is at the heart of all scientific investigations. Parametric modeling, either analytically or qualitatively, either explicitly or implicitly, is intrinsic to human intelligence (i.e., it is essentially the only way we learn about the outside world). Analogously, statistical analysis based on parametric modeling is also essential to our scientific understanding of the data.

At a conceptual level, probabilistic models in statistical analyses serve as a mechanism through which one connects observed data with a scientific premise or hypothesis about the real-world phenomena. No model can completely represents every detail of reality. The goal of modeling is to abstract the key features of the underlying scientific problem into a workable mathematical form with which the scientific premise may be examined. Families of probability distributions characterized by a small number of parameters are most useful for this purpose.

Let $\mathbf{y}$ denote the observed data. In *parametric inference*, we assume that the observation follows a probabilistic law that belongs to a given distribution family; that is, $\mathbf{y}$ is a realization of a random process (i.e., sampling) whose probability law has a particular form (e.g., Gaussian, multinomial, Dirichlet etc.), $f(\mathbf{y} \mid \boldsymbol{\theta})$, which is completely known other than $\boldsymbol{\theta}$. Here, $\boldsymbol{\theta}$ is called a (population) *parameter* and it often corresponds to a scientific premise for our understanding of a natural process. Finding a value of $\boldsymbol{\theta}$ that is most compatible with the observation $\mathbf{y}$ is termed *model fitting* or *estimation*. We make scientific progresses by iterating between fitting the data to the posited model and proposing an improved model to accommodate important features of the data that are not accounted for by the previous model. When the model is given, an efficient method should be used to make inference on the parameters. Both the maximum likelihood estimation method and the Bayes method use the likelihood function to extract information from data and are efficient; these methods will be the main focus of the remaining part of this chapter.

A.2.2 Frequentist approach to statistical inference

Frequentist approach, or sometimes simply referred to as the classical statistics procedure, arrives at its inferential statements by using a point estimate of the unknown parameter and addressing the estimation uncertainty by the *frequency behavior* of the estimator. Among all the estimation methods, the method of *maximum likelihood estimate* (MLE) is most popular.

The MLE of $\boldsymbol{\theta}$ is defined as an argument $\hat{\boldsymbol{\theta}}$ that maximizes the likelihood function; that is,

$$\hat{\boldsymbol{\theta}} = \arg\max_{\text{all } \boldsymbol{\theta}} L(\boldsymbol{\theta} \mid \mathbf{y}),$$

where the *likelihood function* $L(\boldsymbol{\theta} \mid \mathbf{y})$ is defined to be *any* function that is proportional to the probability density $f(\mathbf{y} \mid \boldsymbol{\theta})$. Clearly, $\hat{\boldsymbol{\theta}}$ is a function of $\mathbf{y}$ and its form is determined completely by the parametric model $f(\ )$. Hence, we can write $\hat{\boldsymbol{\theta}}$ as $\hat{\boldsymbol{\theta}}(\mathbf{y})$ to explicate this connection. Any deterministic function of the data $\mathbf{y}$, such as $\hat{\boldsymbol{\theta}}(\mathbf{y})$, is called an *estimator*. Although there are many possible ways of producing an estimator, it can be shown under regularity conditions that the MLE $\hat{\boldsymbol{\theta}}(\mathbf{y})$ is asymptotically the most efficient among all potential estimators. In other words, no other ways of using $\mathbf{y}$ can outperform the MLE procedure in estimating $\boldsymbol{\theta}$ in an asymptotic sense. Some inferior methods, such as the method of moments (MOM), can be used as alternatives when the MLE is difficult to obtain.

For example, if $\mathbf{y} = (y_1, \ldots, y_n)$ are i.i.d. observations from $N(\theta, 1)$, a Normal distribution with mean θ and variance 1, then the MLE of θ is $\hat{\theta}(\mathbf{y}) = \bar{y}$ which is simply a linear combination of the $\mathbf{y}$. Another often-used estimator is the median, which is also easy to compute. If $\mathbf{y} = (y_1, \ldots, y_n)$ are i.i.d. observations from a Cauchy(θ) distribution, which has a density form

$$f(x;\ \theta) = \frac{1}{\pi[1 + (x-\theta)^2]},$$

then the MLE is no longer easy to compute, whereas the median is often a good substitute.

Uncertainty in the estimation is addressed by the *principle of repeated sampling*. Imagine that the same stochastic process which "generates" our observation $\mathbf{y}$ can be repeated indefinitely under identical conditions. A frequentist studies what the "typical" behavior of an estimator [e.g., $\hat{\theta}(\mathbf{y}_{\text{rep}})$] is. Here, $\mathbf{y}_{\text{rep}}$ denotes a hypothetical dataset generated by a replication of the *same* process that generates $\mathbf{y}$ and is, therefore, a random variable that has $\mathbf{y}$'s characteristics. The distribution of $\hat{\theta}(\mathbf{y}_{\text{rep}})$ is called the *frequency behavior* of estimator $\hat{\theta}$. For the normal distribution example, the frequency distribution of $\bar{y}_{\text{rep}}$ is $N(\theta, 1/n)$. With this distribution available, we can calibrate the observed $\hat{\theta}(\mathbf{y})$ with the "typical" behavior of $\hat{\theta}(\mathbf{y}_{\text{rep}})$ [e.g., $N(\theta, 1/n)$] to quantify uncertainty in the estimation.

We want to emphasize that the concepts of an "estimator" and its uncertainty only make sense if a generative model is contemplated. For example, the statement that "$\hat{\theta}_a$ estimates the true frequency of A" only makes sense if we imagine that an i.i.d. model (or another similar model) was used to generate the data. If this model is not really what we have in mind, then the meaning of $\hat{\theta}_a$ is no longer clear. A imaginary random process for the data generation is crucial for deriving a valid statistical statement.

A $(1-\alpha)100\%$ confidence interval (or region) for $\boldsymbol{\theta}$, for instance, is of the form $(\underline{\boldsymbol{\theta}}(\mathbf{y}_{\text{rep}}), \overline{\boldsymbol{\theta}}(\mathbf{y}_{\text{rep}}))$, meaning that under repeated sampling, the probability that the interval (the interval is random under repeated sampling) covers the true $\boldsymbol{\theta}$ is $1-\alpha$. To the contrary of what most people have hoped for, this interval statement *does not* mean that "$\boldsymbol{\theta}$ is in $(\underline{\boldsymbol{\theta}}(\mathbf{y}), \overline{\boldsymbol{\theta}}(\mathbf{y}))$ with probability $1-\alpha$." With observed $\mathbf{y}$, the true $\boldsymbol{\theta}$ is either in or out of the interval and no meaningful probability statement can be given unless $\boldsymbol{\theta}$ can be treated as a random variable.

When finding the analytical form of the frequency distribution of an estimator $\hat{\boldsymbol{\theta}}$ is difficult, some advanced techniques such as the *jackknife* and *bootstrap* method can be applied to numerically simulate the "typical" behavior of an estimator (Efron 1979). Suppose $\mathbf{y} = (y_1, \ldots, y_n)$ and each y_i follows an i.i.d. model. In the bootstrap method, one treats the empirical distribution of $\mathbf{y}$ (the distribution that gives a probability mass of $1/n$ to each y_i and 0 to all other points in the space) as the "true underlying distribution" and repeatedly generates new datasets, $\mathbf{y}_{\text{rep},1}, \ldots, \mathbf{y}_{\text{rep},B}$ from this distribution. Operationally, each $\mathbf{y}_{\text{rep},b}$ consists of n data points; that is, $\mathbf{y}_{\text{rep},b} = (y_{b,1}, \ldots, y_{b,n})$, where each $y_{b,i}$ is a simple random sample (with replacement) from the set of the observed data points $\{y_1, \ldots, y_n\}$. With the bootstrap samples, we can calculate $\hat{\theta}(\mathbf{y}_{\text{rep},b})$, for $b = 1, \ldots, B$, whose histogram tells us how $\hat{\theta}$ varies from sample to sample.

In a sense, the classical inferential statements are *pre-data* statements because they are concerned with the repeated sampling properties of a procedure and do not have to refer to the actual observed data (except in the bootstrap method, where the observed data are used in the approximation of the "true underlying distribution"). A major difficulty in the frequentist approach, besides its awkwardness in quantifying estimation uncertainty, is its difficulty in dealing with nuisance parameters. Suppose $\boldsymbol{\theta} = (\theta_1, \theta_2)$. In a problem where we are only interested in one component, θ_1 say, the other component θ_2 becomes a *nuisance parameter*. No clear principles exist in classical statistics that enables us to get rid of θ_2 in an optimal way. One of the most popular practices in statistical analysis is the so-called *profile likelihood* method, in which one treats the nuisance parameter θ_2 as known and fixes it at its MLE. This method, however, underestimates the involved uncertainty (because it treats unknown θ_2 as if it were known) and can lead to incorrect inference when the dimensionality of θ_2 is high. More sophisticated methods based on orthogonality, similarity, and average likelihood have also been proposed, but they all have their own problems and limitations.

A.2.3 Bayesian methodology

Bayesian statistics seeks a more ambitious goal by modeling all sorts of related information and uncertainty (e.g., physical randomness, subjective opinions, prior knowledge from different sources, etc.) with a joint probability distribution and treating *all quantities* involved in the model, be they observations, missing data, or unknown parameters, as random variables. It uses the calculus of probability as the guiding principle in manipulating data and derives its inferential statements based purely on an appropriate conditional distribution of unknown variables.

Instead of treating $\boldsymbol{\theta}$ as an unknown constant as in a frequentist approach, Bayesian treats $\boldsymbol{\theta}$ as a realized value of a random variable that follows a *prior distribution* $f_0(\boldsymbol{\theta})$. This prior is typically regarded as known to the researcher independently of the data under analysis. The Bayesian approach has at least two advantages. First, through the prior distribution, we can inject prior knowledge and information on the value of $\boldsymbol{\theta}$. This is especially important in bioinformatics since biologists often have substantial knowledge on the subject under study. To the extent that this information is correct, it will sharpen the inference about $\boldsymbol{\theta}$. Second, treating all the variables in the system as random variables greatly clarifies the methods of analysis. It follows from the basic probability theory that information about the realized value of any random variable, $\boldsymbol{\theta}$, say, based on observation of related random variables, $\mathbf{y}$, say, is summarized in the conditional distribution of $\boldsymbol{\theta}$ given $\mathbf{y}$, the so-called *posterior distribution*. Hence, if we are interested only in a component of $\boldsymbol{\theta}$, we have just to integrate out the remaining components of $\boldsymbol{\theta}$ (i.e., nuisance parameters) from the posterior distribution. Furthermore, if we are interested in the prediction of a future observation $\mathbf{y}^+$ depending on $\boldsymbol{\theta}$, we can obtain the posterior distribution of $\mathbf{y}^+$ given $\mathbf{y}$ by completely integrating out $\boldsymbol{\theta}$. However, an associated problem with the Bayesian approach is the specification of an appropriate prior for the unknowns. This task can be daunting in some high-dimensional problems.

The use of probability distributions to describe unknown quantities is also supported by the fact that probability theory is the only known coherent system for quantifying objective and subjective uncertainties. Furthermore, probabilistic models have been accepted as appropriate in almost all information-based technologies, including artificial intelligence, control theory, communication and signal processing, information theory, statistics, system science, etc. When the system under study is modeled properly, the Bayesian approach is always among the most coherent, consistent, and efficient statistical methods.

The theorem that combines the prior and the data to form the posterior distribution (Section A.3) is a simple mathematical result first given by Thomas Bayes in 1763. The statistical procedure based on the systematic use of this theorem appears much later (some people believe that Laplace was the first *Bayesian*) and is also named after Bayes. The adjective *Bayesian* is often used for approaches in which subjective probabilities are emphasized. In this sense, Thomas Bayes is not really a Bayesian.

A main controversial aspect of the Bayesian approach is the use of the prior distribution, to which three interpretations can be given: (a) as frequency distributions; (b) as objective representations of a rational belief of the parameter, usually in a state of ignorance; and (c) as a subjective measure of what a particular individual believes (Cox and Hinkley 1974). Interpretation (a) refers to the case when $\boldsymbol{\theta}$ indeed follows a stochastic process and, therefore, is uncontroversial. But this scenario is of limited applicability. Interpretation (b) is theoretically interesting but is often untenable in real applications. The emotive words "subjective" and "objective" should not be taken too seriously. (Many people regard the frequentist approach as a more "objective" one.) There are considerable subjective elements and personal judgments injected into all phases of scientific investigations. Claiming that someone's procedure is "more objective" based on how the procedure is derived is nearly meaningless. A truly objective evaluation of any procedure is by how well it attains its stated goals. In bioinformatics, we are fortunate to have a lot of known biological facts to serve as objective judges.

In most of our applications, we employ the Bayesian method mainly because of its internal consistency in modeling and analysis and its capability to combine various sources of information. Thus, we often take a combination of (a) and (c) for deriving a "reasonable" prior for our data analysis. We advocate the use of a suitable sensitivity analysis (i.e., an analysis on how our inferential statements are influenced by a change in the prior) to validate our statistical conclusions.

A.3 Bayes Procedure and Missing Data Formalism

A.3.1 The joint and posterior distributions

The full process of a typical Bayesian analysis can be described as consisting of three main steps (Gelman, Roberts and Gilks 1995): (a) setting up a full probability model, the *joint distribution*, that captures the relationship among *all* the variables (e.g., observed data, missing data, unknown parameters) in consideration; (b) summarizing the findings for particular quantities of interest by appropriate posterior distributions, which is typically a conditional distribution of the quantities of interest given the observed data; (c) evaluating the appropriateness of the model and suggesting improvements (model criticizing and selection).

A standard procedure for carrying out step (a) is to formulate the scientific question of interest through the use of a probabilistic model, based on which we can write down the *likelihood function* of $\boldsymbol{\theta}$. Then, a prior distribution $f_0(\boldsymbol{\theta})$ is contemplated which should be both mathematically tractable and scientifically meaningful. The joint probability distribution can then be represented as $Joint = likelihood \times prior$:

$$p(\mathbf{y}, \boldsymbol{\theta}) = p(\mathbf{y} \mid \boldsymbol{\theta}) f_0(\boldsymbol{\theta}) \qquad (A.3)$$

For notational simplicity, we use $p(\mathbf{y} \mid \boldsymbol{\theta})$, hereafter, interchangeably with $f(\mathbf{y} \mid \boldsymbol{\theta})$ to denote the likelihood. From a Bayesian's point of view, this is simply a conditional distribution.

Step (b) is completed by obtaining the *posterior distribution* through the application of the Bayes theorem:

$$p(\boldsymbol{\theta} \mid \mathbf{y}) = \frac{p(\mathbf{y}, \boldsymbol{\theta})}{p(\mathbf{y})} = \frac{p(\mathbf{y} \mid \boldsymbol{\theta}) f_0(\boldsymbol{\theta})}{\int p(\mathbf{y} \mid \boldsymbol{\theta}) f_0(\boldsymbol{\theta}) d\boldsymbol{\theta}} \propto p(\mathbf{y} \mid \boldsymbol{\theta}) f_0(\boldsymbol{\theta}). \qquad (A.4)$$

A.3 Bayes Procedure and Missing Data Formalism

When $\boldsymbol{\theta}$ is discrete, the integral is replaced by summation. The denominator $p(\mathbf{y})$, which is a normalizing constant for the function, is sometimes called the *marginal likelihood* of the model and can be used to conduct model selection (Kass and Raftery 1995). Although evaluating $p(\mathbf{y})$ analytically is infeasible in many applications, Markov chain Monte Carlo methods often can be employed for its estimation.

For example, suppose we toss a loaded k-sided die n times and wish to estimate the frequency vector $\boldsymbol{\theta} = (\theta_1, \ldots, \theta_k)$ for the occurrence frequency of each side. A commonly used model is the multinomial distribution. A mathematically convenient prior for the multinomial families is the *Dirichlet distributions*, Dirichlet($\boldsymbol{\alpha}$), whose density form is given in (A.2). Here, $\boldsymbol{\alpha} = (\alpha_1, \ldots, \alpha_k)$ is the hyperparameter for the Dirichlet distribution and is sometimes referred to as the "pseudo-counts," which can be understood heuristically as the "worth" of one's prior opinion (relative to the number of actual observations). When a simple i.i.d. model is imposed on an observed sequence of letters, $\mathbf{y} = (y_1, \ldots, y_n)$, its likelihood function is

$$p(\mathbf{y} \mid \boldsymbol{\theta}) = \prod_{i=1}^{n} \theta_{y_i} = \prod_{j=1}^{k} \theta_j^{n_j},$$

where n_j is the number of counts of residual type j in $\mathbf{y}$. If a Dirichlet ($\boldsymbol{\alpha}$) prior is used for its parameter $\boldsymbol{\theta}$, the posterior distribution for $\boldsymbol{\theta}$ is simply another Dirichlet distribution with hyperparameter $(\alpha_1 + n_1, \ldots, \alpha_k + n_k)$. The posterior mean of, say, θ_j is $(n_j + \alpha_j)/(n + \alpha)$.

Suppose the parameter vector has more than one component [i.e., $\boldsymbol{\theta} = (\theta_1, \boldsymbol{\theta}_{[-1]})$, where $\boldsymbol{\theta}_{[-1]}$ denotes all but the first component]. One may be interested only in one of components, θ_1, say. The other components that are not of immediate interest but are needed by the model, *nuisance parameters*, can be removed by integration:

$$p(\theta_1 \mid \mathbf{y}) = \frac{p(\mathbf{y}, \theta_1)}{p(\mathbf{y})} = \frac{\int p(\mathbf{y} \mid \theta_1, \boldsymbol{\theta}_{[-1]}) f_0(\theta_1, \boldsymbol{\theta}_{[-1]}) d\boldsymbol{\theta}_{[-1]}}{\iint p(\mathbf{y} \mid \theta_1, \boldsymbol{\theta}_{[-1]}) f_0(\theta_1, \boldsymbol{\theta}_{[-1]}) d\theta_1 d\boldsymbol{\theta}_{[-1]}}. \quad (A.5)$$

Note that computations required for completing a Bayesian inference are the integrations (or summations for discrete parameters) over all unknowns in the joint distribution to obtain the marginal likelihood and over all but those of interest to remove nuisance parameters. Despite the deceptively simple-looking form of (A.5), the challenging aspects of Bayesian statistics are as follows: (i) the development of a model, $p(\mathbf{y} \mid \boldsymbol{\theta}) f_0(\boldsymbol{\theta})$, which must effectively capture the key features of the underlying scientific problem and (ii) the necessary computation for deriving the posterior distributions. For aspect (i), the *missing data* formulation is an important tool to help one formulate a scientific problem; for (ii), the Monte Carlo techniques described in this book are essential.

A.3.2 The missing data problem

The missing data formulation is an important methodology for modeling complex data structures and for designing computational strategies (Little and Rubin 1987). This general framework was motivated in the early 1970s (and maybe earlier) by the need of a proper statistical analysis of certain survey data where parts of the data were missing (Rubin 1976). For example, a large survey of families was conducted in 1967, in which many socioeconomic variables were recorded. Then, a follow-up study of the same families were done in 1970. Naturally, the 1967 data had a large amount of missing values due to either recording errors or some families's refusal to answer certain questions. The 1970 data had an even more severe kind of missing data caused by the fact that many families studied in 1967 could not be located in 1970.

The first important question for a missing data problem is under what conditions one can ignore the "missing mechanism" in the analysis; that is, does the fact that an observation is missing tell us anything about the quantities we are interested in estimating? For example, the fact that many families moved out of a particular region may indicate that the region's economy was having a problem. Thus, if our interested estimand is a certain "consumer confidence" measure of the region, the standard estimate resulting only from the observed families might be biased. Rubin's (1976) pioneering work provides a general guidance on how to judge the ignorability. Since everything in a Bayes model is a random variable, it is especially convenient and transparent in dealing with these ignorability problems in a Bayesian framework. The second important question is how one should conduct computations, such as finding the MLE or the posterior distribution of the estimands. This question has motivated statisticians to develop several important algorithms: the EM algorithm (Dempster et al. 1977), data augmentation (Tanner and Wong 1987), and the Gibbs sampler (Gelfand and Smith 1990)

In the late 1970s and early 1980s, people started to realize that many other problems can be treated as a missing data problem. One typical example is the so-called *latent-class* model, which is most easily explained by the following example (Tanner and Wong 1987). In the 1972-74 General Social Surveys, a sample of 3181 participants were asked to answer the following questions. Question A: whether or not you think it should be possible for a pregnant woman to obtain a legal abortion *if she is married and does not want any more children.* Question B: the italicized phrase in A is replaced with "if she is not married and does not want to marry the man." A latent-class model assumes that a person's answers to A and B are conditionally independent given the value of a dichotomous latent variable Z (either 0 or 1). Intuitively, this model asserts that the population consists of two "types" of persons (e.g., conservative and liberal) and Z is the unobserved "label" of each person. If you know the person's label, then

A.3 Bayes Procedure and Missing Data Formalism

his/her answer to question A will not help you to predict his/her answer to question B. Clearly, variable Z can be thought of as a "missing data" although it is not really "missing" in a standard sense.

The importance of the missing data formulation stems from the following two main considerations. Conceptually, this framework helps in making model assumptions explicit (e.g., ignorable versus nonignorable missing mechanism), in defining precise estimand of interest, and in providing a logical framework for causal inference (Rubin, 1976). Computationally, the missing data formulation inspired the invention of several important statistical algorithms. Mathematically, however, the missing data formulation is not well defined. In real life, what we can observe is always partial (incomplete) information and there is no absolute distinction between parameters and missing data (i.e., some unknown parameters can also be thought of as missing data and vice versa). In the most general and abstract form, the "missing data" can refer to any augmented part of the probabilistic system under consideration. When missing data $\mathbf{y}_{\text{mis}}$ is present, a proper inference on parameters of interest can be achieved by using the "observed-data likelihood," $L_{\text{obs}}(\boldsymbol{\theta}; \mathbf{y}_{\text{obs}}) = p(\mathbf{y}_{\text{obs}} \mid \boldsymbol{\theta})$, which can be obtained by integration:

$$L_{\text{obs}}(\boldsymbol{\theta}; \mathbf{y}_{\text{obs}}) \propto \int p(\mathbf{y}_{\text{obs}}, \mathbf{y}_{\text{mis}} \mid \boldsymbol{\theta}) d\mathbf{y}_{\text{mis}}.$$

Since it is often difficult to complete this integral analytically, one needs advanced computational methods such as the EM algorithm (Dempster et al. 1977) to compute the MLE.

Bayesian analysis for missing data problems can be achieved coherently through integration. Let $\boldsymbol{\theta} = (\theta_1, \boldsymbol{\theta}_{[-1]})$ and suppose we are interested only in θ_1. Then,

$$p(\theta_1 \mid \mathbf{y}_{\text{obs}}) \propto \iint p(\mathbf{y}_{\text{obs}}, \mathbf{y}_{\text{mis}} \mid \theta_1, \boldsymbol{\theta}_{[-1]}) p(\theta_1, \boldsymbol{\theta}_{[-1]}) d\mathbf{y}_{\text{mis}} d\boldsymbol{\theta}_{[-1]}.$$

Since all quantities in a Bayesian model are treated as random variables, the integration for eliminating the missing data is no different than that for eliminating nuisance parameters.

Our main use of the missing data formulation is to construct proper statistical models for bioinformatics problems. As will be shown in the later sections, this framework frees us from being afraid of introducing meaningful but perhaps high-dimensional variables into our model, which is often necessary for a satisfactory description of the underlying scientific knowledge. The extra variables introduced this way, when treated as missing data, can be integrated out in the analysis stage so as to result in a proper inference for the parameter of interest. Although a conceptually simple procedure, the computation involved in integrating out missing data can be very difficult.

A.4 The Expectation-Maximization Algorithm

The expectation-maximization (EM) algorithm is perhaps one of the most well-known statistical algorithms for finding the mode of a *marginal* likelihood or posterior distribution function; that is, the EM algorithm enables one to find the mode of

$$F(\boldsymbol{\theta}) = \int f(\mathbf{y}_{\text{mis}}, \boldsymbol{\theta}) d\mathbf{y}_{\text{mis}}, \tag{A.6}$$

where $f(\mathbf{y}_{\text{mis}}, \boldsymbol{\theta}) \geq 0$ and $F(\boldsymbol{\theta}) < \infty$ for all $\boldsymbol{\theta}$. When $\mathbf{y}_{\text{mis}}$ is discrete, we simply replace the integral in (A.6) by summation. In the traditional description of the EM algorithm, the f function is usually written as a *complete data likelihood*, $p(\mathbf{y}_{\text{obs}}, \mathbf{y}_{\text{mis}}|\boldsymbol{\theta})$. Here we purposely omit $\mathbf{y}_{\text{obs}}$ for a simpler presentation. Although trivial, it is worthwhile to note that the problem setting in (A.6) is very general and can refer to *any* marginal optimization problem. The EM algorithm starts with an initial guess $\boldsymbol{\theta}^{(0)}$ and iterates the following two steps:

- **E-step.** Compute

$$\begin{aligned} Q(\boldsymbol{\theta}|\boldsymbol{\theta}^{(t)}) &= E_t[\log f(\mathbf{y}_{\text{mis}}, \boldsymbol{\theta}) \mid \mathbf{y}_{\text{obs}}] \\ &\equiv \int \log f(\mathbf{y}_{\text{mis}}, \boldsymbol{\theta}) f(\mathbf{y}_{\text{mis}} \mid \boldsymbol{\theta}^{(t)}) d\mathbf{y}_{\text{mis}}, \end{aligned}$$

where $f(\mathbf{y}_{\text{mis}} \mid \boldsymbol{\theta}) = f(\mathbf{y}_{\text{mis}}, \boldsymbol{\theta})/F(\boldsymbol{\theta})$ can be seen as the conditional distribution of $\mathbf{y}_{\text{mis}}$.

- **M-step.** Find $\boldsymbol{\theta}^{(t+1)}$ to maximize $Q(\boldsymbol{\theta}|\boldsymbol{\theta}^{(t)})$.

The E-step is derived from an "imputation" heuristic. Since we assume that the log-likelihood function is easy to compute once the missing data $\mathbf{y}_{\text{mis}}$ are given, it is appealing to simply "fill in" a set of missing data and conduct a complete-data analysis. However, the simple "fill-in" idea is incorrect because it underestimates the variability caused by the missing information. The correct approach is to average over all the missing data. In general, the E-step considers all possible ways of filling in the missing data, computes the corresponding complete-data log-likelihood function, and then obtains $Q(\boldsymbol{\theta}|\boldsymbol{\theta}^{(t)})$ by averaging these functions according to the current "predictive density" of the missing data. The M-step then finds the maximum of the Q-function.

It is shown (Dempster et al. 1977) that this iteration *always* increases the likelihood value. To see this, we note that

$$f(\mathbf{y}_{\text{mis}} \mid \boldsymbol{\theta}) = f(\mathbf{y}_{\text{mis}}, \boldsymbol{\theta})/F(\boldsymbol{\theta}),$$

which is equivalent to

$$\log F(\boldsymbol{\theta}) = \log f(\mathbf{y}_{\text{mis}}, \boldsymbol{\theta}) - \log f(\mathbf{y}_{\text{mis}} \mid \boldsymbol{\theta}).$$

A.4 The Expectation-Maximization Algorithm

Thus, taking expectation with respect to $f(\mathbf{y}_{\text{mis}} \mid \boldsymbol{\theta}^{(t)})$ for both sides, we have

$$\log F(\boldsymbol{\theta}) = E_t[\log f(\mathbf{y}_{\text{mis}}, \boldsymbol{\theta})] - E_t[\log f(\mathbf{y}_{\text{mis}} \mid \boldsymbol{\theta})]$$
$$\equiv Q(\boldsymbol{\theta}|\boldsymbol{\theta}^{(t)}) - S(\boldsymbol{\theta}|\boldsymbol{\theta}^{(t)}). \tag{A.7}$$

If we replace $\boldsymbol{\theta}$ by $\boldsymbol{\theta}^{(t)}$ in (A.7), we have that

$$\log F(\boldsymbol{\theta}^{(t)}) = Q(\boldsymbol{\theta}^{(t)}|\boldsymbol{\theta}^{(t)}) - S(\boldsymbol{\theta}^{(t)}|\boldsymbol{\theta}^{(t)}),$$

and if replaced by $\boldsymbol{\theta}^{(t+1)}$, we have that

$$\log F(\boldsymbol{\theta}^{(t+1)}) = Q(\boldsymbol{\theta}^{(t+1)}|\boldsymbol{\theta}^{(t)}) - S(\boldsymbol{\theta}^{(t+1)}|\boldsymbol{\theta}^{(t)}).$$

By the definition of the EM iteration, we note that

$$Q(\boldsymbol{\theta}^{(t+1)}|\boldsymbol{\theta}^{(t)}) \geq Q(\boldsymbol{\theta}^{(t)}|\boldsymbol{\theta}^{(t)}). \tag{A.8}$$

By the Jensen's inequality, we see that for any two probability distributions $\pi_1(\mathbf{x})$ and $\pi_2(\mathbf{x})$,

$$\int \log\left[\frac{\pi_2(\mathbf{x})}{\pi_1(\mathbf{x})}\right] \pi_1(\mathbf{x}) d\mathbf{x} \leq \log \int \frac{\pi_2(\mathbf{x})}{\pi_1(\mathbf{x})} \pi_1(\mathbf{x}) d\mathbf{x} = 0.$$

Hence,

$$S(\boldsymbol{\theta}^{(t)}|\boldsymbol{\theta}^{(t)}) \geq S(\boldsymbol{\theta}^{(t+1)}|\boldsymbol{\theta}^{(t)}). \tag{A.9}$$

Putting (A.8) and (A.9) together, we have proven that $F(\boldsymbol{\theta}^{(t+1)}) \geq F(\boldsymbol{\theta}^{(t)})$. From the proof, we can see that *any* $\boldsymbol{\theta}^{(t+1)}$ that increases the Q-function will increase $F(\boldsymbol{\theta})$. See Meng and van Dyk (1997) and discussions therein for a recent overview of the methods related to the EM algorithm.

It is instructive to consider the EM algorithm for the latent-class model of Section A.3.2. The observed values are $\mathbf{y}_{\text{obs}} = (y_1, \ldots, y_n)$, where $y_i = (y_{i1}, y_{i2})$ and y_{ij} is the ith person's answer to the jth question. The missing data are $\mathbf{y}_{\text{mis}} = (z_1, \ldots, z_n)$, where z_i is the latent-class label of person i. Let $\boldsymbol{\theta} = (\theta_{0,1}, \theta_{1,1}, \theta_{0,2}, \theta_{1,2}, \gamma)$, where γ is the frequency of $z_i = 1$ in the population and $\theta_{k,l}$ is the probability of a type-k person saying "yes" to the lth question. Then, the complete-data likelihood is

$$f(\mathbf{y}_{\text{mis}}, \boldsymbol{\theta}) = p(\mathbf{y}_{\text{obs}} \mid \mathbf{y}_{\text{mis}}, \boldsymbol{\theta}) p(\mathbf{y}_{\text{mis}} \mid \boldsymbol{\theta})$$
$$= \prod_{i=1}^{n} \left[\prod_{k=1}^{2} \left\{ \theta_{z_i,k}^{y_{ik}} (1 - \theta_{z_i,k})^{1-y_{ik}} \right\} \gamma^{z_i} (1-\gamma)^{1-z_i} \right].$$

The E-step requires us to average over all label imputations. Thus,

$$Q(\boldsymbol{\theta}|\boldsymbol{\theta}^{(t)}) = E_t\left[\sum_{i=1}^{n}\sum_{k=1}^{2} \{y_{ik}\log\theta_{z_i,k} + (1-y_{ik})\log(1-\theta_{z_i,k})\}\right]$$
$$+ E_t\left[\sum_{i=1}^{n} \{z_i\log\gamma + (1-z_i)\log(1-\gamma)\}\right],$$

where the expectation is taken to average out all the z_i according to their "current" predictive probability distribution

$$\tau_i \equiv p(z_i = 1 \mid \mathbf{y}_{\text{obs}}, \boldsymbol{\theta}^{(t)}) = \frac{\gamma^{(t)} \theta_{1y_i}^{(t)}}{\gamma^{(t)} \theta_{1y_i}^{(t)} + (1-\gamma^{(t)})\theta_{0y_i}^{(t)}}.$$

Hence, we simply "fill in" a probabilistic label for each person in the E-step, which gives us

$$Q(\boldsymbol{\theta}|\boldsymbol{\theta}^{(t)}) = \sum_{m=0}^{1}\sum_{k=1}^{2}\sum_{i:\, y_{ik}=1} \tau_i^m(1-\tau_i)^{1-m}\log\theta_{m,k}$$

$$+ \sum_{i:\, y_{ik}=0} \tau_i^m(1-\tau_i)^{1-m}\log(1-\theta_{m,k})$$

$$+ \left(\sum_{i=1}^{n}\tau_i\right)\log\gamma + \left(\sum_{i=1}^{n}(1-\tau_i)\right)\log(1-\gamma).$$

Although the above expression looks overwhelming, the computation is actually quite simple. The M-step simply updates the parameters as $\gamma^{(t+1)} = \sum_{i=1}^{n} \tau_i/n$ and

$$\theta_{m,k}^{(t+1)} = \frac{\sum_{i:\, y_{ik}=1} \tau_i^m(1-\tau_i)^{1-m}}{\sum_{i:\, y_{ik}=1} \tau_i^m(1-\tau_i)^{1-m} + \sum_{i:\, y_{ik}=0} \tau_i^m(1-\tau_i)^{1-m}}.$$

The EM algorithm can only guarantee to converge to a local mode, $\hat{\boldsymbol{\theta}}$, of the observed data likelihood $F(\boldsymbol{\theta})$. Since the algorithm is deterministic, there is no principled way of getting out of such a local mode. Furthermore, although one can use a similar iterative method to compute the *observed Fisher information*, which is defined as the curvature of $\log f(\hat{\boldsymbol{\theta}})$, it is still not as desirable as having a full posterior distribution on $\boldsymbol{\theta}$.

Alternatively, one can take the Bayesian approach via the data augmentation algorithm (Section 6.4), which is conceptually much simpler. More precisely, with a prior distribution $p_0(\boldsymbol{\theta})$, one can iterate the following two Monte Carlo sampling steps:

- Draw $\mathbf{y}_{\text{mis}}^{(t+1)}$ from $p(\mathbf{y}_{\text{mis}} \mid \boldsymbol{\theta}^{(t)}, \mathbf{y}_{\text{obs}})$;
- Draw $\boldsymbol{\theta}^{(t+1)}$ from $p(\boldsymbol{\theta} \mid \mathbf{y}_{\text{mis}}^{(t+1)}, \mathbf{y}_{\text{obs}})$.

Both steps are straightforward to implement provided that the prior distribution $p_0(\ )$ is a standard one (e.g., a product of Beta distributions). Otherwise, means such as the rejection method or a Metropolis-Hastings step may be needed.

References

Aarts, E. H. L. and Korst, J. (1989). *Simulated Annealing and Boltzmann Machines: A Stochastic Approach to Combinatorial Optimization and Neural Computing*, Wiley, Chichester.

Ackley, D. H., Hinton, G. R. and Sejnowski, T. J. (1985). A learning algorithm for Boltzmann machines, *Cognitive Science* **9**(1): 147–169.

Albert, J. H. and Chib, S. (1993). Bayesian analysis of binary and polychotomous response data, *Journal of the American Statistical Association* **88**: 669–679.

Alberts, B., Bray, D., Lewis, J., Raff, M., Roberts, K. and Watson, J. D. (1994). *Molecular Biology of the Cell*, 3rd edn, Garland Publishing, New York.

Alder, B. and Wainwright, T. (1959). Studies in molecular dynamics I. General method, *Journal of Chemical Physics* **31**(2): 459–466.

Aldous, D. J. and Diaconis, P. (1987). Strong uniform times and finite random walks, *Advances in Applied Mathematics* **8**: 69–97.

Anderson, B. D. O. and Moore, J. B. (1979). *Optimal Filtering*, Prentice-Hall, Englewood Cliffs, NJ.

Antoniak, C. E. (1974). Mixtures of Dirichlet processes with applications to Bayesian nonparametric problems, *Annals of Statistics* **2**(6): 1152–1174.

Arnold, V. I. (1989). *Mathematical Methods of Classical Mechanics*, 2nd edn, Springer-Verlag, New York.

Asmussen, S. (1987). *Applied Probability and Queues*, Wiley, Chichester.

Athreya, K. B. and Ney, P. (1978). New approach to the limit theory of recurrent Markov chains, *Transactions of the American Mathematical Society* **245**: 493–501.

Avitzour, D. (1995). Stochastic simulation Bayesian approach to multi-target tracking, *IEE Proceedings-Radar Sonar and Navigation* **142**(2): 41–44.

Bar-Shalom, Y. and Fortmann, T. E. (1988). *Tracking and Data Association*, Academic Press, Boston.

Barker, A. A. (1965). Monte Carlo calculations of radial distribution functions for a proton-electron plasma, *Australian Journal of Physics* **18**(2): 119–133.

Bastolla, U., Frauenkron, H., Gerstner, E., Grassberger, P. and Nadler, W. (1998). Testing a new Monte Carlo algorithm for protein folding, *Proteins-Structure Function and Genetics* **32**(1): 52–66.

Batoulis, J. and Kremer, K. (1988). Statistical properties of biased sampling methods for long polymer chains, *Journal of Physics A (Mathematical and General)* **21**(1): 127–46.

Beckett, L. and Diaconis, P. (1994). Spectral analysis for discrete longitudinal data, *Advances in Mathematics* **103**(1): 107–128.

Beichl, I. and Sullivan, F. (1999). Approximating the permanent via importance sampling with application to the dimer covering problem, *Journal of Computational Physics* **149**(1): 128–147.

Bellman, R. E. (1957). *Dynamic Programming. Rand Corporation Research Study*, Princeton University Press, Princeton.

Berg, B. A. and Neuhaus, T. (1991). Multicanonical algorithms for 1st order phase-transitions, *Physics Letters B* **267**(2): 249–253.

Berman, A. and Plemmons, R. J. (1994). *Nonnegative Matrices in the Mathematical Sciences*, Classics in applied mathematics 9, Society for Industrial and Applied Mathematics, Philadelphia.

Berzuini, C., Best, N. G., Gilks, W. R. and Larizza, C. (1997). Dynamic conditional independence models and Markov chain Monte Carlo methods, *Journal of the American Statistical Association* **92**: 1403–1412.

Besag, J. and Green, P. J. (1993). Spatial statistics and Bayesian computation, *Journal of the Royal Statistical Society, Series B* **55**(1): 25–37.

Besag, J., Green, P. J., Higdon, D. and Mengersen, K. (1995). Bayesian computation and stochastic systems (with discussion), *Statistical Science* **10**(1): 3–41.

Bickel, P. J. and Doksum, K. A. (2000). *Mathematical Statistics: Basic Ideas and Selected Topics*, Vol. I, 2nd edn, Prentice Hall, Englewood Cliffs, NJ.

Box, G. E. P. and Tiao, G. C. (1973). *Bayesian Inference in Statistical Analysis*, Addison-Wesley Publishing Company, Reading, Massachusetts.

Breiman, L. and Friedman, J. H. (1985). Estimating optimal transformations for multiple regression and correlation (with discussion), *Journal of the Americal Statistical Association* **80**: 580–619.

Buntine, W. L. and Weigend, A. S. (1991). Bayesian back-propagation, *Complex Systems* **5**: 603–643.

Campbell, M. K. (1999). *Biochemistry*, 3rd edn, Saunders College Pub., Philadelphia.

Cannings, C., Thompson, E. A. and Skolnick, M. H. (1978). Probability functions on complex pedigrees, *Advances in Applied Probability* **10**(1): 26–61.

Casella, G. and George, E. I. (1992). Explaining the Gibbs sampler, *American Statistician* **46**(3): 167–174.

Casella, G. and Robert, C. P. (1996). Rao-Blackwellisation of sampling schemes, *Biometrika* **83**(1): 81–94.

Ceperley, D. M. (1995). Path integrals in the theory of condensed helium, *Reviews of Modern Physics* **67**(2): 279–355.

Chakraborty, A., Chen, Y., Diaconis, P., Holmes, S. and Liu, J. S. (2001). Counting 0-1 tables and related problems, *Technical report*, Department of Statistics, Stanford University.

Chen, L. Y., Qin, Z. and Liu, J. S. (2001). Exploring hybrid Monte Carlo in Bayesian computation, *Bayesian Methods with Applications to Science, Policy and Official Statistics* pp. 71–80.

Chen, M. H. and Schmeiser, B. W. (1993). Performance of the Gibbs, hit-and-run, and Metropolis samplers, *Journal of Computational and Graphical Statistics* **2**: 251–272.

Chen, M. H. and Shao, Q. M. (1997). On Monte Carlo methods for estimating ratios of normalizing constants, *Annals of Statistics* **25**(4): 1563–1594.

Chen, M. H., Shao, Q. M. and Ibrahim, J. G. (2000). *Monte Carlo Methods in Bayesian Computation*, Springer Series in Statistics, Springer-Verlag, New York.

Chen, R. and Liu, J. S. (1996). Predictive updating methods with application to Bayesian classification, *Journal of the Royal Statistical Society, Series B* **58**(2): 397–415.

Chen, R. and Liu, J. S. (2000a). Mixture Kalman filters, *Journal of the Royal Statistical Society, Series B* **62**: 493–508.

Chen, R., Wang, X. D. and Liu, J. S. (2000). Adaptive joint detection and decoding in flat-fading channels via mixture Kalman filtering, *IEEE Transactions on Information Theory* **46**(6): 2079–2094.

Chen, Y. (2001). *Sequential Importance Sampling and Its Applications*, Ph.d., Stanford University.

Chen, Y. and Liu, J. S. (2000b). Discussion of the paper by Stephens and Donnelly, *Journal of the Royal Statistical Society, Series B* **62**: 644–645.

Chen, Y. and Liu, J. S. (2001). Approximating permanents with sequential importance sampling, *Technical report*, Department of Statistics, Harvard University.

Chib, S., Nardari, F. and Shephard, N. (2002). Markov chain Monte Carlo methods for stochastic volatility models, *Journal of Econometrics* **108**(2): 281–316.

Chung, K. L. (1974). *A Course in Probability Theory*, Academic Press, New York.

Cong, J., Kong, T., Xu, D., Liang, F., Liu, J. S. and Wong, W. H. (1999). Simulated tempering for VLSI floorplan designs, *Asia and South Pacific Design Automation Conference*, Tokyo, pp. 13–16.

Cowles, M. K. and Carlin, B. P. (1996). Markov chain Monte Carlo convergence diagnostics: A comparative review, *Journal of the American Statistical Association* **91**(434): 883–904.

Cox, D. R. and Hinkley, D. V. (1974). *Theoretical Statistics*, Chapman & Hall, New York.

Creighton, T. E. (1993). *Proteins: Structures and Molecular Properties*, 2nd edn, W.H. Freeman, New York.

Csáki, P. and Fischer, J. H. (1960). Contributions to the problem of maximal correlation, *Matematikao Kotato Intezet, Kozlemenyei* **5**: 325–337.

Cybenko, G. (1989). Approximations by superpositions of a signodial function, *Mathematics of Control, Signals, and Systems* **2**: 303–314.

Damien, P., Wakefield, J. and Walker, S. (1999). Gibbs sampling for Bayesian non-conjugate and hierarchical models by using auxiliary variables, *Journal of the Royal Statistical Society, Series B* **61**(pt.2)): 331–344.

Dempster, A. P., Laird, N. M. and Rubin, D. B. (1977). Maximum likelihood from incomplete data via EM algorithm, *Journal of the Royal Statistical Society, Series B* **39**(1): 1–38.

Devroye, L. (1986). *Non-Uniform Random Variate Generation*, Springer-Verlag, New York.

Diaconis, P. (1988). *Group Representations in Probability and Statistics*, Institute of Mathematical Statistics, Hayward, California.

Diaconis, P. and Stroock, D. (1991). Geometric bounds for eigenvalues for Markov chains, *Annals of Applied Probability* **1**: 36–61.

Diaconis, P., Graham, R. and Holmes, S. P. (2001). Statistical problems involving permutations with restricted positions, *Technical report*, Department of Statistics, Stanford University.

Dörrie, H. (1965). *100 Great Problems of Elementary Mathematics; Their History and Solution*, Dover Publications, New York,.

Doss, H. (1994). Bayesian nonparametric estimation for incomplete data via successive substitution sampling, *Annals of Statistics* **22**(4): 1763–1786.

Doucet, A., Godsill, S. J. and Andrieu, C. (2000). On sequential Monte Carlo sampling methods for Bayesian filtering, *Statistics and Computing* **10**(3): 197–208.

Duane, S., Kennedy, A. D., Pendleton, B. J. and Roweth, D. (1987). Hybrid Monte Carlo, *Physics Letters B* **195**(2): 216–222.

Durbin, R. L., Eddy, S. R., Krogh, A. and Mitchison, G. (1998). *Biological Sequence Analysis: Probabilistic Models of Proteins and Nucleic Acids*, Cambridge University Press, Cambridge, UK.

Edwards, R. G. and Sokal, A. D. (1988). Generalization of the Fortuin-Kasteleyn-Swendsen-Wang representation and Monte Carlo algorithm, *Physical Review D* **38**(6): 2009–12.

Efron, B. (1979). Bootstrap methods: Another look at the jackknife, *Annals of Statistics* **7**(1): 1–26.

Efron, B. and Morris, C. (1975). Data analysis using Stein's estimator and its generalizations, *Journal of the American Statistical Association* **70**: 311–319.

Efron, B. and Petrosian, V. (1999). Nonparametric methods for doubly truncated data, *Journal of the American Statistical Association* **94**(447): 824–834.

Elerian, O., Chib, S. and Shephard, N. (2001). Likelihood inference for discretely observed nonlinear diffusions, *Econometrika* **69**(4): 959–993.

Escobar, M. D. (1994). Estimating Normal means with a Dirichlet process prior, *Journal of the American Statistical Association* **89**: 268–277.

Ethier, S. N. and Kurtz, T. G. (1986). *Markov Processes: Characterization and Convergence*, Wiley series in probability and mathematical statistics, Wiley, New York.

Ferguson, T. S. (1974). Prior distributions on spaces of probability measures, *Annals of Statistics* **2**(4): 615–629.

Fill, J. A. (1991). Eigenvalue bounds on convergence to stationarity for nonreversible Markov chains, with application to the exclusion process., *Annals of Applied Probability* **1**: 62–87.

Fill, J. A. (1998). An interruptible algorithm for perfect sampling via Markov chains, *Annals of Applied Probability* **8**: 131–162.

Frenkel, D. and Smit, B. (1996). *Understanding Molecular Simulation: From Algorithms to Applications*, Academic Press, San Diego.

Frigessi, A., Distefano, P., Hwang, C. R. and Sheu, S. J. (1993). Convergence rates of the Gibbs sampler; the Metropolis algorithm and other single-site updating dynamics, *Journal of the Royal Statistical Society, Series B* **55**(1): 205–219.

Gelb, A. (1974). *Applied Optimal Estimation*, M.I.T. Press, Cambridge, Mass.

Gelfand, A. E. and Kuo, L. (1991). Nonparametric Bayesian bioassay including ordered polytomous response, *Biometrika* **78**(3): 657–666.

Gelfand, A. E. and Smith, A. F. M. (1990). Sampling-based approaches to calculating marginal densities, *Journal of the American Statistical Association* **85**: 398–409.

Gelfand, A. E., Sahu, S. K. and Carlin, B. P. (1995). Efficient parametrizations for Normal linear mixed models, *Biometrika* **82**(3): 479–488.

Gelman, A. and Meng, X. L. (1998). Simulating normalizing constants: From importance sampling to bridge sampling to path sampling, *Statistical Science* **13**(2): 163–185.

Gelman, A. and Rubin, D. B. (1992). Inference from iterative simulation using multiple sequences (with discussion), *Statistical Science* **7**: 457–472.

Gelman, A., Carlin, J. B., Stern, H. S. and Rubin, D. B. (1995). *Bayesian Data Analysis*, reprinted 1997. edn, Chapman & Hall, London.

Gelman, A., Roberts, G. O. and Gilks, W. R. (1995). Efficient Metropolis jumping rules, in J. Bernardo, J. Berger, A. Dawid and A. Smith (eds), *Bayesian Statistics*, Vol. 5, Oxford University Press, Oxford.

Geman, S. and Geman, D. (1984). Stochastic relaxation, Gibbs distributions and the Bayesian restoration of images, *IEEE Transactions on Pattern Analysis and Machine Intelligence* **6**: 721–741.

George, E. I. and McCulloch, R. E. (1997). Approaches for Bayesian variable selection, *Statistica Sinica* **7**(2): 339–373.

Geyer, C. (1992). Practical Monte Carlo Markov chain (with discussion), *Statistical Science* **7**: 473–511.

Geyer, C. and Thompson, E. (1995). Annealing Markov chain Monte Carlo with applications to ancestral inference, *Journal of the American Statistical Association* **90**: 909–920.

Geyer, C. J. (1991). Markov chain Monte Carlo maximum likelihood, in E. Keramigas (ed.), *Computing Science and Statistics: The 23rd symposium on the interface*, Interface Foundation, Fairfax, pp. 156–163.

Gilks, W. R., Richardson, S. and Spiegelhalter, D. J. (1998). *Markov Chain Monte Carlo in Practice*, Chapman & Hall, London.

Gilks, W. R., Roberts, G. O. and George, E. I. (1994). Adaptive direction sampling, *Statistician* **43**(1): 179–189.

Gilks, W. R., Roberts, G. O. and Sahu, S. K. (1998). Adaptive Markov chain Monte Carlo through regeneration, *Journal of the American Statistical Association* **93**(443): 1045–1054.

Goodman, J. and Sokal, A. (1989). Multigrid Monte Carlo method: Conceptual foundations., *Physical Review D* **40**(6): 2035–2071.

Gordon, N. J., Salmond, D. J. and Ewing, C. (1995). Bayesian state estimation for tracking and guidance using the bootstrap filter, *Journal of Guidance Control and Dynamics* **18**(6): 1434–1443.

Gordon, N. J., Salmond, D. J. and Smith, A. F. M. (1993). Novel approach to nonlinear non-Gaussian Bayesian state estimation., *IEE Proceedings on Radar and Signal Processing* **140**: 107–113.

Gouriérourx, C. and Monfort, A. (1997). *Simulation Based Econometric Methods*, Oxford University Press, Oxford.

Grassberger, P. (1997). Pruned-enriched Rosenbluth method: Simulations of θ polymers of chain length up to 1,000,000, *Physical Review E* **56**: 3682–3693.

Green, P. J. (1995). Reversible jump Markov chain Monte Carlo computation and Bayesian model determination, *Biometrika* **82**(4): 711–732.

Green, P. J. and Murdoch, D. J. (1998). Exact sampling for Bayesian inference: Towards general purpose algorithms, *in* J. Bernardo, J. Berger, A. Dawid and A. Smith (eds), *Bayesian Statistics*, Vol. 6, Oxford University Press, Oxford, pp. 301–321.

Green, P. J. and Silverman, B. W. (1994). *Nonparametric Regression and Generalized Linear Models: A Roughness Penalty Approach*, Vol. 58 of *Monographs on statistics and applied probability*, Chapman & Hall, London.

Grenander, U. and Miller, M. I. (1994). Representations of knowledge in complex systems, *Journal of the Royal Statistical Society, Series B* **56**(4): 549–603.

Griffiths, R. C. and Tavare, S. (1994). Simulating probability distributions in the coalescent, *Theoretical Population Biology* **46**(2): 131–159.

Gustafson, P. (1998). A guided walk Metropolis algorithm, *Statistics and Computing* **8**(4): 357–364.

Hammersley, J. M. and Handscomb, D. C. (1964). *Monte Carlo Methods*, Methuen's monographs on applied probability and statistics, Methuen; Wiley, London.

Hammersley, J. M. and Morton, K. W. (1954). Poor man's Monte Carlo., *Journal of the Royal Statistical Society , Series B* **16**(1): 23–38.

Hammersley, J. M. and Morton, K. W. (1956). A new Monte Carlo technique: Antithetic variates, *Proceedings of the Cambridge Philosophical Society* **52**: 449–475.

Hansmann, U. H. E. and Okamoto, Y. (1997). Numerical comparisons of three recently proposed algorithms in the protein folding problem, *Journal of Computational Chemistry* **18**(7): 920–933.

Harvey, A. C. (1990). *The Econometric Analysis of Time Series*, 2nd edn, MIT Press, Cambridge.

Hastings, W. K. (1970). Monte Carlo sampling methods using Markov chains and their applications, *Biometrika* **57**(1): 97–109.

Hesselbo, B. and Stinchcombe, R. B. (1995). Monte Carlo simulation and global optimization without parameters, *Physics Review Letters* **74**(12): 2151–2155.

Higdon, D. M. (1998). Auxiliary variable methods for Markov chain Monte Carlo with applications, *Journal of the American Statistical Association* **93**: 585–595.

Hockney, R. W. (1970). The potential calculation and some applications, *Methods in Computational Physics* **9**: 136–211.

Holley, R. A. and Stroock, D. (1988). Simulated annealing via Sobolev inequalities, *Communications in Mathematical Physics* **115**(4): 553–569.

Holley, R. A., Kusuoka, S. and Stroock, D. (1989). Asymptotics of the spectral gap with applications to the theory of simulated annealing, *Journal of Functional Analysis* **83**(2): 333–347.

Hopfield, J. J. (1982). Neural networks and physical systems with emergent collective computational abilities, *Proceedings of the National Academy of Sciences of USA* **79**(8): 2554–2558.

Horn, R. A. and Johnson, C. R. (1985). *Matrix Analysis*, Cambridge University Press, New York.

Hornik, K., Stinchcombe, M. and White, H. (1989). Multilayer feed-forward networks are universal approximators, *Neural Networks* **2**(5): 359–366.

Hukushima, K. and Nemoto, K. (1996). Exchange Monte Carlo method and application to spin glass simulations, *Journal of the Physical Society of Japan* **65**(4): 1604–1608.

Hull, J. and White, A. (1987). The pricing of options on assets with stochastic volatility, *Journal of Finance* **42**: 281–300.

Hurzeler, M. and Kunsch, H. R. (1998). Monte Carlo approximations for general state-space models, *Journal of Computational and Graphical Statistics* **7**(2): 175–193.

Jazwinski, A. H. (1970). *Stochastic Processes and Filtering Theory*, Mathematics in science and engineering v. 64, Academic Press, New York,.

Jerrum, M. and Sinclair, A. (1989). Approximating the permanent, *SIAM Journal On Computing* **18**(6): 1149–1178.

Johnson, N. L. and Kotz, S. (1972). *Distributions in Statistics: Continuous Multivariate Distributions*, Wiley, New York,.

Kalman, R. (1960). A new approach to linear filtering and prediction problems, *Journal of Basic Engineering* **82**: 35–45.

Karlin, S. and Taylor, H. M. (1998). *An Introduction to Stochastic Modeling*, 3rd edn, Academic Press, Orlando.

Karplus, M. and Petsko, G. A. (1990). Molecular dynamics simulations in biology, *Nature* **347**: 631–639.

Kass, R. E. and Raftery, A. E. (1995). Bayes factors, *Journal of the American Statistical Association* **90**(430): 773–795.

Kesten, H. (1974). Renewal theory for functionals of a Markov chain with general state space, *The Annals of Probability* **2**(3): 355–387.

Kirkpatrick, S., Gelatt, C. D. and Vecchi, M. P. (1983). Optimization by simulated annealing, *Science* **220**: 671–680.

Kitagawa, G. (1996). Monte Carlo filter and smoother for non-Gaussian nonlinear state space models, *Journal of Computational and Graphical Statistics* **5**: 1–25.

Knuth, D. E. (1997). *The Art of Computer Programming*, 3rd edn, Addison-Wesley, Reading, Mass.

Kong, A., Cox, N., Frigge, M. and Irwin, M. (1993). Sequential imputation and multipoint linkage analysis, *Genetic Epidemiology* **10**(6): 483–488.

Kong, A., Liu, J. S. and Wong, W. H. (1994). Sequential imputations and Bayesian missing data problems, *Journal of the American Statistical Association* **89**(425): 278–288.

Kremer, K. and Binder, K. (1988). Monte Carlo simulation of lattice models for macromolecules, *Computer Physics Reports* **7**(6): 259–310.

Krogh, A., Brown, M., Mian, I. S., Sjolander, K. and Haussler, D. (1994). Hidden Markov models in computational biology: Applications to protein modeling, *Journal of Molecular Biology* **235**(5): 1501–1531.

Kuznetsov, N. Y. (1996). Computing the permanent by importance sampling method, *Cybernetics and Systems Analysis* **32**(6): 749–755.

Lancaster, H. O. (1958). The structure of bivariate distributions, *Annals of Mathematical Statistics* **29**(3): 719–736.

Lang, K. J. and Witbrock, M. J. (1988). Learning to tell two spirals apart, *Connectionist Models Summer School*, pp. 52–59.

Lauritzen, S. L. and Spiegelhalter, D. J. (1988). Local computations with probabilities on graphical structures and their application to expert systems, *Journal of the Royal Statistical Society, Series B* **50**(2): 157–224.

Lawrence, C. E., Altschul, S. F., Boguski, M. S., Liu, J. S., Neuwald, A. F. and Wootton, J. C. (1993). Detecting subtle sequence signals: a Gibbs sampling strategy for multiple alignment, *Science* **262**(5131): 208–214.

Lawrence, C. E. and Reilly, A. A. (1990). An expectation-maximization (EM) algorithm for the idenification and characterization of common sites in unaligned biopolymer sequences, *Proteins* **7**: 41–51.

Leach, A. R. (1996). *Molecular Modelling: Principles and Applications*, Longman, Harlow, England.

Li, K.-H. (1988). Imputation using Markov chains, *Journal of Statistical Computation and Simulation* **30**: 57–79.

Li, X. J. and Sokal, A. D. (1989). Rigorous lower bound on the dynamic critical exponents of the Swendsen-Wang algorithm, *Physical Review Letters* **63**(8): 827–30.

Liang, F. (1997). *Weighted Markov Chain Monte Carlo and Optimization*, Ph.D. Thesis, The Chinese University of Hong Kong.

Liang, F. and Wong, W. H. (1999). Dynamic weighting in simulations of spin systems, *Physics Letters A* **252**(5): 257–262.

Liang, F. and Wong, W. H. (2000). Evolutionary Monte Carlo: Applications to c_p model sampling and change point problem, *Statistica Sinica* **10**(2): 317–342.

Liang, F. and Wong, W. H. (2001). Real parameter evolutionary Monte Carlo and Bayesian neural network forecasting, *Journal of the American Statistical Association* **96**(454): 653–666.

Liang, F., Truong, Y. K. and Wong, W. H. (2001). Automatic Bayesian model averaging for linear regression and applications in Bayesian curve fitting, *Statistica Sinica* **11**: 1005–1029.

Little, R. J. A. and Rubin, D. B. (1987). *Statistical Analysis with Missing Data*, Wiley series in probability and mathematical statistics. Applied probability and statistics, Wiley, New York.

Liu, C. H., Rubin, D. B. and Wu, Y. N. (1998). Parameter expansion to accelerate EM: The PX-EM algorithm, *Biometrika* **85**(4): 755–770.

Liu, J. and West, M. (2000). Combined parameter and state estimation in simulation-based filtering, *in* A. Doucet, J. F. G. de Freitas and N. J. Gordon (eds), *Sequential Monte Carlo in Practice*, Springer-Verlag, New York.

Liu, J. S. (1991). *Correlation Structure and Convergence Rate of the Gibbs Sampler*, Ph.d. thesis, The University of Chicago.

Liu, J. S. (1994a). The collapsed Gibbs sampler in Bayesian computations with applications to a gene regulation problem, *Journal of the American Statistical Association* **89**(427): 958–966.

Liu, J. S. (1994b). Fration of missing information and convergence rate of data augmentation, *in* J. Small and A. Lehman (eds), *Computationally Intensive Statistical Methods: Proceedings of the 26th symposium on the Interface*, Vol. 26 of *Computing Science and Statistics*, Interface Foundation of North America, North Carolina, pp. 490–497.

Liu, J. S. (1996a). Metropolized independent sampling with comparisons to rejection sampling and importance sampling, *Statistics and Computing* **6**(2): 113–119.

Liu, J. S. (1996b). Nonparametric hierarchical Bayes via sequential imputations, *Annals of Statistics* **24**(3): 911–930.

Liu, J. S. (1996c). Peskun's theorem and a modified discrete-state Gibbs sampler, *Biometrika* **83**(3): 681–682.

Liu, J. S. and Chen, R. (1995). Blind deconvolution via sequential imputations, *Journal of the American Statistical Association* **90**(430): 567–576.

Liu, J. S. and Chen, R. (1998). Sequential Monte Carlo methods for dynamic systems, *Journal of the American Statistical Association* **93**(443): 1032–1044.

Liu, J. S. and Lawrence, C. E. (1999). Bayesian inference on biopolymer models, *Bioinformatics* **15**(1): 38–52.

Liu, J. S. and Sabatti, C. (1998). Simulated sintering: Markov chain Monte Carlo with spaces of varying dimensions (with discussion), *in* J. Bernardo, J. Berger, A. Dawid and A. Smith (eds), *Bayesian Statistics*, Vol. 6, Oxford University Press, New York, pp. 386–413.

Liu, J. S. and Sabatti, C. (2000). Generalised Gibbs sampler and multigrid Monte Carlo for Bayesian computation, *Biometrika* **87**(2): 353–369.

Liu, J. S. and Wu, Y. N. (1999). Parameter expansion for data augmentation, *Journal of the American Statistical Association* **94**: 1264–1274.

Liu, J. S., Chen, R. and Logvinenko, T. (2001). A theoretical framework for sequential importance sampling and resampling, *in* A. Doucet, J. F. G. de Freitas and N. J. Gordon (eds), *Sequential Monte Carlo in Practice*, Springer-Verlag, New York.

Liu, J. S., Chen, R. and Wong, W. H. (1998). Rejection control and sequential importance sampling, *Journal of the American Statistical Association* **93**(443): 1022–1031.

Liu, J. S., Liang, F. and Wong, W. H. (2000). The use of multiple-try method and local optimization in Metropolis sampling, *Journal of the American Statistical Association* **95**: 121–134.

Liu, J. S., Liang, F. and Wong, W. H. (2001). A theory for dynamic weighting in Monte Carlo computation, *Journal of the American Statistical Association* **96**(454): 561–573.

Liu, J. S., Neuwald, A. F. and Lawrence, C. E. (1995). Bayesian models for multiple local sequence alignment and Gibbs sampling strategies, *Journal of the American Statistical Association* **90**(432): 1156–1170.

Liu, J. S., Wong, W. H. and Kong, A. (1994). Covariance structure of the Gibbs sampler with applications to the comparisons of estimators and augmentation schemes, *Biometrika* **81**(1): 27–40.

Liu, J. S., Wong, W. H. and Kong, A. (1995). Covariance structure and convergence rate of the Gibbs sampler with various scans, *Journal of the Royal Statistical Society, Series B* **57**(1): 157–169.

Liu, X., Brutlag, D. L. and Liu, J. S. (2001). BioProspector: Discovering conserved DNA motifs in upstream regulatory regions of co-expressed genes, *Procceedings of the Pacific Symposium on Bioinformatics* **6**: 127–138.

Lyklema, J. W. and Kremer, K. (1986). Monte Carlo series analysis of irreversible self-avoiding walks 2. the growing self-avoiding walk, *Journal of Physics A-Mathematical and General* **19**(2): 279–289.

Maceachern, S. N. (1994). Estimating Normal means with a conjugate style Dirichlet process prior, *Communications in Statistics-Simulation and Computation* **23**(3): 727–741.

MacEachern, S. N., Clyde, M. and Liu, J. S. (1999). Sequential importance sampling for nonparametric Bayes models: The next generation, *Canadian Journal of Statistics* **27**(2): 251–267.

Mallows, C. L. (1973). Some comments on c_p, *Technometrics* **15**: 661–675.

Marinari, E. and Parisi, G. (1992). Simulated tempering: a new Monte Carlo scheme., *Europhysics Letters* **19**(6): 451–458.

Marsaglia, G. and Zaman, A. (1993). The KISS generator, *Technical report*, Department of Statistics, Florida State University.

Marshall, A. (1956). The use of multi-stage sampling schemes in Monte Carlo computations, *in* M. Meyer (ed.), *Symposium on Monte Carlo Methods*, Wiley, New York, pp. 123–140.

Maung, K. (1941). Measurement of association in a contingency table with special reference to the pigmentation of hair and eye colours of Scottish school children, *Annals of Eugenics, London* **11**: 189.

McCue, L. A., Thompson, W., Carmack, C. S., Ryan, M. P., Liu, J. S., Derbyshire, V. and Lawrence, C. E. (2001). Phylogenetic footprinting of transcription factor binding sites in proteobacterial genomes, *Nucleic Acids Research* **29**: 774–782.

McFadden, D. (1989). A method of simulated moments for estimation of discrete response models without numerical integration, *Econometrica* **57**: 995–1026.

Meirovitch, H. (1982). A new method for simulation of real chains: Scanning future steps, *Journal of Physics A-Mathematical and General* **15**(12): L735–L741.

Meirovitch, H. (1985). Scanning method as an unbiased simulation technique and its application to the study of self-attracting random-walks, *Physical Review A* **32**(6): 3699–3708.

Meng, X. L. and van Dyk, D. (1997). The EM algorithm: An old folk-song sung to a fast new tune, *Journal of the Royal Statistical Society, Series B* **59**(3): 511–540.

Meng, X. L. and van Dyk, D. (1999). Seeking efficient data augmentation schemes via conditional and marginal augmentation, *Biometrika* **86**(2): 301–320.

Meng, X. L. and Wong, W. H. (1996). Simulating ratios of normalizing constants via a simple identity: A theoretical exploration, *Statistica Sinica* **6**(4): 831–860.

Mengersen, K. L., Robert, C. P. and Cuihenneuc-Jouyaux, C. (1999). MCMC convergence diagnostics: A reviewww, *in* J. M. Bernardo, J. O. Berger, A. P. Dawid and A. F. M. Smith (eds), *Bayesian Statistics*, Vol. 6, Clarendon Press, Oxford, pp. 415–440.

Metropolis, N., Rosenbluth, A. W., Rosenbluth, M. N., Teller, A. H. and Teller, E. (1953). Equations of state calculations by fast computing machines, *Journal of Chemical Physics* **21**(6): 1087–1091.

Meyn, S. P. and Tweedie, R. L. (1994). Computable bounds for convergence rates of Markov chains, *Annals of Applied Probability* **4**: 981–1011.

Müller, P. and Insua, D. R. (1998). Issues in Bayesian analysis of neural network models, *Neural Computation* **10**(3): 749–770.

Murdoch, D. J. and Green, P. J. (1998). Exact sampling from a continuous state space, *Scandinavian Journal of Statistics* **25**(3): 483–502.

Murray, G. (1977). Comment on "Maximum likelihood from incomplete data via the EM algorithm" by A.P. Dempster, N.M. Laird, and D.B. Rubin, *Journal of the Royal Statistical Society, B.* **39**: 27–28.

Mykland, P., Tierney, L. and Yu, B. (1995). Regeneration in Markov chain samplers, *Journal of the American Statistical Association* **90**(429): 233–241.

Neal, R. M. (1993). A theoretical analysis of Monte Carlo algorithms for the simulation of Gibbs random field images: Comments, *IEEE Transactions On Information Theory* **39**(1): 310–310.

Neal, R. M. (1994). An improved acceptance procedure for the hybrid Monte Carlo algorithm, *Journal of Computational Physics* **111**: 194–203.

Neal, R. M. (1996). *Bayesian Learning for Neural Networks*, Springer-Verlag, New York.

Neuwald, A. F., Liu, J. S. and Lawrence, C. E. (1995). Gibbs motif sampling: Detection of bacterial outer-membrane protein repeats, *Protein Science* **4**(8): 1618–1632.

Newman, M. E. J. and Barkema, G. T. (1999). *Monte Carlo Methods in Statistical Physics*, Oxford University Press, Oxford.

Niedermayer, F. (1988). General cluster updating method for Monte Carlo simulations, *Physical Review Letters* **61**(18): 2026–2029.

Nienhuis, B. (1982). Exact critical-point and critical exponents of o(n) models in 2 dimensions, *Physical Review Letters* **49**(15): 1062–1065.

Nummelin, E. (1984). *General Irreducible Markov Chains and Nonnegative Operators*, Cambridge University Press, Cambridge.

Nummelin, E. and Tweedie, R. L. (1978). Geometric ergodicity and r-positivity for general Markov chains, *Annals of Probability* **6**: 404–420.

Odell, P. L. and Feiveson, A. H. (1966). A numerical procedure to generate a sample covariance matrix, *Journal of the American Statistical Association* **61**: 199–203.

Oh, M.-S. and Berger, J. O. (1992). Adaptive importance sampling in Monte Carlo integration, *Journal of Statistical Computation and Simulation* **41**: 143–168.

Oh, M.-S. and Berger, J. O. (1993). Integration of multimodal functions by Monte Carlo importance sampling, *Journal of the American Statistical Association* **88**: 450–456.

Onsager, L. (1949). Statistical hydrodynamics, *Nuovo Cimento (suppl.)* **6**: 261.

Pederson, A. R. (1995). A new approach to maximum likelihood estimation for stochastic differential equations based on discrete observations, *Scandinavian Journal of Statistics* **22**: 55–71.

Peskun, P. H. (1973). Optimum Monte Carlo sampling using Markov chains, *Biometrika* **60**(3): 607–612.

Pitman, J. (1993). *Probability*, Springer-Verlag, New York.

Pitt, M. K. and Shephard, N. (1999). Filtering via simulation: Auxiliary particle filters, *Journal of the American Statistical Association* **94**(446): 590–599.

Press, W. H. and Vetterling, W. T. (1995). *Numerical Recipes in C*, Cambridge University Press, Cambridge.

Propp, J. G. and Wilson, D. B. (1996). Exact sampling with coupled Markov chains and applications to statistical mechanics, *Random Structures and Algorithms* **9**(1-2): 223–252.

Propp, J. G. and Wilson, D. B. (1998). How to get a perfectly random sample from a generic Markov chain and generate a random spanning tree of a directed graph, *Journal of Algorithms* **27**(2): 170–217.

Qin, Z. and Liu, J. S. (2001). Multi-point Metropolis method with application to hybrid Monte Carlo, *Journal of Computational Physics* **172**: 827–840.

Rabiner, L. R. (1989). A tutorial on hidden Markov models and selected applications in speech recognition, *Proceedings of the IEEE* **77**(2): 257–286.

Rao, M. M. (1987). *Measure Theory and Integration*, Wiley, New York.

Reeds, J. (1981). Theory of riffle shuffling, *Unpublished manuscript*.

Renyi (1959). On measure of dependence, *Acta Mathematica Academiae Scientiarum Hungaricae* **10**: 441–451.

Rice, J. A. (1994). *Mathematical Statistics and Data Analysis*, 2nd edn, Duxbury Press, Belmont, California.

Ripley, B. (1996). *Pattern Recognition and Neural Networks*, Cambridge University Press, Cambridge.

Ripley, B. D. (1987). *Stochastic Simulation*, Wiley, New York.

Ritter, C. and Tanner, M. A. (1992). Facilitating the Gibbs sampler: the Gibbs stopper and the griddy-Gibbs sampler, *Journal of the American Statistical Association* **87**(419): 861–868.

Roberts, G. O. and Gilks, W. R. (1994). Convergence of adaptive direction sampling, *Journal of Multivariate Analysis* **49**(2): 287–298.

Roberts, G. O. and Rosenthal, J. S. (1998). Markov chain Monte Carlo: Some practical implications of theoretical results, *Canadian Journal of Statistics* **26**(1): 5–20.

Roberts, G. O. and Sahu, S. K. (1997). Updating schemes; correlation structure; blocking and parameterization for the Gibbs sampler, *Journal of the Royal Statistical Society, Series B* **59**(2): 291–317.

Roberts, G. O., Gelman, A. and Gilks, W. R. (1997). Weak convergence and optimal scaling of random walk Metropolis algorithms, *Annals of Applied Probability* **7**: 110–120.

Rosenbluth, M. N. and Rosenbluth, A. W. (1955). Monte Carlo calculation of the average extension of molecular chains, *Journal of Chemical Physics* **23**(2): 356–359.

Rosenthal, J. S. (1995). Minorization conditions and convergence rates for Markov chain Monte cCrlo, *Journal of the American Statistical Association* **90**(430): 558–566.

Rosenthal, J. S. (1996). Analysis of the Gibbs sampler for a model related to James-Stein estimators, *Statistics and Computing* **6**(3): 269–275.

Rubin, D. B. (1976). Inference and missing data, *Biometrika* **63**(3): 581–590.

Rubin, D. B. (1980). Using empirical Bayes techniques in the law-school validity studies, *Journal of the American Statistical Association* **75**(372): 801–816.

Rubin, D. B. (1987). A noniterative sampling/importance resampling alternative to the data augmentation algorithm for creating a few imputations when fractions of missing information are modest: the SIR algorithm, *Journal of the American Statistical Association* **52**: 543–546.

Rubinstein, R. Y. (1981). *Simulation and the Monte Carlo Method*, Wiley series in probability and mathematical statistics, Wiley, New York.

Rumelhart, D. E. and McClelland, J. (1986). *Parallel Distributed Processing: Exploitations in the Micro-Structure of Cognition*, Vol. 1 and 2, MIT Press, Cambridge.

Rumelhart, D. E., Hinton, G. E. and Williams, R. J. (1986). Learning representations by back-propagating errors, *Nature* **323**: 533–536.

Sanderson, J. G. (2000). Testing ecological patterns, *American Scientist* **88**(4): 332–339.

Sarmanov, O. V. (1958). The maximal correlation coefficient (symmetrical case), *Doklady Akademii Nauk USSR* **120**: 715–718.

Schervish, M. and Carlin, B. (1992). On the convergence of successive substitution sampling, *Journal of Computational and Graphical Statistics* **1**: 111–127.

Schmidler, S. C., Liu, J. S. and Brutlag, D. L. (2000). Bayesian segmentation of protein secondary structure, *Journal of Computational Biology* **7**(1-2): 233–248.

Shephard, N. and Pitt, M. K. (1997). Likelihood analysis of non-Gaussian measurement time series, *Biometrika* **84**(3): 653–667.

Siepmann, J. I. and Frenkel, D. (1992). Configurational bias Monte Carlo: a new sampling scheme for flexible chains, *Molecular Physics* **75**(1): 59–70.

Smith, M. C. and Winter, E. M. (1978). On the detection of target trajectories in a multi-target environment, *The 17th IEEE conference on Decision and Control*, San Diego, California.

Stephens, M. and Donnelly, P. (2000). Inference in molecular population genetics, *Journal of the Royal Statistical Society, Series B* **62**(4): 605–635.

Stormo, G. D. and Hartzell, G. W. (1989). Identifying protein binding sites from unaligned DNA fragments, *Proceedings of the Nathional Academy of Science of USA* **86**: 1183–1187.

Swendsen, R. H. and Wang, J. S. (1987). Nonuniversal critical dynamics in Monte Carlo simulations, *Physical Review Letters* **58**(2): 86–88.

Tanner, M. A. and Wong, W. H. (1987). The calculation of posterior distributions by data augmentation (with discussion), *Journal of the American Statistical Association* **82**(398): 528–540.

Thisted, R. A. (1988). *Elements of Statistical Computing*, Chapman & Hall, New York.

Tierney, L. (1994). Markov chains for exploring posterior distributions, *Annals of Statistics* **22**(4): 1701–1728.

Tierney, L. (1998). A note on Metropolis-Hastings kernels for general state spaces, *The Annals of Applied Probability* **8**(1): 1–9.

Torrie, G. M. and Valleau, J. P. (1977). Non-physical sampling distributions in Monte Carlo free energy estimation: Umbrella sampling, *Journal of Computational Physics* **23**(2): 187–199.

Unger, R. and Moult, J. (1993). Genetic algorithms for protein folding simulations, *Journal of Molecular Biology* **231**(1): 75–81.

Valleau, J. P. (1999). Thermodynamic scaling methods in Monte Carlo and their application to phase equilibria, *Advances in Chemical Physics* **105**: 369–404.

Verlet, L. (1967). Computer "experiments" on classical fluids. I. Thermodynamical properties of lennard-jones molecules, *Physical Review* **159**: 98–103.

von Neumann, J. (1951). Various techniques used in connection with random digits, *National Bureau of Standards Applied Mathematics Series* **12**: 36–38.

Wall, F. T. and Erpenbeck, J. J. (1959). New method for the statistical computation of polymer dimensions, *Journal of Chemical Physics* **30**(3): 634–37.

Wall, F. T., Rubin, R. J. and Isaacson, L. M. (1957). Improved statistical method for computing mean dimension of polymer molecules, *Journal of Chemical Physics* **27**: 186–188.

Weisberg, S. (1985). *Applied Linear Regression*, 2nd edn, Wiley, New York.

West, M. and Harrison, J. (1989). *Bayesian Forecasting and Dynamic Models*, Springer-Verlag, New York.

White, H. (1992). *Artificial Neural Networks: Approximation and Learning Theory*, Blackwell Publishers, Cambridge, MA.

Wolff, U. (1989). Collective Monte Carlo updating for spin systems, *Physical Review Letters* **62**(4): 361–364.

Wong, W. H. and Liang, F. M. (1997). Dynamic weighting in Monte Carlo and optimization, *Proceedings of the National Academy of Sciences of USA* **94**(26): 14220–14224.

Wu, Y. N., Zhu, S. C. and Liu, X. W. (1999). The equivalence of Julesz and Gibbs ensembles, *International Conference on Computer Vision*, Corfu, Greece.

Yang, C. N. (1952). The spontaneous magnetization of a 2-dimensional Ising model, *Physics Review* **85**(5): 808–815.

Zhu, S. C., Liu, X. W. and Wu, Y. N. (2000). Exploring texture ensembles by efficient Markov chain Monte Carlo: Toward a "trichromacy" theory of texture, *IEEE Transactions On Pattern Analysis and Machine Intelligence* **22**(6): 554–569.

Zhu, S. C., Wu, Y. N. and Mumford, D. (1997). Minimax entropy principle and its application to texture modeling, *Neural Computation* **9**(8): 1627–1660.

Zhu, S. C., Wu, Y. N. and Mumford, D. (1998). Filters; random fields and maximum entropy (frame): Towards a unified theory for texture modeling, *International Journal of Computer Vision* **27**(2): 107–126.

Author Index

Aarts, E., 210
Ackley, D. H., 223
Albert, J. H., 176
Alberts, B., 84
Alder, B. J., viii, 183
Aldous, D., 251
Altschul, S. F., viii
Anderson, B., 16
Andrieu, C., 72
Antoniak, C. E., 96
Asmussen, S., 247, 254
Athreya, K., 254
Avitzour, D., 65, 98, 100

Bar-Shalom, Y., 98
Barkema, G. T., 154
Barker, A. A., 112, 113, 134, 277
Bastolla, U., 80
Batoulis, J., 76
Beckett, L., 95
Beichl, I., 92
Bellman, R. E., 28
Berg, B. A., 207, 216
Berger, J. O., 42
Berman, A., 248

Besag, J., 133, 134, 280
Bickel, P., 28
Binder, K., 55, 59, 71, 74
Boguski, M. S., viii
Box, G. E. P., 40
Breiman, L., 267
Brown, M., 86
Brutlag, D. L., 13, 65
Buntine, W. L., 223, 240

Campbell, M. K., 11
Cannings, C., 31
Carlin, B. P., 130, 139, 272, 275
Carlin, J. B., 40
Casella, G., 28, 132
Ceperley, L., 117
Chakraborty, A., 92
Chen, L., 199
Chen, M. H., 8, 134
Chen, R., 54, 59, 64, 67, 71, 74, 102, 141
Chen, Y., 39, 70, 74, 81, 92
Chib, S., 176, 178, 201
Chung, K. L., 291, 300
Clyde, M., 38, 97

Cowles, M. K., 272
Cox, D. R., 305
Cox, N., 62
Creighton, T. E., 9
Csáki, P., 267
Cybenko, G., 222

Dörrie, H., vii
Damien, P., 133
Dempster, A. P., 40, 308
Devroye, L., 24
Diaconis, P., 3, 18, 90, 92, 95, 147, 246, 249, 251, 256
Doksum, K., 28
Donnelly, P., 48–51, 74, 83
Doss, H., 97
Doucet, A., 72
Duane, S., 189
Durbin, R. L., 86

Eddy, S. R., 86
Edwards, R. G., 133, 154
Efron, B., viii, 16, 138, 303
Elerian, O., 178
Erpenbeck, J. J., 59, 67, 71, 73
Escobar, M. D., 97, 150
Ethier, S., 247
Ewing, C., 98

Feiveson, A. H., 41, 95
Ferguson, T. S., 96
Fill, J. A., 147, 148, 259, 287
Fischer, J. H., 267
Fortmann, T. E., 98
Frenkel, D., viii, 4, 47, 55, 116, 118–120
Friedman, J. H., 267
Frigessi, A., 112, 280
Frigge, M., 62

Gelatt, C. D., viii, 3, 209
Gelb, A., 16
Gelfand, A., viii, 19, 97, 131, 139, 146, 162, 308
Gelman, A., 40, 62, 68, 115, 127, 135, 272

Geman, D., viii, 129, 210
Geman, S., viii, 129, 210
George, E., 132, 135, 238
Geyer, C. J., 4, 210, 212, 272
Gilks, W. R., 21, 115, 127, 135, 226, 235
Godsill, S. J., 72
Goodman, J., viii, 4, 125, 161, 162
Gordon, N. J., 4, 54, 59, 66, 71, 74, 98
Gouriéroux, C., viii
Graham, R., 18, 90
Grassberger, P., 55, 59, 67, 71, 73
Green, P. J., 122, 133, 240, 287
Grenander, U., 122
Griffiths, R. C., 38, 48–51, 81
Gustafson, P., 193

Hammersley, J. M., 4, 26, 27, 38, 54–58, 76
Handscomb, D. C., 26, 38, 76
Hansmann, U. H. E., 218
Harrison, J., 14, 65
Hartzell, G. W., 13, 85–88
Harvey, A. C., 65
Hastings, W. K., 4, 111, 112, 115
Haussler, D., 86
Hesselbo, B., 207, 217
Higdon, D., 133, 156
Hinkley, D. V., 305
Hinton, G. E., 223
Hockney, R. W., 186
Holley, R. A., 210, 260
Holmes, S., 18, 90, 92
Hopfield, J., 221
Horn, R. A., 257
Hornik, K., 222
Hukushima, K., 4
Hull, J., 201
Hurzeler, M., 75

Insua, D. R., 241
Irwin, M., 62
Isaacson, L. M., 69

Jazwinski, A. H., 16

Jerrum, M., 90, 260
Johnson, C. R., 257
Johnson, N. L., 40
Jones, C. S., 178

Kalman, R., 16, 65
Karlin, S., 106
Karplus, M., viii
Kass, R., 306
Kennedy, A. D., 184, 189–191
Kesten, H., 291
Kirkpatrick, S., viii, 3, 209
Kitagawa, G., 74
Knuth, D., 23
Kong, A., 4, 35, 54, 59, 61, 62, 64, 74, 149
Korst, J., 210
Kotz, S., 40
Kremer, K., 55, 56, 59, 71, 74, 76
Krogh, A., 86
Kunsch, H. R., 75
Kuo, L., 97
Kurtz, T., 247
Kusuoka, S., 210
Kuznetsov, N. Y., 92

Laird, N., 40, 308
Lancaster, H. O., 267
Lang, K. J., 223
Lauritzen, S., 31
Lawrence, C. E., viii, 13, 14, 86, 142, 299
Leach, A. R., viii, 10, 68
Li, K.-H., 131
Li, X. J., 156
Liang, F., 4, 118–120, 124, 221, 241
Little, R. J. A., 176, 274, 307
Liu, C., 168
Liu, J., 65
Liu, J. S., viii, 3, 4, 13, 14, 35, 38, 39, 54, 59, 67, 71, 74, 81, 112, 118–120, 124, 130, 131, 135, 141, 142, 150, 161, 199

Liu, X., 13
Liu, X. W., 165
Logvinenko, T., 89
Lyklema, J. W., 56

Müller, P., 241
MacEachern, S. N., 38, 76, 97, 150
Mallow, C. L., 238
Marinari, E., viii, 210
Marsaglia, G., 23
Marshall, A., 3, 32
Maung, K., 267
McClelland, J., 221
McCue, L. A., 13
McCulloch, R., 238
McFadden, D., 199
Meirovitch, H., 69, 76
Meng, X. L., 8, 70, 168, 275, 311
Mengersen, K. L., 272
Metropolis, N., viii, 3, 34, 105
Meyn, S. P., 254, 255
Mian, I. S., 86
Miller, M. I., 122
Mitchison, G., 86
Monfort, A., viii
Moore, J., 16
Morris, C., 138
Morton, K. W., 4, 27, 54–58
Moult, J., 80
Mumford, D., 165
Murdoch, D. J., 287
Murray, G., 40
Mykland, P., 254

Nardari, F., 201
Neal, R., 191, 195, 223, 240
Nemoto, K., 4
Neuhaus, T., 207, 216
Neuwald, A. F., viii, 14, 86, 142
Newman, M. E. J., 154
Ney, P., 254
Niedermayer, F., 157
Nienhuis, B., 59
Nummelin, E., 247, 254, 271

Odell, P. L., 41, 95

Oh, M.-S., 43
Okamoto, Y., 218

Parisi, G., viii, 210
Pederson, A. R., 178
Pendleton, B. J., 184, 189–191
Peskun, P. H., 112, 114, 127, 134, 277
Petrosian, V., 16
Petsko, G., viii
Pitman, J., ix
Pitt, M. K., 64, 73, 201
Plemmons, R., 248
Press, W. H., 228
Propp, J. G., 284

Qin, Z., 118, 200

Rényi, A., 267
Rabiner, L. R., 64
Raftery, A. E., 307
Rao, M. M., 172
Reeds, J., 251
Reilly, A. A., 86
Rice, J. A., ix
Richardson, S., 21
Ripley, B., 38, 39, 222
Ritter, C., 128
Robert, C. P., 272
Roberts, G. O., 115, 127, 131, 135, 235, 254
Rosenbluth, A., viii, 4, 34, 54–58
Rosenbluth, M., viii, 4, 34, 54–58
Rosenthal, J., 254, 256, 263
Roweth, D., 184, 189–191
Rubin, D. B., 3, 18, 19, 40, 66, 73, 115, 135, 136, 138, 143, 176, 272, 274, 308
Rubin, R. J., 69
Rubinstein, R. Y., 26, 76
Rumelhart, D. E., 222, 223

Sabatti, C., 124, 139, 161, 162, 202
Sahu, S. K., 131, 139, 235

Salmond, D. J., 4, 54, 98
Sanderson, J. G., 92
Sarmanov, O. V., 267
Schervish, M., 130, 275
Schmeiser, B. W., 134
Schmidler, S., 64
Sejnowski, T. J., 223
Shao, Q. M., 8
Shephard, N., 64, 73, 179, 201
Siepmann, J. I., 4, 48, 55, 116
Silverman, B., 240
Sinclair, A., 90, 261
Sjolander, K., 86
Skolnick, M. H., 31
Smit, B., viii, 4, 118–120
Smith, A. F. M., viii, 4, 19, 54, 131, 146, 308
Smith, M. C., 100
Sokal, A., viii, 4, 126, 133, 154, 156, 161, 162
Spiegelhalter, D. J., 21, 31
Stephens, M., 49–51, 74, 83
Stern, H. S., 40
Stinchcombe, M., 222
Stinchcombe, R. B., 207, 217
Stormo, G. D., 13, 85–88
Stroock, D., 147, 210, 249, 256, 261
Sullivan, F., 92
Swendsen, R. H., 133, 154–156

Tanner, M. A., viii, 20, 128, 131, 154, 308
Tavare, S., 39, 49–51, 81
Taylor, H., 106
Teller, A., viii, 4, 34
Teller, E., viii, 4, 34
Thisted, R., 2
Thompson, E. A., 31, 210
Tiao, G., 40
Tierney, L., 114, 115, 135, 254, 271, 280
Torrie, G. M., 206
Tweedie, R., 254, 255

Unger, R., 80

Valleau, J. P., 206
van Dyk, D., 169, 275, 311
Vecchi, M. P., viii, 3, 209
Verlet, L., 186
Vetterling, W. T., 228
von Neumann, J., 3, 24

Wainwright, T. E., viii, 183
Wakefield, J., 133
Walker, S., 133
Wall, F. T., 59, 67, 69, 71, 73
Wang, J. S., 133, 154–156
Weigend, A. S., 223, 240
Weisberg, S., 239
West, M., 14, 65
White, A., 201

White, H., 222
Williams, R. J., 223
Wilson, D. B., 284
Winter, E. M., 100
Witbrock, M. J., 223
Wolff, U., 154, 157
Wong, W. H., viii, 4, 8, 20, 54, 70, 118–120, 124, 131, 154, 221, 241, 308
Wooton, J. C., viii
Wu, Y. N., 3, 131, 165, 172, 275

Yang, C. N., 221
Yu, B., 254

Zaman, A., 23
Zhu, S., 165, 167

Subject Index

$1/k$-ensemble sampling, 215, 217
5' UTR, 13
χ^2 distance, 249
cv^2, *see* coefficient of variation

adaptive direction sampling, 226
ADS, *see* adaptive direction sampling
algorithm
    adaptive direction sampling, 226
    conjugate gradient Monte Carlo, 228
    data augmentation, 4
    exchange Monte Carlo, 212
    hybrid Monte Carlo, 4
    Metropolis, 4, 106
    Metropolis-Hastings, 111
    multicanonical sampling, 4
    multigrid Monte Carlo, 4
    multiple-try Metropolis, 4, 118
    orientational bias Monte Carlo, 119
    parallel tempering, 212
    parameter expansion, 4
    random-walk Metropolis, 114
    sequential imputation, 60–64
    simulated tempering, 4
amino acid, 11
annealing, 209
    simulated, *see* simulated annealing
antithetic variates, 27
aperiodic, 114, 249
autocorrelation, 110
autocorrelation time
    exponential, 126
    integrated (IAT), 126, 215, 236, 273
auxiliary distributions, 116

back-propagation, 223
backward operator, 264
Barker's scheme, 112
Bayes
    method, 19–21, 135, 136, 304–309
    sequential imputation for, 62
Bayes inference, *see* Bayes method
Bayesian method, 3

Subject Index

Beta distribution, 297, 299
binding motif, 13
binding site, 13
binomial distribution, 297
bioinformatics, 10–14
Black-Scholes formula, 201
block-motif model, 13, 140
Boltzmann distribution, 7
bond angle, 10
bootstrap, viii, 304
bootstrap filter, 4, 66, 67, 71, 74, 81, 98, 99, 101
Buffon's needle, vii

canonical ensemble, 215
card shuffling, 246
CBMC, *see* configurational bias Monte Carlo
cdf, *see* cumulative distribution function
central limit theorem, 1, 300
CFTP, *see* coupling from the past
CGMC, *see* conjugate gradient Monte Carlo
chain-structured model, 28
Chapman-Kolmogorov equation, 247
classification, 222
clique, 31
CLT, *see* central limit theorem
coalescence, 49, 50, 81
codon, 11
coefficient of variation, 35, 74, 75, 82, 93, 100
conditional probability, 296
configurational bias Monte Carlo, 4, 116
conjugate gradient Monte Carlo, 227
control variates, 26
convergence
 almost sure, 300
 in distribution, 300
 in probability, 300
coupling from the past, 272, 284
coupling method, 250, 272

covariance, 299
critical point, 221
crossover, 231
cumulative distribution function, 24, 296

data augmentation, 4, 131, 135–139, 154, 174
demographic model, 49
density of the states, 216
density-scaling Monte Carlo, 4
detailed balance, 113, 117, 119
Dirichlet distribution, 299, 307
Dirichlet process, 96
distance
 χ^2, 249
 L^1-, 249
 variation, 249
DNA, 10–14
dynamic linear model, 14
dynamic programming, 29–30
dynamic system, 14
dynamic weighting, 124–125, 220, 287–293

effective sample size, 126
EM algorithm, *see* expectation-maximization algorithm
empirical Bayes, 138
encoder problem, 223
energy
 free, 8
 interacting, 68
 internal, 8
 potential, 7
entropy, 8
equilibrium distribution, 106
ergodicity theorem, 184
error rate
 Monte Carlo, 2
 Riemann integral, 2
Euclidean space, 114
evolution process, 49
exact simulation, 28, 30, 109
exchange Monte Carlo, 4, 212, 231

expectation-maximization algorithm, 310–312
exponential distribution, 297

fading channel, 102
fiber, 162–164, 170, 173, 176
forward operator, 147, 148, 264

Gamma distribution, 297
Gaussian sum filter, 16
gene, 11
gene regulation, 10, 13
genealogy, 49
GeneBank, 12
genetic code, 11
genome, 11–14
geometric distribution, 298
Gibbs motif sampler, 12
Gibbs sampler, 14, 129–158
    blocking, 131
    collapsing, 146–151
    generalized, 171
    grouping, 130, 146–151
    Metropolized, 133
    random-scan, 130
    systematic scan, 130
graphical model, 31
Griffiths-Tavare algorithm, 51
group, 171
    Haar measure for, 172
    topological, 171
    locally compact, 171
growth method, 56–60, 67

Haar measure, 172
    invariant, 172
    left-invariant, 172
    right invariant, 172
hard-shell ball model, 107
helix-turn-helix motif, 12
hidden Markov model, 29, 64–65
hierarchical Bayes, 19, 138–139
hierarchical model, 95
Hilbert space, 264
    inner product in, 264

hit-and-run algorithm, 134
HMC, see hybrid Monte Carlo, see hybrid Monte Carlo
HMM, see hidden Markov model
HR, see hit-and-run algorithm
human genome project, 11
hybrid Monte Carlo, 4, 5, 183–203

i.i.d., see independent and identically distributed
IAT, see autocorrelation time
importance sampling, 3, 31–42
    adaptive, 42
    efficiency of, 34
    marginalization, 37
    properly weighted sample of, 36
    Rao-Blackwellization, 37
    rule of thumb, 34
    sequential, 4, 46–48
    with rejection control, 43
independent, 299
independent and identically distributed, 1, 53, 301
indirect observation model, 199
integrated autocorrelation time, 269
invariance with respect to importance weighting, 125, 288
inversely restricted sampling, see growth method
inversion method, 24
irreducible, 114, 249
Ising model, 7, 107, 153–158, 221
IWIW, see invariance with respect to importance weighting

Kalman filter, 16
    extended, 16
    iterated extended, 16
kinetic energy, 185

L^1-distance, 249
Laplacian, 257
law of large numbers, 1

strong, 300
weak, 291, 300
leap-frog method, 186
Lennard-Jones potential, 9
likelihood, 20, 60
limiting distribution, 106
linear equations, 38
Liouville's theorem, 187
liquid model, 9
locally compact group, 171

macromolecule, 9
magnetization, 8, 110
marginalization, 37
Markov chain, 245
    aperiodic, 249
    forward operator of, *see* forward operator
    interleaving, 132
    irreducible, 249
    periodic, 246
    reversible, 112, 113, 257
Markov chain Monte Carlo, viii, 4, 105
Markov process, 64, 105
Markov property, 245
Markov random field, 163
Markovian structure, 71
mathematical expectation, 296
maximal correlation, 130, 132
maximum likelihood estimate, 20, 50, 60, 302
MCMC, *see* Markov chain Monte Carlo
mean squared error, 35
Metropolis algorithm, 4, 105, 106
Metropolized Gibbs sampler, 133
Metropolized independence sampler, 115, 116, 119, 281
    multiple-trial, 120
MGMC, *see* multigrid Monte Carlo
MIH, *see* Metropolized independence sampler
missing data problem, 19–21, 60–64, 135–136, 308–312

mixture Kalman filter, 98, 100
MKF, *see* mxture Kalman filter100, 102
MLE, *see* maximum likelihood estimate
MLP, *see* multi-layer perceptrons
mobile communication, 102
molecular dynamics, 183
molecular structure, 3, 9
Monte Carlo
    Markov chain, viii
    sequential, 4
Monte Carlo filter, 81–104
MTM, *see* multiple-try Metropolis, *see* multiple-try Metropolis
MTMIS, *see* Metropolized independence sampler
multi-layer perceptrons, 222
multicanonical sampling, 4, 216
multigrid Monte Carlo, 4
multinomial distribution, 298, 307
multiple alignment, 12, 139–143
multiple-try Metropolis, 4, 118, 135, 227
multipoint
    hybrid Monte Carlo, 197
multipoint method, 120

neural network training, 222
Newton's law of motion, 184
nonlinear filtering, 14, 64
nonlinear state-space model, 201
nonparametric Bayes, 95
normal distribution, 296
normalizing constant, 3, 7, 24, 37, 45, 48, 56, 59, 69, 70, 72, 77, 80, 97, 105, 109, 122, 181, 206–208, 307
nucleotide, 11

OBMC, *see* orientational bias Monte Carlo
observation equation, 64
on-line estimation, 16

Onsager, L., 221
optimization, 3
orbit, 170
orientational bias Monte Carlo, 119

parallel tempering, 4, 211, 212, 225, 229, 231
parameter expanded
    data augmentation, 175
parameter expansion, 4, 174
parity problem, 223
partial rejection control, 75
partial resampling, 4
partial sample, 54, 58, 72
particle filter, *see* bootstrap filter
partition function, 3, 7, 8, 59
PDS, *see* probabilistic dynamic system (PDS)
peeling algorithm, 31
perect simulation, *see* coupling from the past
PERM, *see* prune-enriched Rosenbluth method
permanent, 90
permutation group, 247
permutation test, 17
phase flow, 188
phase space, 185
phylogeny, 81
Pitman, D., 298
Poincaré inequality, 257
Poisson Distribution, 298
polymer model, 55, 56
    lattice, 55
population genetics, 49, 81
posterior distribution, 68
Potts model, 153–158
probabilistic dynamic system (PDS), 68
probit regression, 176
propagation method, 31
properly weighted sample, 36
proposal function, 106, 161
proposal transition, 129
protein, 10–14

protein folding, 79
prune-enriched Rosenbluth method, 71
pruning and enrichment method, 71
pseudo-random number, 23
PT, *see* parallel tempering
PX-DA, *see* parameter expansion

quadratic form, 257

random-grid Monte Carlo, 121
random-ray Monte Carlo, 134
random-walk Metropolis, 114
Rao-Blackwellization, 27, 28
rejection control, 43–46, 48, 75
    partial, *see* partial rejection control
rejection method, 3, 24–25
resampling, 72–75
    residual, 72
    simple random, 72
reversible, 112, 113
reversible jumping rule, 122–124
Robert, C. P., 28
Rosenbluth method, *see* growth method

SA, *see* simulated annealing
sampling importance-resampling, 3, 66
SAW, *see* self-avoid walk
self-avoid walk, 55–60, 79–81
    attrition, 56
sequential importance sampling, 4, 46–48, 53–104
    application, 49
    rejection control in, 48
sequential imputation, 4, 60–64
sequential Monte Carlo, 4, 53–104
sequentially symmetric, 120
shrinkage estimate, 139
sigmoidal function, 222
simple random walk
    on a cube, 246

on a line, 246
simulated annealing, 3, 209
simulated tempering, 4, 210
simulated tempering with dynamic weighting, 220
SIR, *see* sampling importance resampling
SIS, *see* sequential importance sampling
site sampler, 14
SKF, *see* split-track filter
slice sampler, 133, 156
snooker algorithm, *see* adaptive direction sampling
specific heat, 8
spectral density, *see* density of the states
spectral radius, 264
split-track filter, 100, 104
stage-wise rejection, 116
state equation, 64
state-space model, 14, 64
stationary distribution, 106, 111, 125
STDW, *see* simulated tempering with dynamic weighting, 221
stochastic volatility model, 201
stratified sampling, 26
stream, 72–75, 101
strong uniform time, 252
Swendsen-Wang algorithm, 153–158

target tracking, 14, 98–102
tempering
    parallel, *see* parallel tempering
    simulated, *see* simulated tempering
topological group, 171
torsion angle, 10
transformation group, 170, 171
transition function, 39, 106, 111, 245

transition rule, *see* transition function
truncated Gaussian, 25
truncated observation, 16

umbrella sampling, 4
unbiased estimator, 28

variance reduction, 26
variation distance, 249
Verlet algorithm, 186

water molecule, 10
Wishart Distribution
    inverse, 40
Wishart distribution, 41
    sampling, 41
Wolff's algorithm, 157

Springer Series in Statistics (continued from p. ii)

Kotz/Johnson (Eds.): Breakthroughs in Statistics Volume I.
Kotz/Johnson (Eds.): Breakthroughs in Statistics Volume II.
Kotz/Johnson (Eds.): Breakthroughs in Statistics Volume III.
Küchler/Sørensen: Exponential Families of Stochastic Processes.
Le Cam: Asymptotic Methods in Statistical Decision Theory.
Le Cam/Yang: Asymptotics in Statistics: Some Basic Concepts, 2nd edition.
Liu: Monte Carlo Strategies in Scientific Computing.
Longford: Models for Uncertainty in Educational Testing.
Mielke/Berry: Permutation Methods: A Distance Function Approach.
Pan/Fang: Growth Curve Models and Statistical Diagnostics.
Parzen/Tanabe/Kitagawa: Selected Papers of Hirotugu Akaike.
Politis/Romano/Wolf: Subsampling.
Ramsay/Silverman: Applied Functional Data Analysis: Methods and Case Studies.
Ramsay/Silverman: Functional Data Analysis.
Rao/Toutenburg: Linear Models: Least Squares and Alternatives.
Reinsel: Elements of Multivariate Time Series Analysis, 2nd edition.
Rosenbaum: Observational Studies, 2nd edition.
Rosenblatt: Gaussian and Non-Gaussian Linear Time Series and Random Fields.
Särndal/Swensson/Wretman: Model Assisted Survey Sampling.
Schervish: Theory of Statistics.
Shao/Tu: The Jackknife and Bootstrap.
Simonoff: Smoothing Methods in Statistics.
Singpurwalla and Wilson: Statistical Methods in Software Engineering: Reliability and Risk.
Small: The Statistical Theory of Shape.
Sprott: Statistical Inference in Science.
Stein: Interpolation of Spatial Data: Some Theory for Kriging.
Taniguchi/Kakizawa: Asymptotic Theory of Statistical Inference for Time Series.
Tanner: Tools for Statistical Inference: Methods for the Exploration of Posterior Distributions and Likelihood Functions, 3rd edition.
van der Laan: Unified Methods for Censored Longitudinal Data and Causality.
van der Vaart/Wellner: Weak Convergence and Empirical Processes: With Applications to Statistics.
Verbeke/Molenberghs: Linear Mixed Models for Longitudinal Data.
Weerahandi: Exact Statistical Methods for Data Analysis.
West/Harrison: Bayesian Forecasting and Dynamic Models, 2nd edition.